recent advances in phytochemistry

volume 20

# The Shikimic Acid Pathway

# RECENT ADVANCES IN PHYTOCHEMISTRY

*Proceedings of the Phytochemical Society of North America*
**General Editor: Eric E. Conn,** *University of California, Davis, California*

*Recent Volumes in the Series*

A Continuation Order Plan is available for this series. A continuation order will bring delivery of each new volume immediately upon publication. Volumes are billed only upon actual shipment. For further information please contact the publisher.

recent advances in phytochemistry

volume 20

# The Shikimic Acid Pathway

Edited by

**Eric E. Conn**
University of California, Davis
Davis, California

PLENUM PRESS • NEW YORK AND LONDON

Library of Congress Cataloging in Publication Data

Phytochemical Society of North America. Meeting (25th: 1985: Pacific Grove, Ca.)
   The shikimic acid pathway.

   (Recent advances in phytochemistry; v. 20)
   "Proceedings of the twenty-fifth Annual Symposium of the Phytochemical Society of
North America . . . held June 12–16, 1985, at Asilomar Convention Center, Pacific
Grove, California.
   Bibliography: p.
   Includes index.
   1.  Shikimic  acid—Metabolism—Congresses.  2.  Amino  acids—Metabolism—
Congresses. 3. Plants—Metabolism—Congresses. I. Conn, Eric E. II. Title. III. Series.
QK861.R38 vol. 20              581.19′2 s [581.1′33]                 86-8885
[QK898.S55]

ISBN 978-1-4684-8058-0        ISBN 978-1-4684-8056-6 (eBook)
DOI 10.1007/978-1-4684-8056-6

Proceedings of the Twenty-fifth Annual Symposium of the Phytochemical Society
of North America, honoring the founders, Ted Geissman and Gestur Johnson,
held June 12–16, 1985, at Asilomar Convention Center, Pacific Grove, California

© 1986 Plenum Press, New York

Softcover reprint of the hardcover 1st edition 1986

A Division of Plenum Publishing Corporation
233 Spring Street, New York, N.Y. 10013

PREFACE


This volume contains the invited papers presented as a symposium of The Phytochemical Society of North America which met for its annual meeting at the Asilomar Conference Center, Pacific Grove, California on June 12-16, 1985. The topic of the symposium, "The Shikimic Acid Pathway - Recent Advances", was especially appropriate for this, the Silver Anniversary of the Society because of the many natural products derived from that pathway. The organizers of the symposium recognized that it would not be possible to cover all groups of compounds derived from shikimic acid and therefore decided to omit any detailed discussion of flavonoid compounds and lignin. Research in these two areas has been the subject of several recent symposiums and/or published volumes. By omitting these topics, it was possible to devote more attention to other, equally interesting products derived from the shikimate pathway.

Each chapter in the volume authoritatively speaks for itself on an important topic. However, the reader is invited to enjoy the lead chapter by Ulrich Weiss who describes his role in the research on the shikimate pathway during 1952/53. We are grateful to Dr. Weiss for this charming account of his work carried out in the laboratory of Dr. B.D. Davis during that period.

Those who attended the Silver Anniversary Meeting were privileged to hear Dr. Gestur Johnson reminisce about the founding of the Society, initially called the Plant Phenolics Group of North America. At the annual banquet R. Horwitz also shared with us some recollections of Dr. Ted Geismann. Geissman and Johnson are considered by many to be the founding fathers of Plant Phenolics Group, renamed the Phytochemical Society of North America, in 1966. Three other members who attended that inaugural meeting and who were also at Asilomar were S.C. Brown, L. Jurd, and G.H.N. Towers.

v

The organizing committee of the Silver Anniversary Meeting consisted of B.G. Chan (Chair), J.A.R. Ladyman, V.L. Singleton and E.E. Conn.  As is so often the case, much of the labor fell on the Chair of the committee.  It is therefore a pleasure to acknowledge that much of the success of the meeting is due to Dr. Chan.  Generous financial support for the Symposium was provided by the Monsanto Chemical Company, FMC Corporation, Shell Development Company, E.I. duPont de Nemours and Company, Cetus Madison Corporation, Dow Chemical Company, Eli Lilly and Company, and Idetek, Incorporated.

As editor, I wish to thank the symposium speakers for their participation and prompt submission of their manuscripts.  Finally, thanks are due to Ms. Billie Gabriel for her skillful preparation of the camera-ready copy.


February, 1986                                          Eric E. Conn

CONTENTS

Chapter One

EARLY RESEARCH ON THE SHIKIMATE PATHWAY: SOME PERSONAL
REMARKS AND REMINISCENCES

ULRICH WEISS

Laboratory of Chemical Physics
National Institute of Arthritis, Diabetes,
and Digestive and Kidney Diseases
Bethesda, Maryland 20205

The 25th Annual Conference at Asilomar, California,
June 12 - 16, 1985, was a most enjoyable and stimulating
event, which has splendidly accomplished its goal:   to
provide an overview of the present state of our knowledge
of the shikimate pathway and its ramifications.  As parti-
cipants, we all are indebted to the organizers for the job
they did; their effort must have been great.

Having taken part, during 1952/1953, in the early
research on this biosynthetic sequence, attendance at this
conference was for me a great experience indeed.  To my
regret, I have never had an opportunity since those ancient
times for making further direct contributions to the elucidation of the shikimate pathway.  I have, however, during
all these years maintained a lively interest in the advances
of research in this field.  It was, therefore, a real joy to
participate in the conference and to get a coherent picture
of the present status of our knowledge through lectures by
many of the colleagues responsible for these advances, and
of course by private discussions with old friends and new
acquaintances, a method so much more satisfying and efficient than having to follow the progress of research
piecemeal from the original literature.

And what progress!  The few remaining gaps in our
understanding of the biosynthesis of the classical five
primary aromatic metabolites are now being closed; all the
known intermediates have been prepared by total synthesis;
the enzymes involved and their regulations have been studied
in great detail.  Our knowledge of this area of biosynthesis
should thus be complete in the foreseeable future.  Fortu-

1

nately, there is no danger of this completion meaning
the end of research on the shikimate pathway.  Side-branches
of this sequence are being found in increasing numbers,
giving it additional significance:  the biosyntheses via
isochorismic acid, the formation of phenylalanine and
tyrosine through arogenate rather than prephenate, and
the origin of the $C_7N$ unit present in the important
ansamycin antibiotics and in the maytansinoids will suffice
to prove the point.  Perhaps the most interesting and novel
development, however, is the discovery of a number of
shikimate-derived metabolites with non-aromatic six-membered
rings; it shows that the function of the shikimate pathway
is by no means limited to the biosynthesis of aromatic
compounds.  Much remains to be explored in these new areas,
and we can expect with confidence that other such novel
sequences branching off from the classical one will be
discovered.  In addition, some features remain unexplored
even in the initial pathway.  The old puzzle of the origin
of quinic acid is still unsolved; interestingly, it was not
discussed at Asilomar.

It is particularly gratifying that the research on
the shikimate pathway has also brought dividends in the
form of important practical applications, such as the
discovery that the widely used herbicide glyphosate acts
by inhibiting one of the enzymes of the common part of the
sequence, the enzyme which catalyzes the conversion of
shikimate 3-phosphate into 5-enolpyruvylshikimate 3-phosphate.
This discovery is one more example supporting my firmly held
conviction that sound exploration of new fields will eventually lead to findings of practical utility, even if these
could not have been anticipated in the beginning and were by
no means the initial goal of the research.  This, in my
opinion, makes the division into "fundamental" or "academic",
and "applied" research artificial, meaningless, and often
harmful.

I have been invited to reminisce a little about the
early days of research on the shikimate pathway, and to
attempt a brief outline of the stages of further development
to its present level.  I joined the small, oddly named
"Tuberculosis Research Laboratory" of the U.S. Public Health
Service on East 69th Street in Manhattan, N.Y., on December
10, 1951, to participate in the research directed by Dr.
B.D. Davis.  This work was devoted to the exploration of
biosynthetic pathways leading to the amino acids and other

metabolites with the help of mutants of bacteria such
as <u>Escherichia</u> <u>coli</u> and <u>Aerobacter</u> <u>aerogenes</u>. Suitable
strains of such bacteria had become readily accessible
through the penicillin method developed by Dr. Davis.
My task, as the organic chemist of the group, was the
isolation and characterization of intermediates in the
shikimate pathway. Initially, I had every reason to
expect a long-term association with this research, which
I found fascinating. However, subsequent developments
limited my active participation in it to only two years.

Besides Dr. Davis and myself, the senior staff
consisted of the biochemist, Henry J. Vogel, with whom I
shared the laboratory, and the geneticist, Werner Maas.
Each one of us senior people had an assistant; my work
during those two years owes very much to the exceptional
skill, competence, and enthusiasm of Mrs. Elizabeth S.
Mingioli, universally known as "Ber". The group also
included several postdoctoral associates: Drs. Chaim
Yaniv, who unfortunately passed away not long after
returning to his native Israel; E.B. Kalan; Susumu
Mitsuhashi from Tokyo, now an internationally known
bacteriologist; and Sam Beiser, likewise a bacteriologist,
whose early death a few years later was a sad loss. The
last two colleagues, in particular, became good friends.
Their helpful advice was very important to me, working as
I was in an entirely new field, and in the unaccustomed
milieu of institutional research. Before accepting Dr.
Davis' invitation to join the group, I had held two ten-
year jobs in the pharmaceutical industry, doing work of an
entirely different kind and purpose. Later on, Dr. Charles
Gilvarg came from the Department of Pharmacology of New
York University; he joined me for some time in the work on
the structure of prephenic acid.

In addition, the laboratory was visited for shorter
periods by several scientists who wished to become familiar
with the preparation and use of bacterial mutants. I
recall Drs. Bruce Ames, Sam Aronoff, Alton Meister, and Dan
Rudman; there may have been others.

The laboratory consisted of a few small, modestly
equipped rooms on the top floor of a four-story building
which served otherwise as a Public Health Center; I
remember that the superintendent of the building was quite
unenthusiastic about the invasion of his realm by scientific

research.  At first, the laboratories were not air-
conditioned and therefore miserably uncomfortable during
the hot humid summer of 1952; later on, air-conditioning
was installed, so that the research during my second
summer was carried out under acceptable conditions.

Before Mrs. Mingioli and I started work on the isolation of a new intermediate, its excretion by suitable
mutants was studied by Dr. Davis, who selected the strain
which showed the best production of the metabolite.  Our
research led to the isolation in pure form, and to the
complete structure proof, of three new intermediates:
3-dehydroquinic acid, shikimic acid 3-phosphate, and
prephenic acid.  Of these, the last one turned out to be
a precursor of phenylalanine and tyrosine only.  In contrast,
the two other compounds were shown to be intermediates in
the biosynthesis of all five primary aromatic metabolites.
It was established, eventually, through work in the
Tuberculosis Research Laboratory and elsewhere, that these
metabolites are formed through a common series of seven
intermediates, with sequences to the individual aromatic
compounds branching off from the last one of these common
stages.  In December, 1951, only two of these substances
had already been discovered.  Shikimic acid itself, the
only previously known one among the intermediates in the
common sequence leading to phenylalanine, tyrosine,
tryptophan, p-aminobenzoic acid, and p-hydroxybenzoic
acid, had been recognized as being involved in their
biosynthesis through the use of a large collection of
bacterial mutants.  Its immediate precursor, 3-dehydro-
shikimic acid (called 5-dehydroshikimic acid at that time)
had been isolated and fully characterized by the excellent
work of my predecessor, I.I. Salamon.  He had introduced
the use of charcoal columns, with elution by aqueous alcohol
of increasing strength, for the isolation of his compound.
I found this method very valuable in my subsequent work on
other intermediates, and was delighted and surprised to
hear at Asilomar from Dr. Amrhein that the technique is
still in daily use in his laboratory as the first step in
the routine isolation of shikimic acid 3-phosphate needed
for his research on glyphosate.

The first intermediate that we isolated was 3-dehydro-
quinic acid, the compound directly preceding 3-dehydroshi-
kimic acid.  Following chromatography on charcoal and
elution with very dilute ethanol (2.5 - 5 percent), further

purification was achieved by precipitation from the concentrated effluent fractions as the insoluble brucine salt.
The question may be asked: what deep chemical deliberations
prompted the choice of brucine from among the many
possible alkaloids? The answer is simple: I happened to
find a supply of brucine in the small collection of chemi-
cals in the laboratory, tried it, and saw that it worked
beautifully. I never knew from what earlier research it was
left, but am glad that it was. Removal of the brucine as
the insoluble picrate, and further routine manipulations
gave a residue which crystallized slowly on contact with
ether. I vividly remember my delight when I inspected the
small test tube a few days after putting it into the refri-
gerator, and found that it contained a lot of pretty
crystals of 3-dehydroquinic acid. I also recall the day
on which this happened: February 29, 1952; 1952 was a leap
year. Proof of structure and stereochemistry of the
compound rests essentially on its reduction to quinic acid,
on the absence of the pair of cis-hydroxyls present in
shikimic acid, and on the conversion into 3-dehydroshikimic
acid on mild treatment with acid.

The structure of 3-dehydroquinic acid as a cyclic aldol
made it appear probable that its six-membered ring is formed
from an acyclic precursor through an intramolecular aldoli-
zation; it would then be the first compound in the pathway
in which the ring of phenylalanine, tyrosine, tryptophan,
etc., is already present, although not yet aromatic. Not
long afterwards, this interpretation was shown to be correct
when Sprinson and his coworkers at Columbia University
isolated DAHP, the acyclic earliest intermediate, and
studied its cyclization to 3-dehydroquinic acid.

While this work was in progress, Mrs. Mingioli and I
also isolated another compound which had been observed in
culture filtrates of certain mutants. This was a compound
which was not utilized by any available bacterial strain,
but which gave shikimic acid on fairly energetic acid hydrolysis. Techniques quite similar to those used for the
isolation of dehydroquinic acid led to a pure, although
non-crystalline K salt. Before submitting this substance
to elementary analysis for C and H, I made qualitative tests
for N, S, and Cl; they were negative. I was about to test
the product also for P, when a very knowledgeable colleague
advised against wasting any of the small amount of available
material on such a test, since no phosphorylated intermediate

in the biosynthesis of amino acids had yet been observed; I
followed his advice. The counsel of this colleague turned
out to be unhelpful; he shall therefore remain anonymous.
On subsequent analysis, the amount of C and H was found to
add up to less than 50 percent. Making the usual assumption
that the balance is oxygen, I tried, completely unsuccess-
fully, to formulate a derivative of shikimic acid which
would contain that much oxygen. The only possible way out
of this dilemma was the hypothesis that the compound does
contain phosphorus after all, and indeed the qualitative
test was strongly positive. Another elementary analysis
now was entirely consistent with a monohydrate of the potassium salt of a monophosphate of shikimic acid; the task of
proving its structure resolved itself to localizing the
phosphate residue. That it is the hydroxyl at C-3 which is
phosphorylated was proved by showing the presence of one
pair of hydroxyls on adjacent carbons (uptake of one mol of
periodate) and by demonstrating that this is not the cis-
oriented pair at C-3 and C-4 of shikimic acid. Furthermore,
the bacterial metabolite was shown to be different from
5-phosphoshikimic acid, readily synthesized for comparison
from the 3,4-acetonide of shikimic acid. Surprisingly, the
5-phosphate is not reported in Chemical Abstracts. In
contrast to the facile synthesis of this compound, conversion of shikimic acid into the 3-phosphate would encounter
great difficulties. As a consequence, it has not been accomplished; shikimate 3-phosphate was one of the last intermediates of the shikimate pathway to be made by total synthesis,
using a method which does not involve shikimic acid (Bartlett
and McQuaid, 1984).

The failure of 3-phosphoshikimic acid to be utilized by
bacteria is in line with the general inability of intact
cells to take up phosphate esters. Its utilization by cell-
free extracts of suitable mutants shows it to be an essential
intermediate in the common portion of the shikimate pathway.

When the entire chain of seven common intermediates had
been elucidated, the surprising fact emerged that this
sequence oscillates back and forth between phosphorylated
and phosphorus-free stages. It begins with 3-deoxy-D-
arabino-heptulosonic acid 7-phosphate (DAHP), continues
with the three phosphorus-free metabolites 3-dehydroquinic
acid, 3-dehydroshikimic acid, and shikimic acid, proceeds
next through the two phosphorylated intermediates shikimic
acid 3-phosphate and 5-enolpyruvylshikimic acid 3-phosphate,
and ends with the phosphorus-free chorismic acid.

During the spring and summer of 1953, we investigated a precursor of phenylalanine. Before we started our work, Dr. Davis had established that this product was not utilized by mutants blocked in the biosynthesis of phenylalanine, but that it was converted with extraordinary ease, on the slightest exposure to acidic conditions, into phenylpyruvic acid, $C_9H_8O_3$; this acid was already known to be transformed into phenylalanine in vivo by transamination. The new metabolite was thus tentatively assumed to be an unusually sensitive derivative or conjugate of phenylpyruvic acid. The name prephenic acid was proposed for this compound by Dr. Davis.

We observed that the effluents from charcoal columns containing the metabolite lacked any specific absorption band in the UV, even if they were quite concentrated. This fact eliminated the most plausible working hypothesis: presence of a derivative of the enol form of phenylpyruvic acid. Such a product would indeed be quite sensitive to acid, but it would show the intense absorption band above 300 nm which is characteristic of the chromophore of enolized phenylpyruvic acid. Alternatively, an acetal of phenylpyruvic acid could be present; while, however, acetals are readily cleaved by acid, their sensitivity is usually much less extreme than that of the new metabolite. The acetals of phenylpyruvic acid prepared later on by Plieninger were indeed much less sensitive than our compound. These two possible interpretations of our metabolite as a derivative of phenylpyruvic acid thus appeared quite improbable, a situation which led me to consider the possibility that the substance might not be an aromatic compound at all, so that its conversion into phenylpyruvic acid would amount to the formation of the benzene ring from an alicyclic precursor. If so, the biosynthetic aromatization might well follow a similar course and would thus, for the first time, become accessible to detailed interpretation.

Theoretical considerations of suitable structures led at once to a formulation embodying a 2,5-cyclohexadienol ring which carries the $C_3$ side-chain of phenylpyruvic acid at position 4; some features of the structure were left open at this point. The cyclohexadienol unit would be expected on mechanistic grounds to undergo very ready acid-catalyzed transformation into a benzene ring, and it would not contain any grouping capable of producing an absorption band in the accessible UV. It would be feasible to test this tentative

interpretation of the few observations on rather crude
eluates as soon as pure material would become available,
since the two double bonds of the ring should be readily
saturated by catalytic hydrogenation.  At this stage,
examination of suitable model compounds would have been
very helpful; unfortunately, none were known.  A number of
2,5-cyclohexadienols had been synthesized earlier, but all
of those had tertiary hydroxyls instead of the secondary
one of my hypothetic formulation.  After publication of our
complete structure of prephenic acid, Professor H.
Plieninger in Heidelberg succeeded in preparing a variety
of 2,5-cyclohexadienols with the required secondary
hydroxyls.  Their chemical properties turned out to be
quite analogous to those we had observed in prephenic acid;
this research of Plieninger thus provided welcome support
for our formulation.  The good personal friendship with
this fine scientist, which developed on the basis of our
shared interest in this class of compound, was a real bonus
from our research; his sudden death last Christmas, during
cardiac surgery, was a deeply felt loss.

These assumptions had thus yielded a working hypothesis
which could be used for interpretation of the available
observations, and which had the additional virtue of
suggesting crucial tests.  In retrospect, I regret that I
stopped speculating at this point.  Had I been willing to
go further with it, and to give more attention to the structural requirements for an intermediate between shikimic acid
and phenylpyruvic acid, I could have written the entire
structure of prephenic acid even then, before the compound
had been isolated.  Evidently, during the transformation of
shikimic acid into phenylpyruvic acid, the carboxyl of the
former is lost (that this loss actually takes place during
the biosynthesis of phenylalanine had been previously
established), while the $C_3$ side-chain must be acquired at
some stage.  I could, therefore, have recognized prephenic
acid as an intermediate where the side-chain was already in
place, but the carboxyl still present.  But I felt,
mistakenly, that I had gone far enough and should not
overdo the speculating.

Prephenic acid was next purified as the barium salt,
which was obtained in fine crystals.  At this stage, Dr.
Gilvarg joined in the research.  The purification steps
leading to the salt were monitored by spectroscopic methods,
which suggested the presence of the barium salt of a

dicarboxylic acid; phenylpyruvic acid is of course a mono-
carboxylic acid. The structure of prephenic acid should
thus embody a second carboxyl, so situated that it would be
lost on conversion into phenylpyruvic acid. This consideration, together with the assumption of the cyclohexadienol
ring, led at once to the complete formulation of prephenic
acid, which was shown to be correct by all further findings:
elementary analysis in surprisingly close agreement with
theory and proving the expected 10:1 ratio of carbon to
barium; liberation of one mol $CO_2$ on acidification; and
uptake of the expected amount of hydrogen on catalytic reduction. In addition, the resulting structure of prephenic
acid also fitted nicely with expectations, just discussed,
of an intermediate between shikimic acid, retaining its
carboxyl, and phenylpyruvic acid, with the three-carbon
side-chain already in place and the ring so structured that
conversion into phenylpyruvic acid becomes understandable.
There is no reason to doubt that the aromatization step
during the biosynthesis of phenylalanine proceeds by essentially the same mechanism. Subsequent research has shown
that prephenic acid in many organisms is also a precursor
of tyrosine. In the alternative arogenate pathway,
discovered later and already mentioned, the transamination
step precedes the aromatization, and prephenic acid is
converted into its amino acid analog.

The 2,5-cyclohexadienol ring, which I had postulated
initially as a working hypothesis capable of rationalizing
a few observations made before prephenic acid had even been
isolated in pure form, has subsequently assumed significance
in the biosynthesis of natural compounds other than prephenic
acid. It occurs in several substances not connected with
the shikimic acid pathway, and plays a role in the biosynthesis of a number of alkaloids, including morphine. Again,
we encounter a relationship that could not have been anticipated in 1953 when the idea of this ring occurred to me.

Clearly, our work on the structure of prephenic acid,
and much other related research not mentioned here, required
quite an expenditure of time and effort. It is interesting
to consider how much easier and faster the structure proof
would have been if techniques such as $^1$H-NMR had become
generally available a few years earlier. The research on
prephenic acid has been published only in the form of a
preliminary note in Science; much to my regret, my manuscript of a full paper has remained unpublished.

The structure of prephenic acid actually represents two stereoisomers. I was no longer able any more to try solving this problem; unpublished mechanistic interpretation of some observations did suggest the stereochemistry later on shown to be the correct one, but at this stage, my active participation in research on the shikimate pathway came to an end. Fortunately, the stereochemistry of prephenic acid was elucidated by Plieninger, who also did much pioneering work on the total synthesis, which he achieved in 1978; independently, Danishefsky and Hirama succeeded in synthesizing prephenic acid.

This brings me to the end of my story. It remains to give a brief account of the most important advances made since then, i.e. since the end of 1953. No complete coverage of the large amount of fine work done is intended.

Research in David Sprinson's laboratory at Columbia University, in part contemporary with the research outlined in the preceding pages, has unraveled the earliest stages of the sequence and the derivation of the first, non-cyclic intermediate, DAHP. In the same laboratory, Judith Levin discovered 5-enolpyruvylshikimate 3-phosphate and predicted the existence and correct structure of the next, and last one, of the common intermediates, the compound from which the pathways to the individual primary aromatic products branch off. Isolation and structure proof of this branch-point intermediate, chorismic acid, by the Australian workers F. Gibson, L.M. Jackman, and J.M. Edwards completed the elucidation of the general pathway.

A few of the significant advances in the later stages of the sequence should be mentioned. Isomerization of chorismic acid to isochorismic acid by an allylic shift has opened a large new area of biosynthesis via shikimic acid; this transformation was studied, again, by Gibson and Jackman in Australia and their co-workers, among whom I wish to mention the late Tom Batterham, who became my first postdoctoral co-worker at NIH, and a close friend. Compounds formed biosynthetically from isochorismic acid include the meta-carboxylated aromatic amino acids of many higher plants, studied by P. Olesen Larsen in Copenhagen, the iron-chelators derived from 2,3-dihydroxybenzoic acid and, at least in some organisms, salicylic acid. The recent observation by Leistner that isochorismic acid is also the parent of ortho-succinyl-benzoic acid and hence of various quinones in higher plants and of Vitamin $K_2$ in bacteria, has added greatly to the importance of the isochorismate branch. The discovery of arogenate, rather than prephenate, as an intermediate in the biosynthesis of phenylalanine and tyrosine has led its discoverer Roy Jensen to interesting comparisons of the two alternative pathways and their enzymology and regulation. The formation of the $C_7N$ unit, already mentioned, has been found by R.W. Rickards to proceed through 3-amino-5-hydroxybenzoic acid, but the details of this pathway are not yet known. The mode of derivation of anthranilic acid and para-amino-benzoic acid from chorismic acid has long remained obscure, since the search for identifiable intermediates has been so far unrewarding. Recently, however, progress has been made by Berchtold and by Ganem through the total synthesis of compounds which could be involved, and proof that they are converted into the aromatic metabolites by enzyme extracts. Excellent, difficult research has shed light on the detailed stereochemistry of the enzymatic interconversions of the various intermediates of the shikimate pathway. The discovery and study of the shikimate-derived metabolites with non-aromatic six-membered rings has brought further exciting enrichment of the shikimate pathway; it has been shown, largely through work by H.G. Floss, that these compounds are not formed through reduction of aromatic compounds, and that the carboxyl of early precursors is retained in their biosynthesis.

In these few pages I have tried to give an idea of the way in which our early research was carried out, and simultaneously to provide an overview of the subsequent developments which led to our present-day knowledge. To me, at least (and I hope not to me only!) it is fascinating to watch the unfolding of the story, and its growth in many directions. As is so often the case, it starts with a methodological advance: the penicillin method for selection of bacterial mutants. Next follows the painstaking, gradual elucidation of the stages through which one of the great, all-important pathways of biosynthesis proceeds: the source of metabolites essential for every living cell, of vital hormones and several vitamins, of indispensable drugs; a pathway operating on a gigantic scale in the biosynthesis of lignin, which constitutes up to about 30% of all wood. During this elucidation, initially unsuspected side-branches to still other substances came to light, practical applications emerged, and the regulatory mechanisms which enable this biosynthetic sequence to function with marvellous efficiency in providing the needed metabolites in exactly the amounts required were revealed.

Many further developments of this story can be expected with confidence.

Chapter Two

THE SHIKIMATE PATHWAY – AN OVERVIEW

HEINZ G. FLOSS

Department of Chemistry
The Ohio State University
Columbus, Ohio 43210

INTRODUCTION

It has been eight years since the Phytochemical Society last held a symposium centered on the theme of Biosynthesis of Aromatic Compounds at its joint meeting with the European Phytochemical Society in Ghent.[1] The intervening years have seen a tremendous revival of interest in the shikimate pathway. This renewed focus was stimulated by the discovery in Amrhein's group[2] that the herbicide glyphosate (Fig. 1) acts by inhibiting the enzyme 5-enolpyruvylshikimate 3-phosphate (EPSP) synthase. Glyphosate is a very successful commercial product for which estimated sales of about $480 million in 1984 have been reported. The discovery of its mode of action has spurred intense efforts in many laboratories to design new or improved compounds of this type.

Figure 2 shows the common features of the shikimate pathway.[3,4] Phosphoenolpyruvate and erythrose 4-phosphate are condensed to form 3-deoxy-arabino-heptulosonic acid 7-phosphate (DAHP), which is then cyclized to give dehydro-quinic acid. Dehydration followed by reduction of the

13

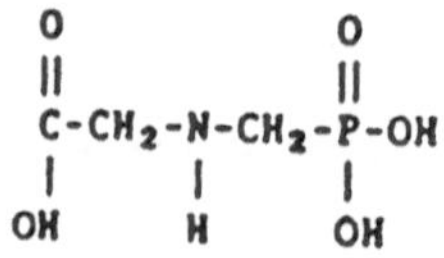

$$\begin{array}{ccc}
O & & O \\
\| & & \| \\
C-CH_2-N-CH_2-P-OH \\
| & | & | \\
OH & H & OH
\end{array}$$

Fig. 1.  Structure of glyphosate [N-(phosphonomethyl)
glycine].

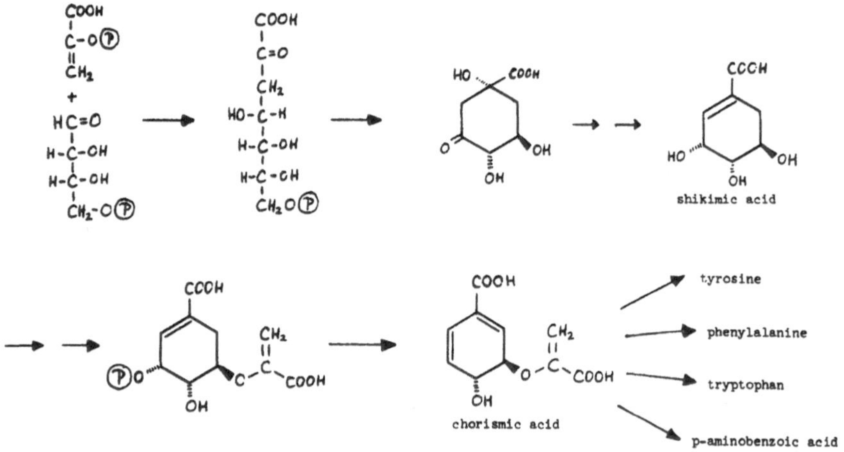

Fig. 2.   The common branch of the shikimate pathway.

carbonyl group then leads to shikimic acid, which is phosphorylated in the 3-position.  Attachment of an enolpyruvyl
side-chain derived from another molecule of phosphoenolpyruvate and 1,4-elimination of phosphoric acid then gives
chorismic acid which is located at a major branch point in
the pathway.  Chorismic acid on the one hand is the
precursor of phenylalanine and tyrosine and on the other
hand gives rise to tryptophan via anthranilic acid and as
well is transformed into p-aminobenzoic acid and a host of
other compounds.  A major development during the last few
years has been the more detailed unraveling of the arogenate
pathway by Jensen and coworkers (Fig. 3).  In most organisms
this is the predominant route leading to the formation of
tyrosine, and in many instances it also is the pathway by
which  phenylalanine is formed (see Chapter 3 by R.A. Jensen
in this volume).

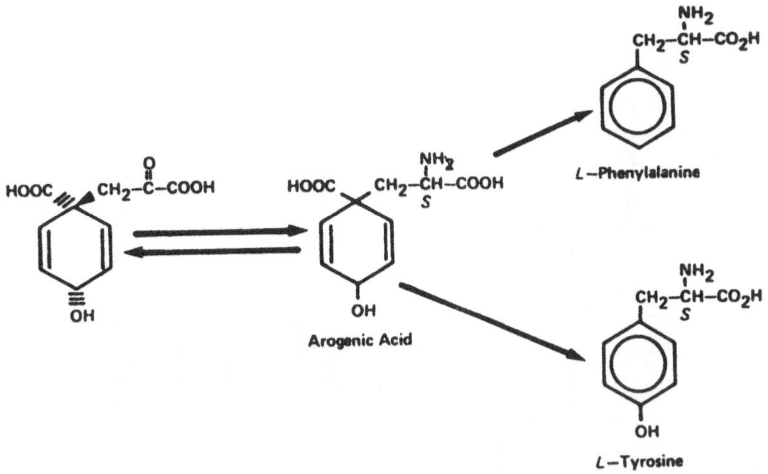

Fig. 3.  The arogenate pathway of tyrosine and phenylalanine biosynthesis.


SYNTHESIS OF PATHWAY INTERMEDIATES

Organic synthesis has made significant contributions to the study of the shikimate pathway, including the total synthesis of arogenate (Fig. 4) by Danishefsky's group.[5] This synthesis relies on the Diels-Alder strategy used earlier in Danishefsky's synthesis of prephenic acid.[6] It uses L-pyroglutamic acid to build up the side-chain array in optically active form.  Of considerable practical utility is a new synthesis of DAHP reported by Knowles' group[7] starting from 2-deoxyglucose (Fig. 5).  Several new syntheses of shikimic acid have been reported in recent years,[8-13] as well as the first synthesis of 3-phosphoshikimic acid by Bartlett's group.[13]  This synthesis, as well as Bartlett's synthesis of EPSP, are described in Chapter 5 in this symposium volume.  The first synthesis of EPSP has just been reported by the group of Ganem.[14]  Both the groups of Berchtold[15,16] and of Ganem[17] also published the first total syntheses of chorismic acid.  An improved version of the Berchtold synthesis[16] is shown in Figure 6. The enolpyruvyl side-chain is elaborated from a malonyloxy side-chain by the Mannich reaction followed by a Hofmann elimination.  The elaboration of the allylic alcohol function proceeded by ring opening of an epoxide with phenylselenide followed by oxidative elimination.

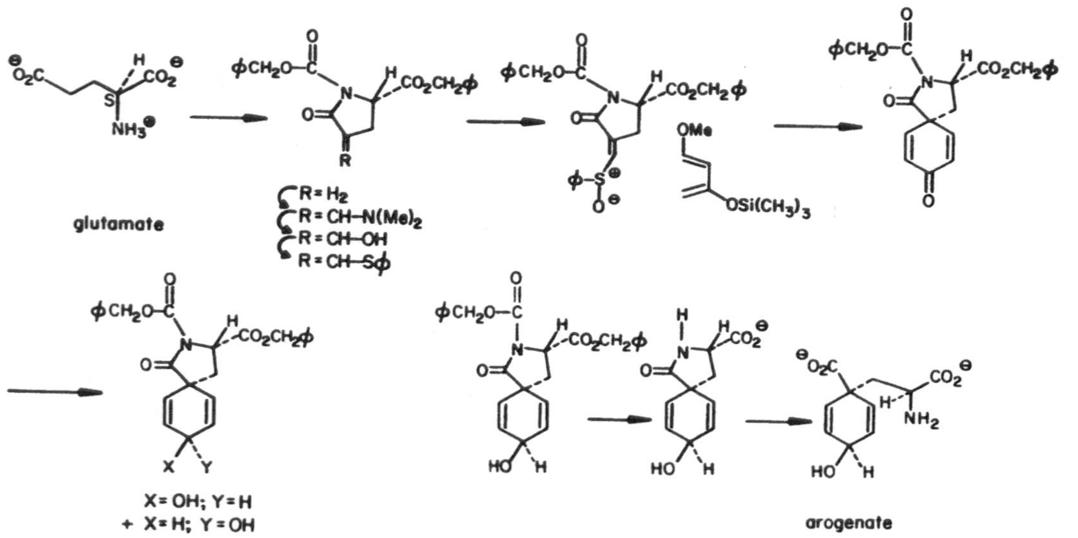

Fig. 4. Total synthesis of arogenate. Reprinted, with modifications, from Reference 5 with permission of the American Chemical Society.

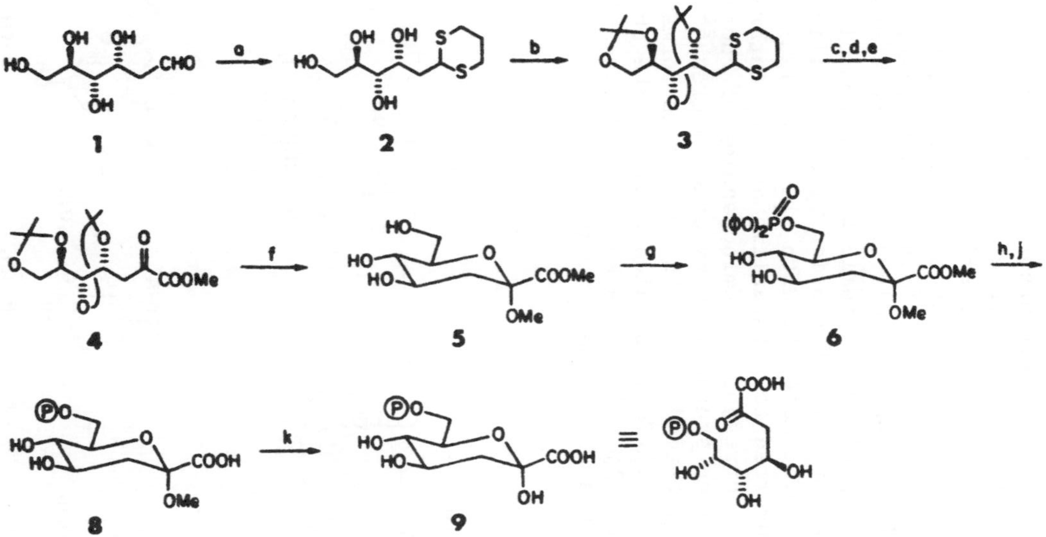

Fig. 5. Synthesis of 3-deoxy-D-arabino-heptulosonic acid 7-phosphate. Reprinted from Reference 7 with permission of the American Chemical Society.

*a, HCl–HS(CH$_2$)$_3$SH–EtOH; b, (CH$_3$)$_2$CO–H$_2$SO$_4$; c, n–BuLi; d, CH$_3$OCOCl; e, N–bromosuccinimide–acetone; f, HBr–MeOH; g, (PhO)$_2$POCl–pyridine; h, PtO$_2$–H$_2$; j, KOH; k, Dowex 50 (H$^+$).

Development of the key methodologies for the construction of the chorismate molecule by the Berchtold and Ganem
groups has paved the way for the synthesis of a number of
other compounds which are known or postulated intermediates
in the shikimate pathway.  One such compound is isochorismic
acid, the precursor of the 2,3-dihydroxybenzoic acid moiety
of enterochelin and also the precursor of the metacarboxy-
phenylamino acids found in a number of higher plant species.
As we shall discuss later, isochorismic acid is also the
precursor of vitamin K.  The synthesis of isochorismic acid
as described by Berchtold's group[18] is shown in Figure 7.
Both Ganem's group[19] and Berchtold's group[20] have synthesized the amino analog of isochorismic acid, a postulated
intermediate in the enzymatic formation of anthranilic acid
(Fig. 8).  The synthetic material with _trans_ configuration
was shown to be converted to anthranilic acid by anthranilate synthase.  The corresponding amino analog of chorismic
acid has also just been synthesized by Ganem's group[21] and
was shown to be an intermediate in the enzymatic formation
of p-aminobenzoic acid.  These studies complete the chemical
synthesis of essentially all of the key intermediates in
the main branch of the shikimate pathway.

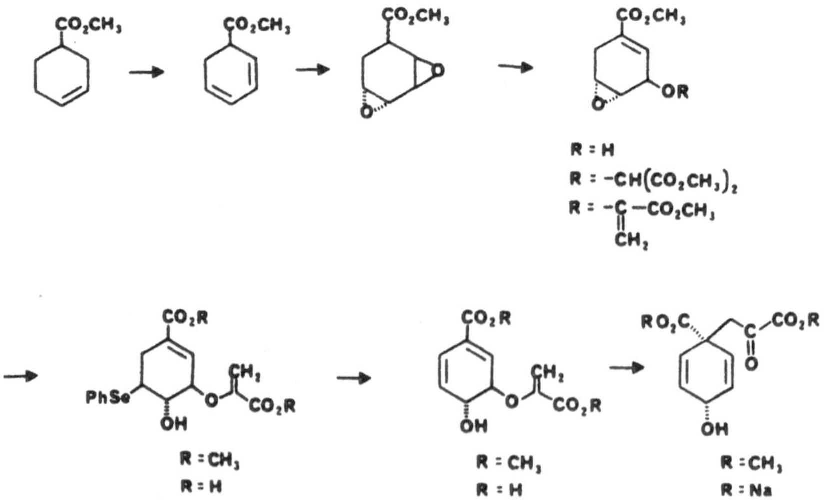

Fig. 6.  Total synthesis of racemic chorismic acid.
Reprinted, with modification, from Reference 16 with permission of the American Chemical Society.

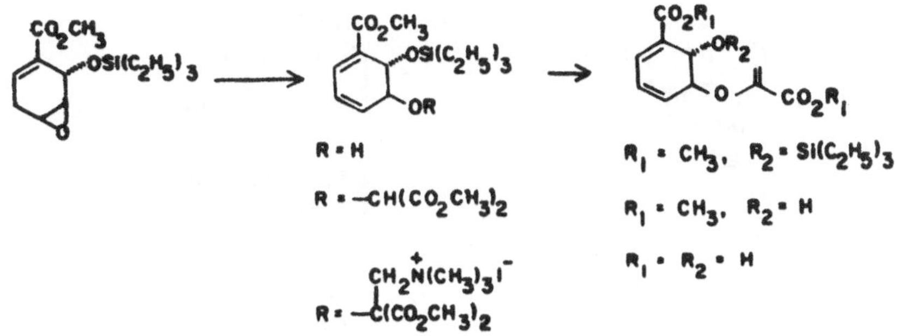

Fig. 7. Total synthesis of isochorismic acid. Reprinted from Reference 18 with permission of the American Chemical Society.

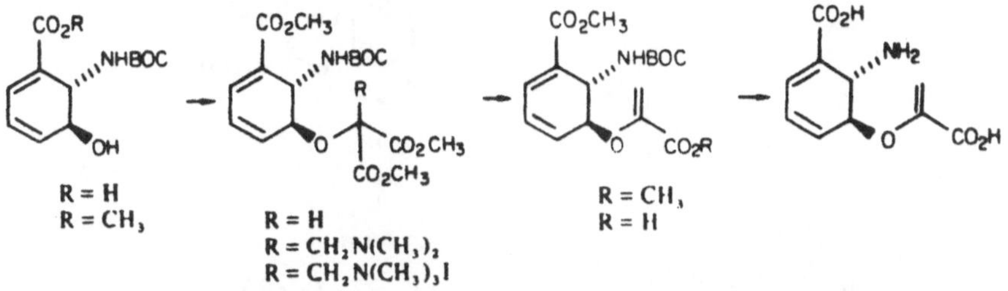

Fig. 8. Synthesis of an intermediate in the enzymatic formation of anthranilic acid. Reprinted, with modifications, from Reference 19 with permission of the American Chemical Society.

A number of new syntheses of labeled intermediates of the shikimate pathway have also been published. These include syntheses of shikimic acid labeled in a variety of different positions.[9,22,23] Particularly elegant is a synthesis of chorismic acid labeled stereospecially with tritium in either the E or the Z position of the enolpyruvyl side-chain.[24] The synthesis is shown in Figure 9. Following elaboration of the enolpyruvyl side chain, bromination and dehydrobromination produces a Z-bromo-enolpyruvyl side chain in which the bromine can be selectively replaced by hydrogen without affecting the stereochemistry by reduction with a Zn/Ag couple. Tritium can be introduced into the Z position by carrying out the reduction in tritiated water and into

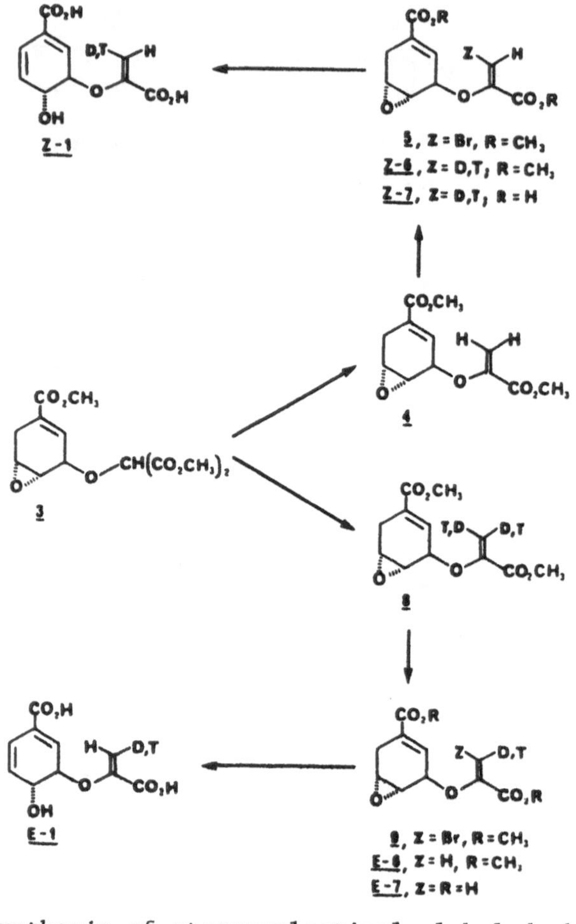

Fig. 9.  Synthesis of stereoselectively labeled chorismic
acid.  Reprinted from Reference 24 with permission of the
American Chemical Society.


the E position by using tritiated formaldehyde in the
Mannich reaction.  This material was used in the stereochemical analysis of the chorismate mutase reaction.[25]


ENZYMES OF THE SHIKIMATE PATHWAY

     Many studies have been carried out during the last few
years on enzymes in the shikimate pathway.  Several of the

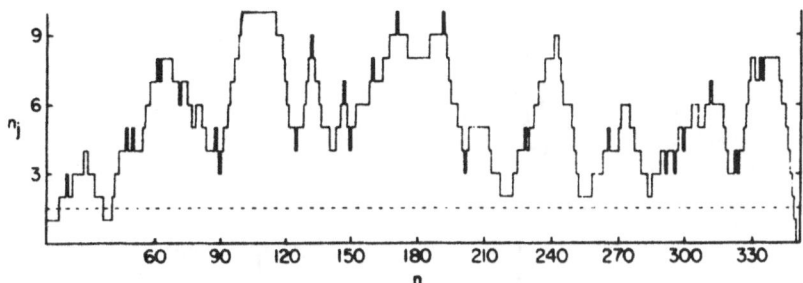

Fig. 10. Amino acid homologies between the tyrosine-
sensitive and the phenylalanine-sensitive isoenzymes of
3-deoxy-D-arabino-heptulosonate 7-phosphate synthase from
Escherichia coli. Reprinted from Reference 28 with permission of the American Society of Biological Chemists.

Comparison of the tyrosine- and the phenylalanine-sensitive
DAHP synthase isoenzymes of E. coli; $n_j$ is the number of
amino acid residues common to decapeptides, n to n + 9,
where n is the residue number in DAHP synthase (Tyr). The
dashed line indicates the $n_j$ value for random decapeptides.


enzymes have been purified to homogeneity, cloned and/or
sequenced. These include the tyrosine- and the phenylalanine-
sensitive DAHP synthases, dehydroquinate synthase and EPSP
synthase from several sources. Some interesting studies
have been carried out on the relatedness of the tyrosine-
and the phenylalanine-sensitive DAHP synthase from
Escherichia coli. Antibodies raised against the native
phenylalanine-sensitive enzyme were found not to cross-
react at all with the tyrosine- or the tryptophan-sensitive
isoenzymes.[26] Comparison of the 40 amino acid residues at
the N-terminal end of the three enzymes also showed very
little relatedness.[27] However, when the entire amino acid
and DNA base sequences of the tyrosine- and the phenylalanine-sensitive enzymes were compared it became evident
(Fig. 10) that the two proteins are much more closely
related than the initial evidence had suggested.[28] Several
regions of high degrees of amino acid conservation are seen
in the protein chains, interspersed with regions of little
homology. This suggests that the aroG and aroF genes may
not be directly derived from a common ancestor but may have
evolved by a combination of different pieces of different
origins. In accord with this hypothesis a high degree of

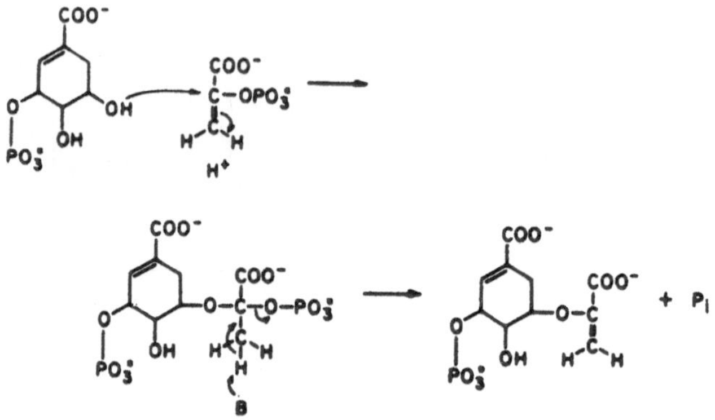

Fig. 11. Reaction mechanism of enolpyruvylshikimate 3-phosphate synthase as proposed by Sprinson and coworkers.[31] Reprinted from Reference 33 with permission of the American Chemical Society.

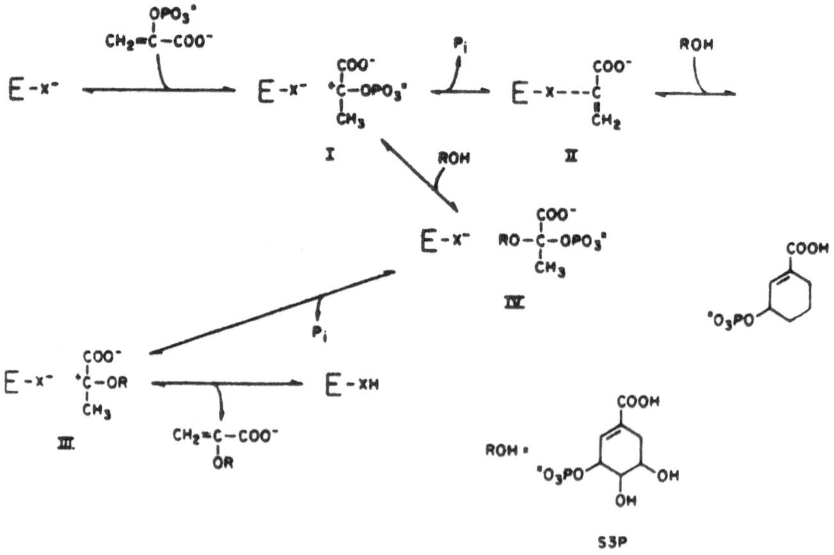

Fig. 12. Reaction mechanism of enolpyruvylshikimate 3-phosphate synthase as proposed by Abeles and coworkers. Reprinted, with modifications, from Reference 33 with permission of the American Chemical Society.

homology was seen between amino acid residues 10 to 18 of
the E. coli DAHP synthase, presumed to be part of the iron
binding region, and residues 54 to 62 of hemerythrin, an
iron binding protein from a seaworm.[29]

A considerable amount of effort has focussed on the
mechanism of action of the enzyme EPSP synthase, spurred by
the finding that glyphosate acts by inhibiting this enzyme.[2]
Amrhein and coworkers have proposed that glyphosate binds
to the enzyme in an orientation largely overlapping the
phosphoenolpyruvate (PEP) binding site[30] (also see Chapter
4 in this volume). Earlier work by the group of Sprinson,[31]
supported by findings by Haslam and coworkers,[32] has led to
postulation of the addition-elimination mechanism for EPSP
synthase shown in Figure 11. Subsequently, Abeles and
coworkers[33] observed that PEP will undergo the enzyme-
catalyzed exchange of the methylene protons in the presence
of an analog of shikimic acid 3-phosphate which lacks the 4
and 5 hydroxyl groups. This observation necessitates a
modification of the mechanism, shown in Figure 12, in which
the addition of the proton to C-3 of PEP is not dependent
on the simultaneous addition of phosphoshikimate to C-2.
Whether compound II (Fig. 12) is an intermediate is not
clear from Abeles' results; in light of stereochemical
studies by Knowles,[34] a path directly from I to IV seems
more likely.

At the meeting of this Society in Ghent[35] in 1978 I
described 3 stereochemical questions in the central part
of the shikimate pathway which at the time were not resolved.
As summarized in Figure 13, these center around the EPSP
synthase reaction, the anthranilate synthase reaction and
the conversion of chorismate into prephenate. These
questions have now been answered through the work of
Knowles, Berchtold and their coworkers[24,25,34] and of our
laboratory.[36,37] Elucidation of these stereochemical
questions is complicated by the fact that the EPSP synthase
reaction proceeds with obligatory formation of a transient
methyl group at C-3 of PEP. Thus, PEP stereospecifically
tritiated in the methylene group will produce EPSP in which
the tritium is evenly distributed between the two methylene
hydrogens. However, if the PEP is labeled stereospecifically with tritium in one hydrogen and deuterium in the
other, such that every tritiated molecule also contains
deuterium, the intermediate methyl group generated will be
chiral. As shown in Figure 14, subsequent abstraction of

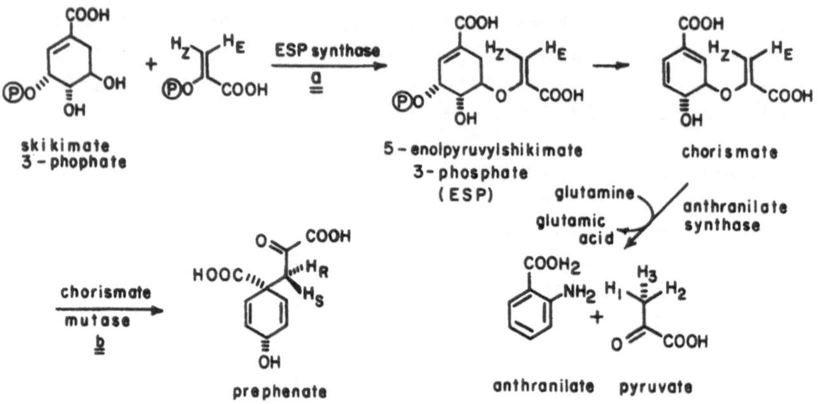

Fig. 13.  Three stereochemical questions in the shikimate
pathway.  Reprinted from Reference 37 with permission of the
American Chemical Society.

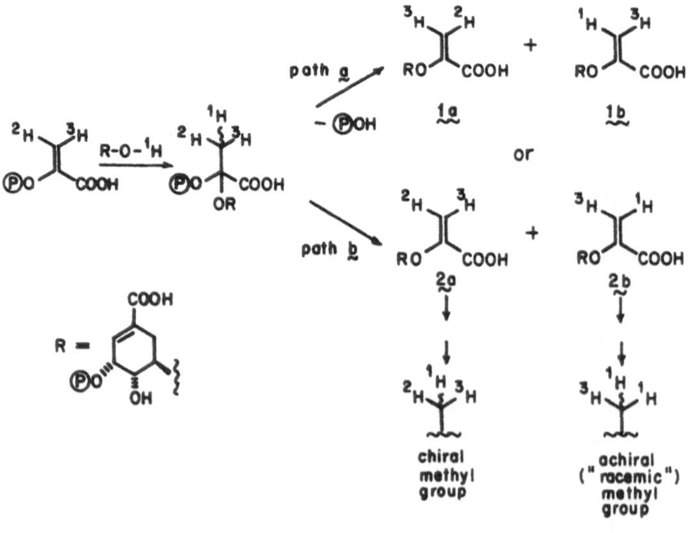

Fig. 14.  Principle of the approach to the analysis of the
steric course of the enolpyruvylshikimate 3-phosphate
synthase reaction.  Reprinted from Reference 37 with permission of the American Chemical Society.

a hydrogen from that methyl group will produce two tritiated species which differ depending on whether the addition
and the elimination step occur with the same stereochemistry, i.e. both syn or both anti, or whether they occur
with opposite stereochemistry, i.e. syn-anti or anti-syn.
The two sets of products can be distinguished if the
methylene group is again converted into a chiral methyl
group by stereospecific addition of a hydrogen.  The
methylene group containing deuterium and tritium will
produce a chiral methyl group upon addition of an unlabeled
hydrogen, whereas the methylene group which is only tritiated will produce an achiral methyl group.  From the
residual chirality of the mixture of tritiated methyl groups
and from the steric course of the hydrogen addition in the
analytical step, the configuration of the deuterated,
tritiated methylene group of EPSP can be determined, and
with that the overall steric course of the enzyme reaction
can be deduced.  This approach was implemented both by
Knowles' group[34] and by our laboratory.[36]

We considered it important to couple the formation
of EPSP tightly to its further conversion to chorismate,
because Haslam's work[32] had shown that EPSP can bind back
to the enzyme and undergo exchange.  Hence we reasoned that
it would be best to carry out the reaction in vivo in a
system of high metabolic flux.  We chose Klebsiella
pneumoniae mutant 62-1 which accumulates large quantities
of chorismic acid for our experimental system.  In view
of the impermeability of bacterial cells to phosphate
esters, this necessitated generating the stereospecifically
labeled PEP within the cells from a suitable precursor.  We
therefore synthesized glycerol labeled stereospecifically
with deuterium and tritium in the pro-R methylene group as
shown in Figure 15.  Ethyl isopropylidene-D-glycerate was
reduced with LiAlD$_4$, followed by the alcohol dehydrogenase-
catalyzed equilibration with [1-$^3$H]ethanol to introduce
tritium into the pro-R position of the hydroxymethyl group.

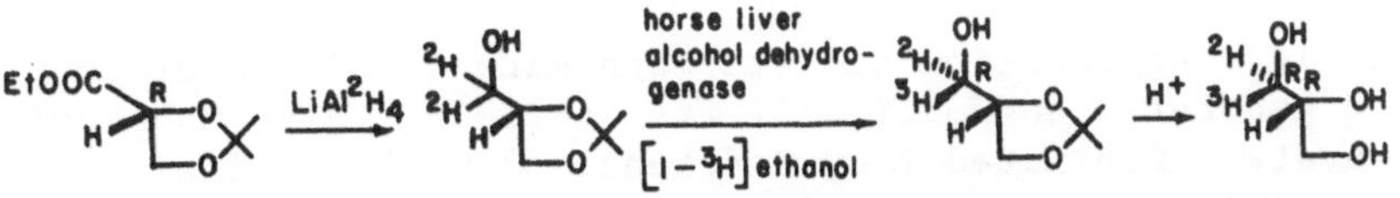

Fig. 15.  Synthesis of (1R,2R)-[$^2$H$_1$,$^3$H]glycerol.

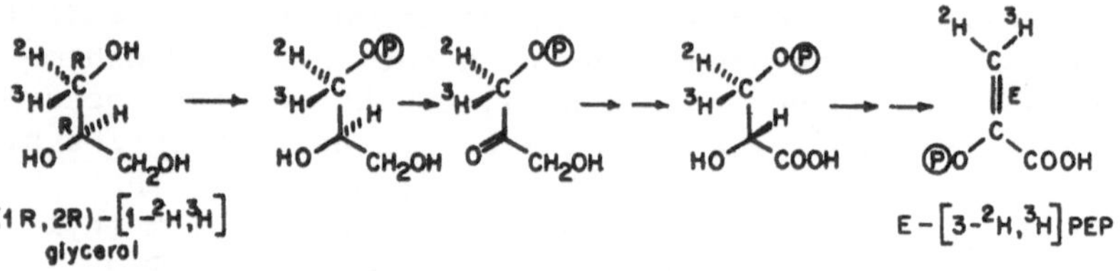

Fig. 16.   Conversion of (1R,2R)-[1-$^2$H$_1$,$^3$H]glycerol into E-[3-$^2$H$_1$,$^3$H]phosphoenolpyruvate.

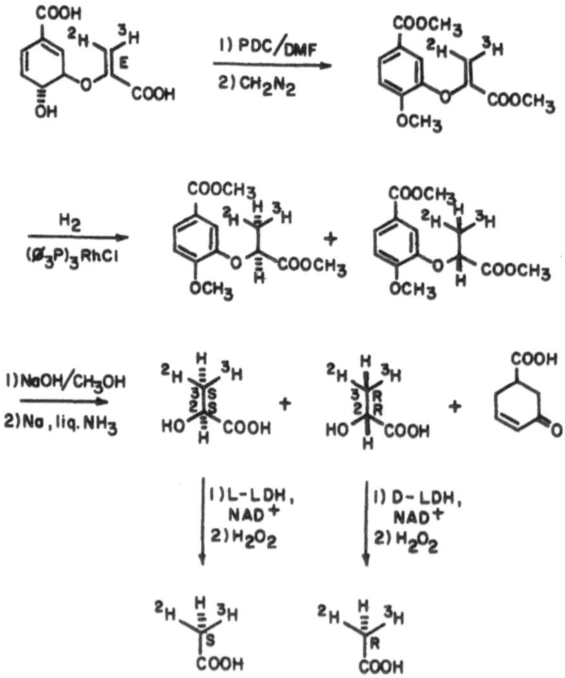

Fig. 17.   Degradation and stereochemical analysis of chorismate generated enzymatically from E-[3-$^2$H$_1$,$^3$H]phosphoenolpyruvate.   Reprinted from Reference 37 with permission of the American Chemical Society.

Deprotection then gave (1R,2R)-[1-$^2$H,$^3$H]glycerol, which
was fed to the cultures of the mutant together with excess
unlabeled shikimic acid.  Unfortunately, it was not
practical to prepare the 1S isomer by the same route, and
an alternate chemical synthesis of both isomers of the
deuterated, tritiated glycerol gave material of low chiral
purity.  Hence, the experiments were carried out only with
the 1R,2R material.  As shown in Figure 16 the labeled
glycerol of 1R,2R configuration will generate deuterated,
tritiated PEP of E configuration within the cells.  This
material is then further converted into chorismic acid
which was isolated and purified.  The degradation of this
chorismate sample to convert the side chain methylene group
stereospecifically into a methyl group is shown in Figure
17.  Aromatization followed by hydrogenation with
Wilkinson's catalyst and final removal of the aromatic
ring gave a sample of D,L-lactate in which the configurations at C2 and C3 are correlated to each other by virtue
of the cis-addition of hydrogen in the hydrogenation step.
Aliquots of this sample were incubated with D-lactate
dehydrogenase and L-lactate dehydrogenase, respectively,
followed by hydrogen peroxide oxidation to give samples of
acetic acid derived either from the L-isomer or the D-
isomer in the racemate.  Chirality analysis indicated F
values of 44.6 (17% e.e. S) from the L-lactate and F = 55.8
(20% e.e. R) from the D-isomer of lactate.  These results
demonstrate that the side chain configuration of the
deuterated, tritiated chorismate formed from E-PEP was
also E.  The EPSP synthase reaction therefore involves
addition and elimination at the 2,3 double bond of PEP
with opposite stereochemistry, i.e., either syn addition –
anti elimination or anti addition – syn elimination.  This
result may seem surprising at first glance, but probably
reflects simply the principle of minimal motion.  As shown
in Figure 18, after the initial addition reaction, a
rotation of 120° is required if the elimination of phosphoric acid is to proceed with the same stereochemistry as
addition of the OH group.  However, if the two steps
proceed with opposite stereochemistry, only a 60° rotation
is necessary.  In the latter case the protonation and
deprotonation could quite conceivably be catalyzed by a
single base which needs to move relatively little in the
catalytic process.  Addition and elimination with the same
stereochemistry more likely require two basic groups to
mediate the proton addition and proton abstraction.  Hence,
addition and elimination with opposite stereochemistry

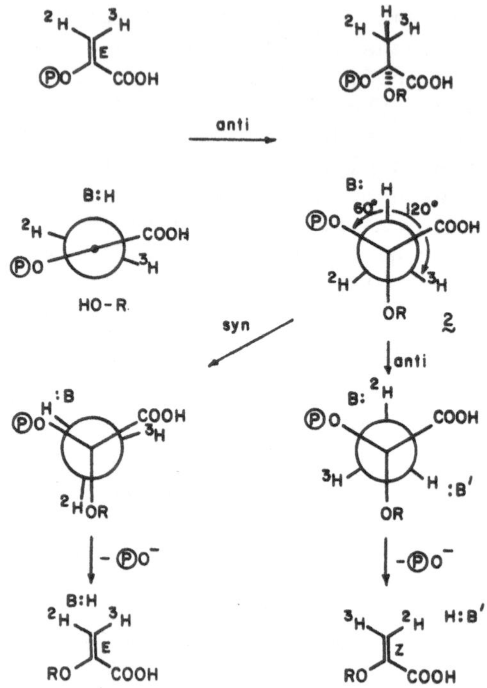

Fig. 18. Stereochemical course of the enolpyruvylshikimate 3-phosphate synthase reaction. Reprinted from Reference 37 with permission of the American Chemical Society.

allows the process to proceed with a minimum of motion and catalysis by a single base, whereas either a major reorientation of functional groups or two bases are required if the two processes occurred with the same stereochemistry.

Having chorismic acid available with a stereospecific label in the side chain methylene group we then proceeded to determine the steric course of the anthranilate synthase reaction. The pyruvate generated in the reaction in the presence of $H_2O$ (Fig. 19) was oxidized to acetate and analyzed for its chirality. An F value of 44.5 (19% e.e. S) indicated that the protonation had occurred on the Re face. This stereochemistry is probably of no mechanistic significance, but it contrasts with that seen in virtually all the reactions in which a proton or other electrophile is added at the methylene carbon of PEP.

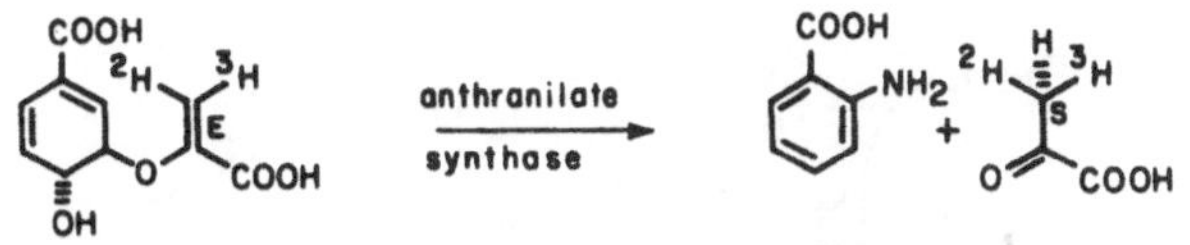

Fig. 19. Steric course of the anthranilate synthase reaction. Reprinted from Reference 37 with permission of the American Chemical Society.

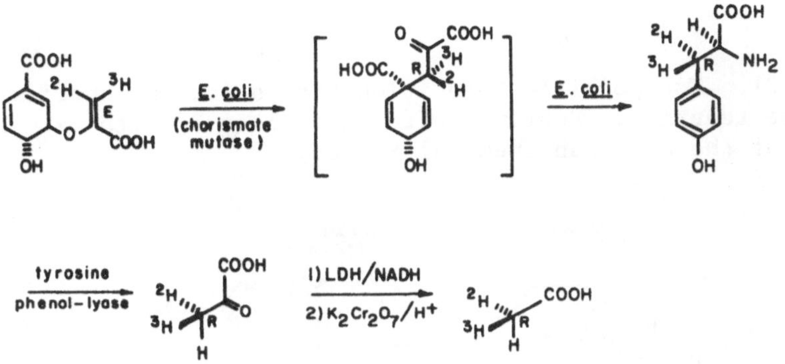

Fig. 20. Stereochemical analysis of the chorismate mutase reaction. Reprinted from Reference 37 with permission of the American Chemical Society.

Finally, to determine the steric course of the chorismate mutase reaction, another portion of the stereospecifically labeled chorismate was fed to a culture of E. coli and the cellular protein was then hydrolyzed to recover phenylalanine and tyrosine. The tyrosine was degraded as shown in Figure 20 to give pyruvate which was further oxidized to acetate for chirality analysis. An F value of 56.3 (22% e.e. R), together with the knowledge that the tyrosine phenol-lyase reaction proceeds with retention of configuration, indicates R configuration at the methylene group of the deuterated, tritiated tyrosine which was formed. The same configuration was deduced for phenylalanine by hydroxylation to tyrosine followed by the same degradation. The results indicate that the two chorismate mutases operating in the phenylalanine and in the tyrosine biosynthetic branch operate with the same stereochemistry. The observed steric course points to a chair-like transition state for the chorismate mutase reaction rather than a boat

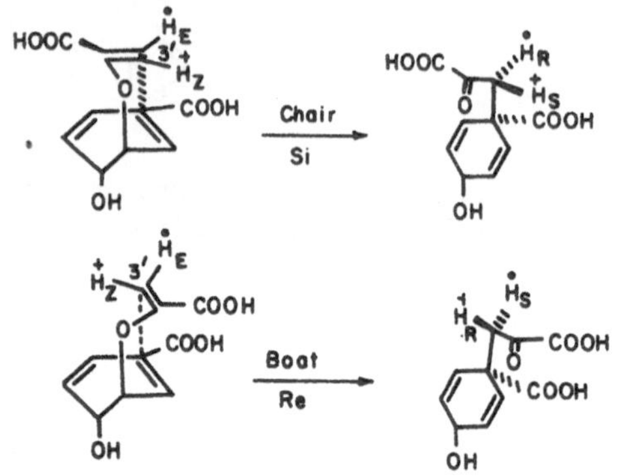

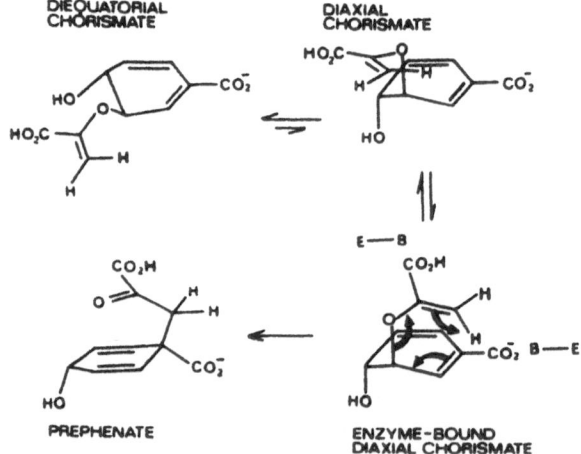

Fig. 21. Two possible transition states for the chorismate mutase reaction. Reprinted from Reference 37 with permission of the American Chemical Society.

Fig. 22. Conformational changes imposed upon chorismate by the enzyme during the chorismate mutase reaction. Reprinted from Reference 67 with permission of Elsevier Biochemical Press.

transition state (Fig. 21). The same transition state geometry had already been suggested earlier based on experiments with transition state analogs.[38] A major role of the enzyme in achieving the observed rate acceleration for the conversion of chorismate to prephenate seems to be the

conformational reorientation of the substrate from the lowest
energy, equatorial conformation to the diaxial one.   This is
a process requiring approximately 7 kcal/mole (Fig. 22).[39]
In addition the enzyme presumably also maintains the proper
side chain orientation for a chair-like transition state.
In accordance with this assumption, it was observed by
Berchtold and coworkers[16] that the non-enzymatic rearrangement of epi-chorismic acid to epi-prephenic acid is
substantially faster than that of chorismic acid to
prephenic acid.

     Another enzyme which has been studied in considerable
detail is tryptophan synthase which catalyzes the conversion
of indoleglycerol 3-phosphate into tryptophan.   The mechanism
of this reaction[40] and that catalyzed by the related enzyme
tryptophanase[41] involves Schiff's base formation with an
enzyme-bound pyridoxal phosphate (PLP).   As shown in Figure
23, initial elimination of the α hydrogen in both reactions
followed by removal of the β substituent generates the PLP
Schiff's base of α-aminoacrylate.   This species can then
undergo either hydrolysis to pyruvate and ammonia or addition
of a different β substituent to generate tryptophan.   In
collaboration with Miles at the NIH we[42] provided the first
chemical evidence for the existence of this PLP-aminoacrylate
Schiff's base by reduction of the enzyme-serine complex with
tritiated sodium borohydride.   The reaction produced phosphopyridoxylalanine with the tritium distribution shown in
Figure 24.   Another postulate regarding the tryptophan
synthase and tryptophanase reactions was that the addition
or removal of the β-indolyl group should proceed via an
indolenine intermediate as shown in Figure 25.   This notion
was tested by Miles and coworkers[43] by preparing and evalu-
ating 2,3-dihydrotryptophan as an inhibitor of these enzymes.
Both tryptophanase and tryptophan synthase were effectively
inhibited by the mixture of diastereomers with L configuration in the side chain.   In subsequent work, these authors[44]
separated the two diastereomers and tested each pure
stereoisomer against the two enzymes.   Remarkably, it was
found that the 3S isomer is an effective inhibitor of tryptophan synthase, but not of tryptophanase, whereas the 3R
isomer inhibits tryptophanase but not tryptophan synthase.
These results indicate that the tryptophan synthase reaction
and the tryptophanase reaction proceed through indolenine
intermediates of opposite stereochemistry, as shown in
Figure 26.

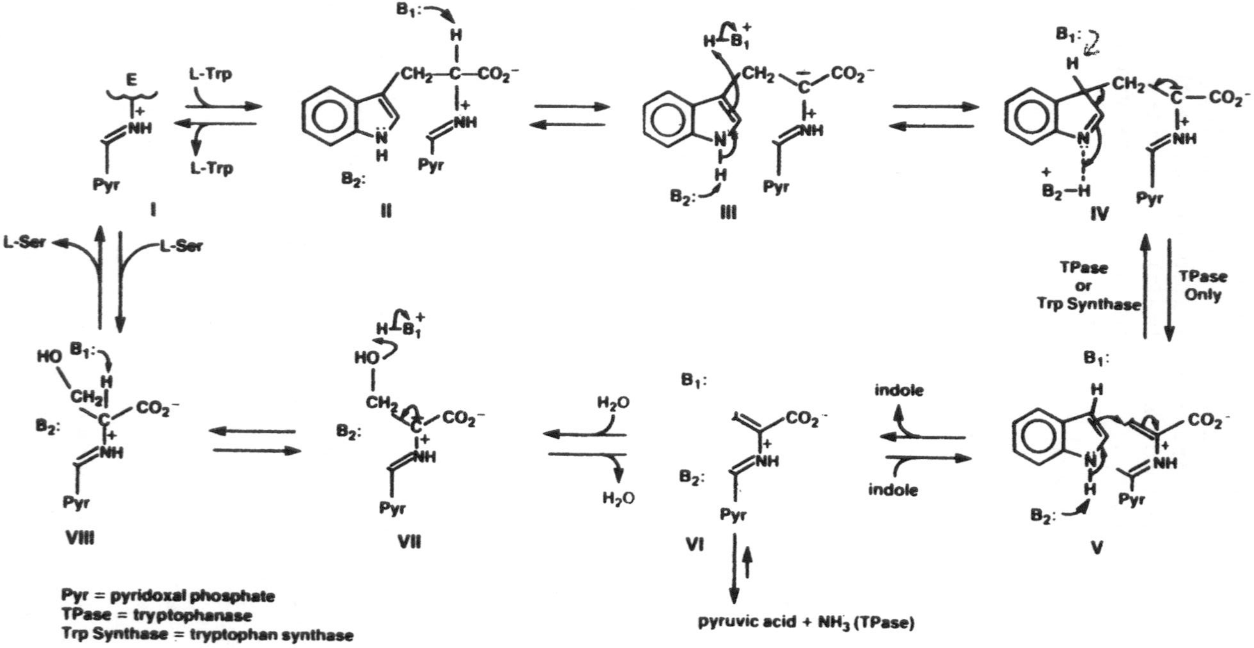

Fig. 23. Mechanism of reaction of tryptophanase and tryptophan synthase. Reprinted from Reference 43 with permission of the American Chemical Society.

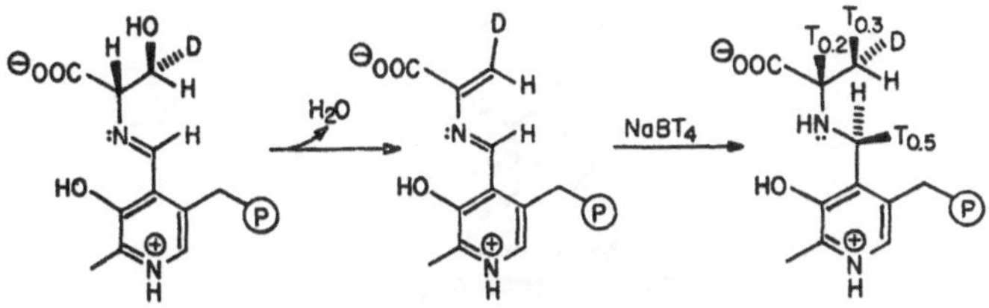

Fig. 24. Tritium distribution in phosphopyridoxylalanine obtained upon [³H]NaBH₄ reduction of the tryptophan synthase-L-serine complex. Reprinted from Reference 42 with permission of the American Society of Biological Chemists.

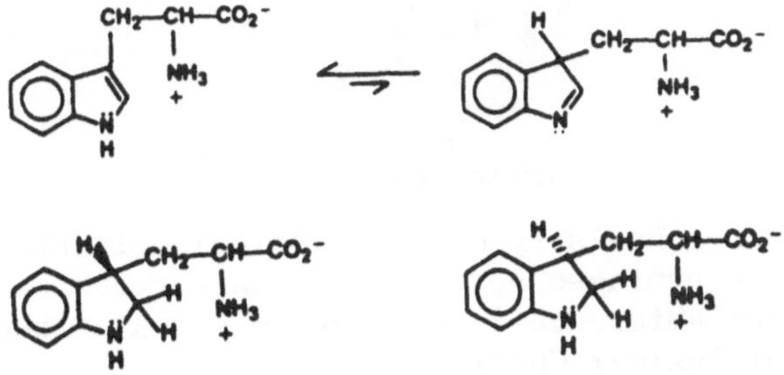

Fig. 25. The indolenine intermediate in the tryptophanase and tryptophan synthase reactions, and the two diastereomers of the inhibitor 2,3-dihydrotryptophan. Reprinted from Reference 43 with permission of the American Chemical Society.

HIGHER PLANT PRODUCTS DERIVED FROM THE SHIKIMATE PATHWAY

A substantial amount of work has been reported in the last few years on products derived from the shikimate pathway in higher plants. Limitations of time and space permit a review of only a few of the interesting results obtained.

Following the earlier isolation and characterization of chalcone synthase, the key enzyme in flavonoid biosynthesis, by Hahlbrock's group, the laboratory of Kindl[45] has now reported the isolation and purification of an analogous enzyme involved in the biosynthesis of stilbenes in higher

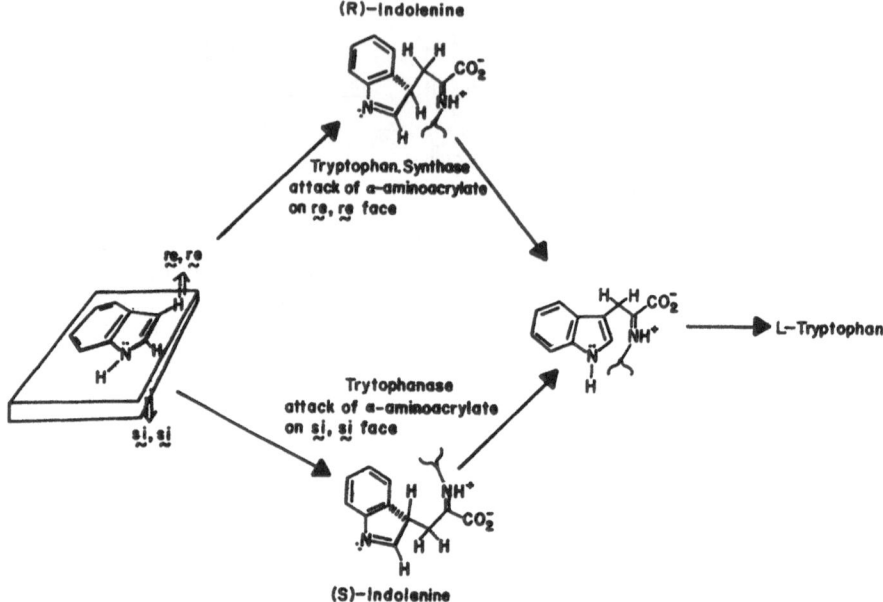

Fig. 26.   Stereochemistry of the indolenine intermediates in
the tryptophan synthase and tryptophanase reactions.
Reprinted from Reference 44 with persmission of the American
Society of Biological Chemists.


plants.   In both cases, the enzyme utilizes the CoA ester of
a cinnamic acid as a starter for a polyketide chain which is
extended by the addition of three molecules of malonyl-CoA
(Fig. 27).   Presumably the conformation imposed upon the
substrate by the particular protein matrix determines that
the cyclization of the initial enzyme-bound polyketide will
produce either a chalcone or a stilbene such as resveratrol.
Stilbene synthase and chalcone synthase are clearly distinct
and different enzymes which show no antigenic crossreacti-
vity.[45,46]   However, their molecular architecture seems to
be somewhat similar in that they both are dimeric proteins
of molecular weight around 90,000.[45]

Hahlbrock and coworkers have continued their pace-
setting studies on the induction of the enzymes of flavonoid
biosynthesis.[47]   Following the earlier demonstration of
coordinate appearance of the enzyme activities in parsley

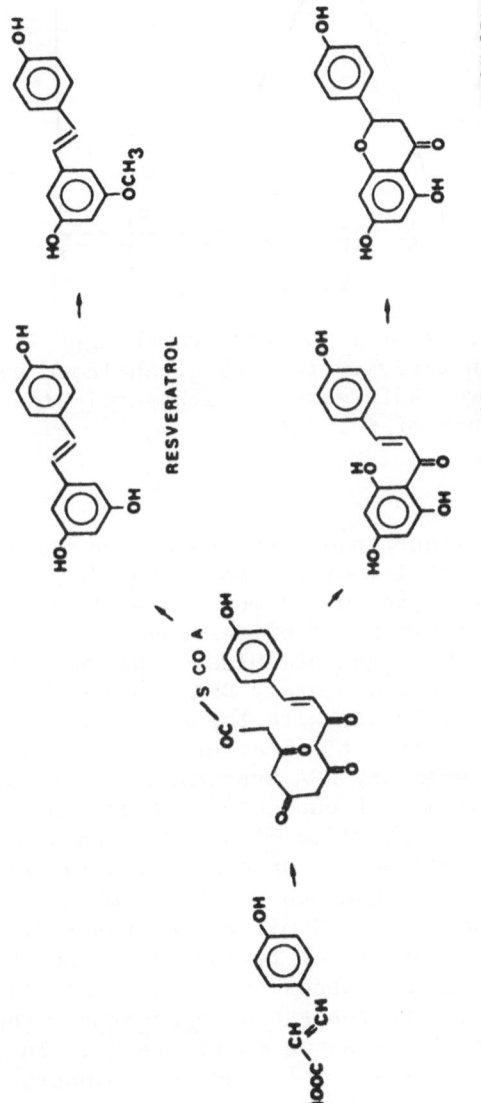

Fig. 27. Reactions catalyzed by stilbene synthase and by chalcone synthase. Reprinted, with modifications, from Reference 46 with permission of the American Society of Plant Physiology.

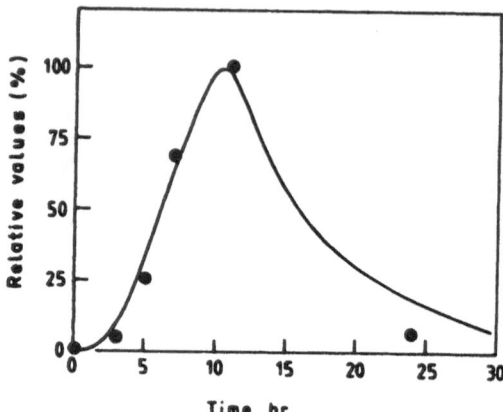

Fig. 28. Time courses of light-induced changes in amount
(·) and translation activity (curve) of chalcone synthase
mRNA. Reprinted from Reference 48 with permission of the
U.S. National Academy of Science.

cell cultures upon light induction and of the increase in
the translational activity of the corresponding messenger
RNAs, these workers[48] have now demonstrated that the process
involves the de novo synthesis of messenger RNA for chalcone
synthase following the light stimulus. This was shown by
preparing a labeled, complementary DNA probe which was used
in hybridization experiments with the mRNA fraction of
parsley cells at different times after onset of illumination.
The amount of complementary DNA hybridized to the messenger
RNA indicates the amount of chalcone synthase mRNA present
at any given time. As shown in Figure 28, the amount of
specific mRNA for chalcone synthase nicely parallels the
translation activity for chalcone synthase mRNA measured
under analogous conditions. Similar mechanisms seem to
operate in the induction of the formation of the phytoalexin
glyceollin in soybean in response to a chemical elicitor, a
glucan from the fungus Phytophthora megasperma. The pathway
of glyceollin synthesis is shown in Figure 29. In vitro
translation of messenger RNA followed by immunoprecipitation
indicated increases in messenger RNA for phenylalanine
ammonia-lyase and chalcone synthase following induction with
the elicitor, which parallel the increases in enzyme activity
and in phytoalexin accumulation.[49]

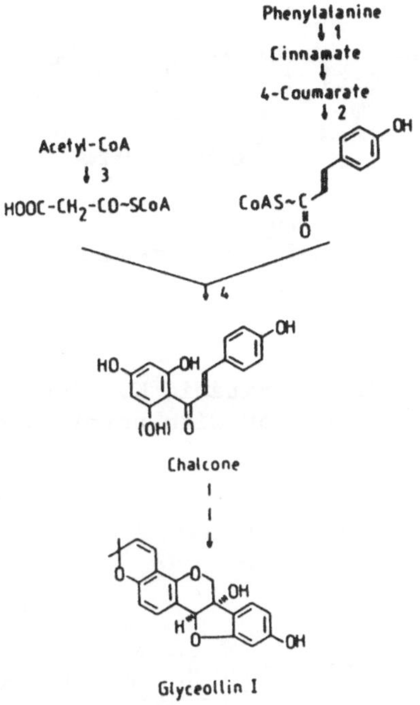

Fig. 29. Proposed biosynthetic pathway of the soybean phytoalexin, glyceollin I. Reprinted from Reference 49 with permission of Academic Press Inc.

BIOSYNTHESIS OF VITAMIN K

A rather unique branch of the shikimate pathway operates in the biosynthesis of naphthoquinones related to vitamin K (menaquinone)[50,51] (Fig. 30). Seven of the ten carbon atoms of the naphthoquinone ring system are derived from the seven carbon atoms of shikimic acid. The remaining three carbons are provided by the three center carbon atoms of α-ketoglutaric acid in a reaction leading to the unique intermediate o-succinylbenzoic acid. Cyclization of the latter produces the intermediate 1,4-dihydroxynaphthoic acid, the substrate for an isoprenylation reaction which occurs with simultaneous loss of $CO_2$. Methylation of the 3 position then completes the reaction sequence. Studies by Meganathan and Bentley[50-52] had indicated that chorismic acid is the substrate for the thiamine pyrophosphate-

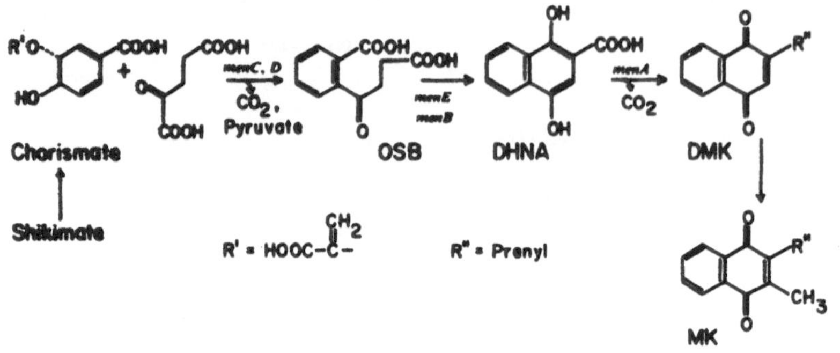

Fig. 30. Biosynthesis of vitamin K. Reprinted, with modifications, from Reference 50 with permission of the American
Society of Microbiology.

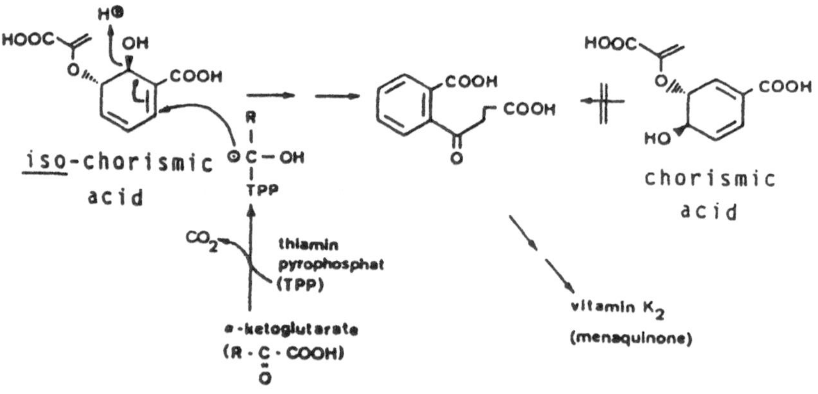

Fig. 31. Formation of o-succinylbenzoic acid from iso-
chorismic acid. Reprinted from Reference 53 with permission
of Pergamon Press Ltd.

catalyzed condensation with α-ketoglutaric acid. However,
recent work by Leistner and coworkers[53] has provided
unequivocal evidence showing that the true substrate for
the reaction is iso-chorismic acid (Fig. 31). The earlier
findings seem to be due to contaminations in the commercial
samples of chorismic acid used. Another question addressed
by Leistner and his group was the activation of the intermediate o-succinylbenzoic acid. Earlier work[54] had

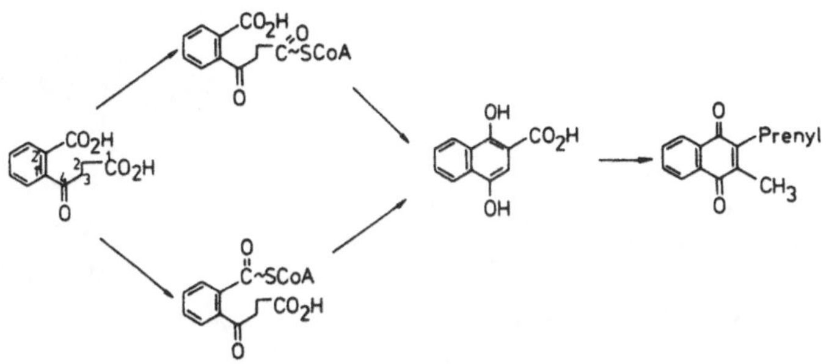

Fig. 32.  Two possible modes of activation of o-succinylbenzoic acid.  Reprinted from Reference 54 with permission of
Federation of European Biochemical Societies.

suggested that the aromatic carboxyl group is activated via
the CoA ester.  However, unequivocal synthesis of both
thioesters and comparison as substrates in the reaction
showed that activation occurs on the aliphatic carboxyl
group (Fig. 32).[55]  These results are discussed in detail
in Chapter 9 by E. Leistner in this volume.

## MICROBIAL METABOLITES OF THE SHIKIMATE PATHWAY

A large number of microbial metabolites are derived in
part or in toto from the shikimate pathway.  Rather promi-
nent among the particular structural moieties found in
antibiotics is a $C_7N$ unit found in most of the ansa
macrolide antibiotics[56] and in other compounds, such as the
mitomycins.[57]  This moiety, consisting of a six-membered
carbocyclic ring with an attached extra carbon and nitrogen
in meta orientation, has been found in many feeding experiments to show a labeling pattern from, for example,
carbohydrate precursors, which clearly matches that of
shikimic acid and shikimate-derived compounds.  On the other
hand, shikimic acid is clearly not incorporated into this
$C_7N$ unit, nor is, in the only case tested, dehydroquinic
acid or dehydroshikimic acid.  The results suggest that the
pathway to the $C_7N$ unit branches off somewhere prior to
shikimic acid.  This conclusion is supported by genetic
experiments with a rifamycin-producing organism.[58]  Based
on a structural comparison of the various $C_7N$ units,

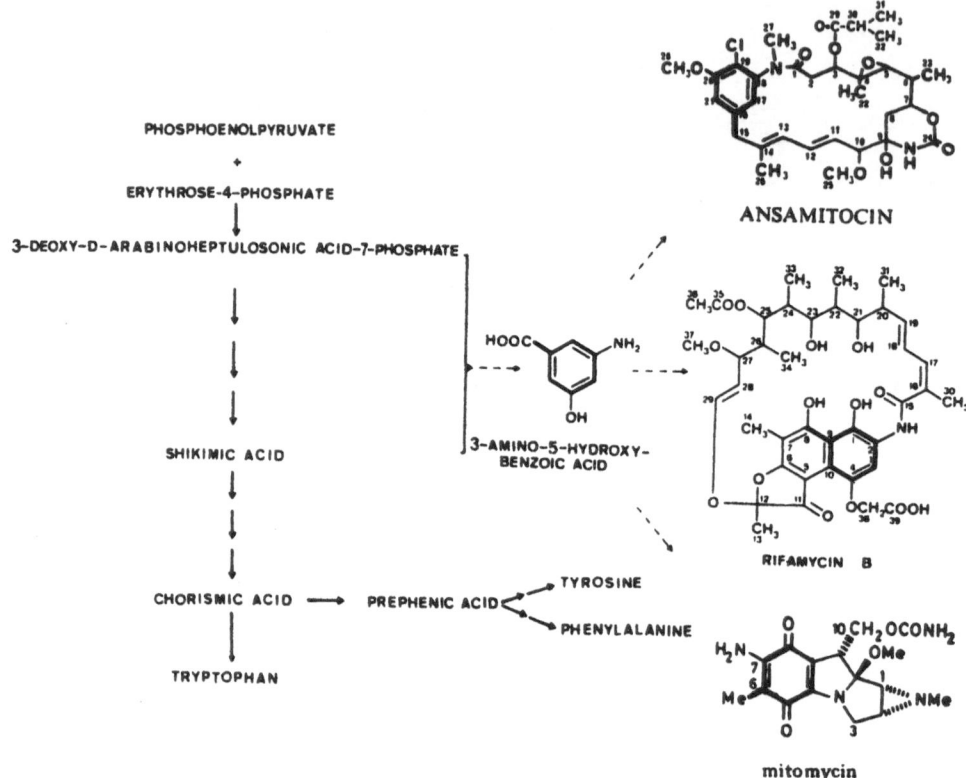

Fig. 33.   The C₇N unit of ansa macrolide and mitomycin
antibiotics and its origin from the shikimate pathway.
Reprinted, with modifications, from Reference 68 with permission of the Federation of European Microbiological Socieites.

Richards and coworkers[50],[60] concluded that the precursor
should be 3-amino-5-hydroxybenzoic acid (Fig. 33).   They
synthesized this compound in labeled form and demonstrated
its efficient and specific incorporation into the major
classes of antibiotics containing a C₇N unit, rifamycin,
mitomycin and the ansamitosins (Fig. 33).   From our own
laboratory,[61] we can add two additional compounds, naphtho-
mycin (Fig. 34) and ansatrienin (Fig. 35), to these
examples.   Both of these compounds, as the spectra show,
are efficiently and specifically labeled from 3-amino-5-
hydroxy-[carboxy-¹³C]benzoic acid.   Studies by Hornemann's

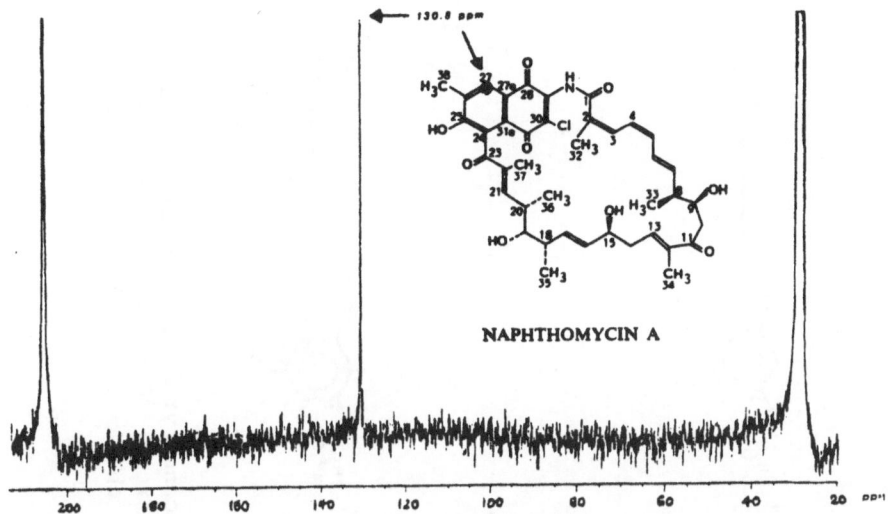

Fig. 34. Proton noise-decoupled $^{13}$C-NMR spectrum of naph-
thomycin biosynthetically formed from 3-amino-5-hydroxy-
[7-$^{13}$C]benzoic acid (from Reference 61).

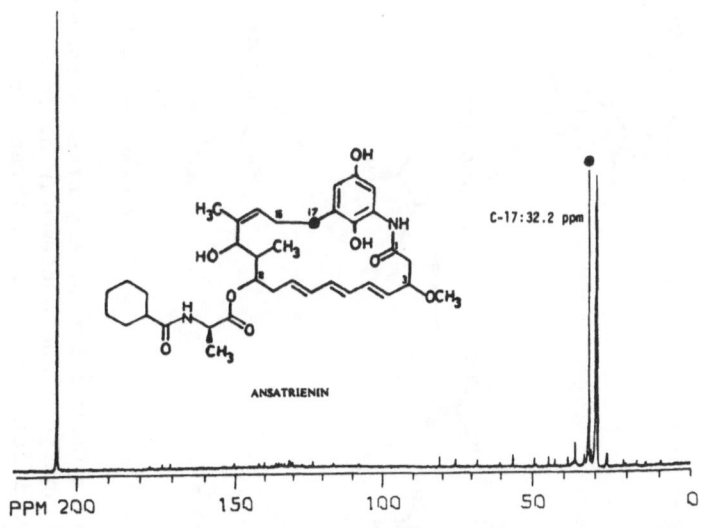

Fig. 35. Proton noise-decoupled $^{13}$C-NMR spectrum of ansatrienin B biosynthetically formed from 3-amino-5-hydroxy-
[7-$^{13}$C]benzoic acid (from Reference 61).

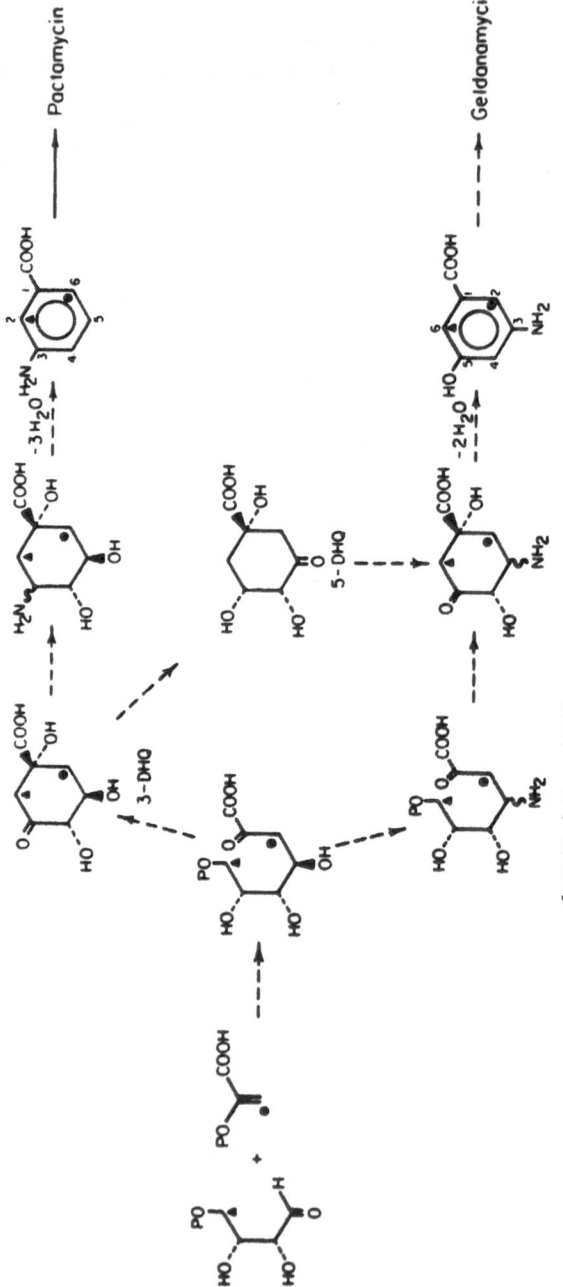

Fig. 36. Conversion of erythrose 4-phosphate and phosphoenolpyruvate into the C₇N units of geldanamycin and pactamycin derived biosynthetically from [¹³C]acetate. Reprinted, with modifications, from Reference 62 with permission of the American Chemical Society.

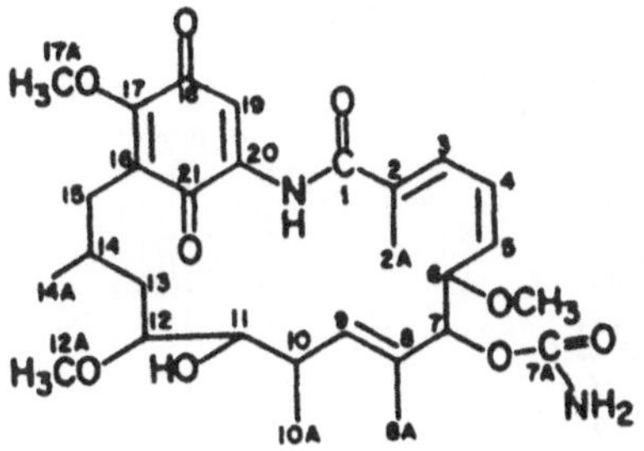

Geldanamycin 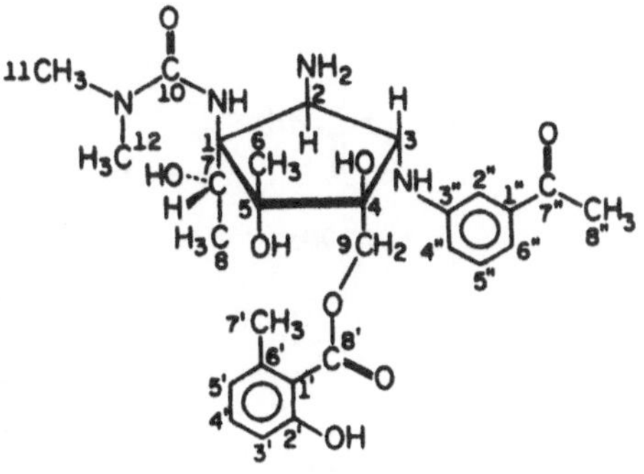                    Pactamycin

Fig. 37.  Structures of geldanamycin and pactamycin.
Reprinted from References 62 and 63 with permission of the
American Chemical Society.

group[57] on mitomycin and by Rinehart's group[62] on geldanamycin have defined the orientation of the nitrogen
substituent relative to the origin of the carbon skeleton
of the $C_7N$ unit.  Surprisingly, the nitrogen is attached to
the carbon which corresponds to carbon 5 of shikimic acid,
not carbon 3 which, after the initial cyclization, is left
at the oxidation state of a carbonyl group (Fig. 36).  This
led Hornemann[57] to suggest a 4-deoxy-4-amino analog of DAHP
as an intermediate.

On the other hand, a $C_7N$ unit found in pactamycin
(Fig. 37), a 3-aminoacetophenone moiety, is derived specifically from 3-aminobenzoic acid,[56] the methyl group being
provided by methionine.  The orientation of the nitrogen in
this case is such that it is attached to the carbon corresponding in its origin to that of C-3 of shikimic acid.[63]
Hence, not all $C_7N$ units are alike.  A pathway to the $C_7N$
units in pactamycin and in geldanamycin as proposed by
Rinehart and coworkers[62] is shown in Figure 36.  To compli-
cate matters further, not all $C_7N$ units are even derived
from the shikimate pathway.  In collaboration with the group
of Omura we have studied the formation of the $C_7N$ unit in
asukamycin and, in collaboration with the group of Zeeck, in
the related manumycin.[64]  In both cases 3-amino-5-hydroxy-
[carboxy-[13]C]benzoic acid is not incorporated at all into
the antibiotic (Fig. 38).  Feeding experiments on asukamycin

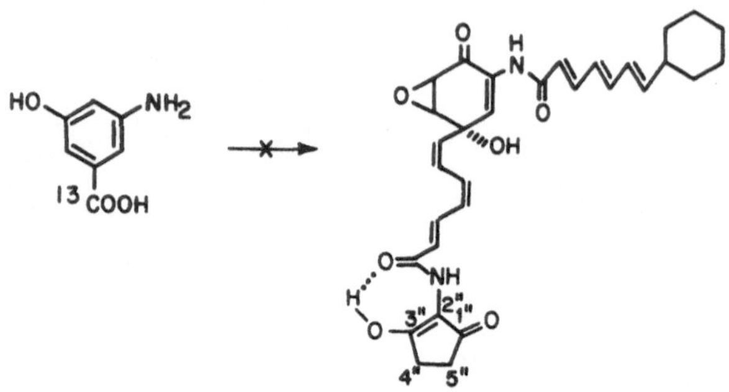

Fig. 38. Non-incorporation of 3-amino-5-hydroxy-[7-$^{13}$C] benzoic acid into the C$_7$N unit of asukamycin.

[U-$^{13}$C] Glycerol - C$_7$N Unit

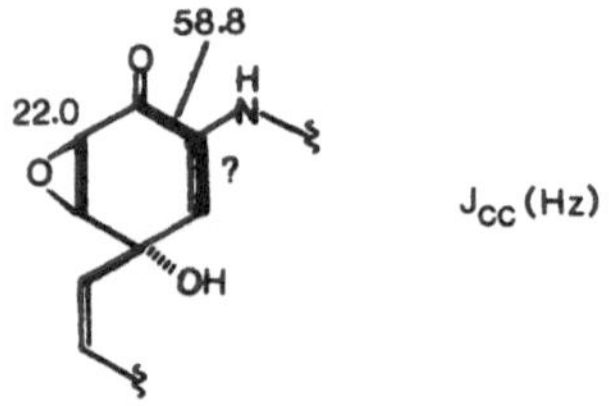

Fig. 39. $^{13}$C-$^{13}$C Coupling pattern in the C$_7$N unit of asukamycin derived biosynthetically from [U-$^{13}$C$_3$]glycerol.

with [U-$^{13}$C$_3$]glycerol (Fig. 39) and with [1-$^{13}$C]- and [1,2-$^{13}$C$_2$]-acetate (Fig. 40) clearly produced labeling and coupling patterns which bear no relationship to the shikimate pathway. Obviously, the mode of formation of this C$_7$N unit requires further study, as do the details of synthesis of 3-amino-5-hydroxybenzoic acid from the shikimate pathway.

Asukamycin contains a second six-membered carbocyclic ring which is indeed derived from shikimic acid. The cyclohexane ring at the end of a short polyketide chain and its attached carbon are labeled by shikimic acid as well as by [carboxy-$^{14}$C]cyclohexanecarboxylic acid (Fig. 41). The same is true for the cyclohexanecarboxylic acid moiety

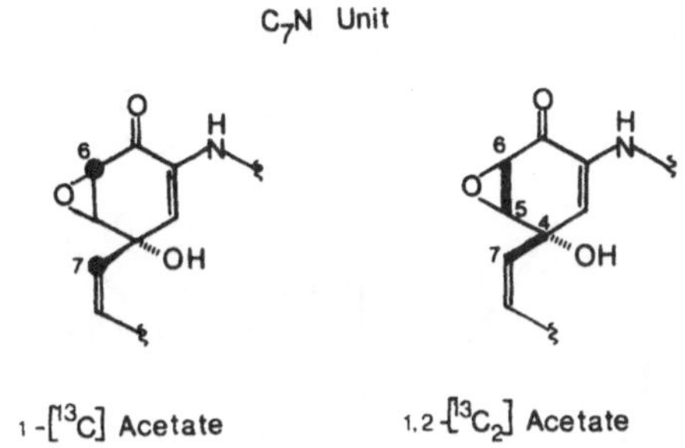

C$_7$N  Unit

1-[$^{13}$C] Acetate                1,2-[$^{13}$C$_2$] Acetate

Fig. 40.   $^{13}$C-Labeling and $^{13}$C-$^{13}$C coupling pattern in the C$_7$N unit of asukamycin derived biosynthetically from [$^{13}$C] acetate.

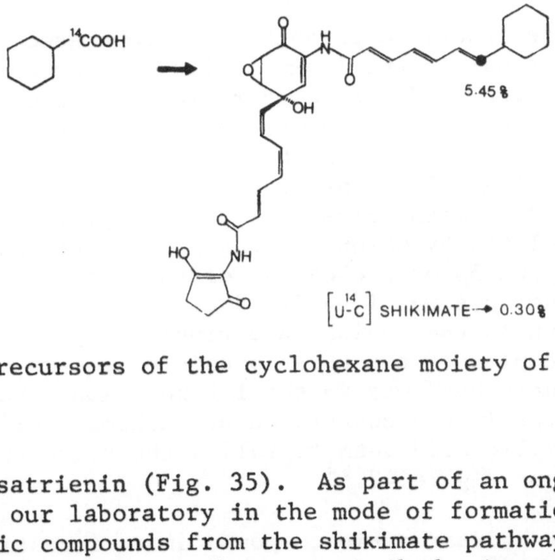

Fig. 41.   Precursors of the cyclohexane moiety of asukamycin.

found in ansatrienin (Fig. 35). As part of an ongoing interest in our laboratory in the mode of formation of hydroaromatic compounds from the shikimate pathway of aromatic biosynthesis, we have examined the incorporation of various cyclohexanecarboxylic acids into ansatrienin (Fig. 42). Interestingly, 2,5-dihydrophenylalanine, a known metabolite of Streptomyces, is not incorporated into ansatrienin, in accordance with the finding that all seven carbon atoms of shikimic acid are retained in the conversion into the cyclohexanecarboxylic acid moiety. Obviously, the pathway requires a number of dehydration and reduction steps. In order to try to establish the sequence of the terminal steps, we fed 2,5-dihydrobenzoic acid and found it to be an

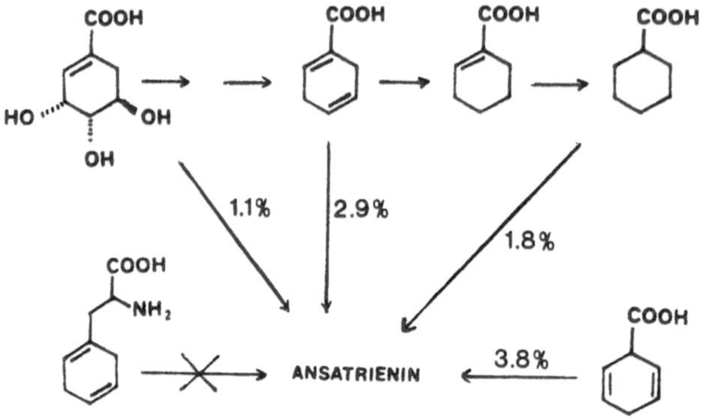

Fig. 42.  Precursors of the cyclohexanecarboxylic acid
moiety of ansatrienin.

efficient precursor, comparable to cyclohexanecarboxylic
acid.  Unfortunately the isomeric 1,4-dihydrobenzoic acid
was also incorporated with comparable efficiency.  Obviously,
at best only one of these compounds can be on the normal
pathway.  Interestingly, in the course of experiments in
which the cyclohexylcarbonylalanine moiety of ansatrienin
was recovered by methanolysis as the methyl ester, which
was then analyzed by GC/MS, it was found that the product
contained about 5% of a closely related compound.  By its
mass spectrum this was identified as the analog containing
a double bond in the cyclohexane ring.  Comparison with
authentic samples of the three possible isomers established
that the double bond was in the 1,2 position.  Hence, the
terminal steps in the conversion of shikimic acid to cyclohexanecarboxylic acid seem to follow the sequence shown in
the top row of Figure 42.[64]

Another hydroaromatic compound under investigation in
our laboratory[65] is the antibiotic ketomycin.  Again,
shikimic acid was efficiently incorporated, but degradation
showed that in this case only the six ring carbons are
derived from shikimic acid (Fig. 43).  In agreement with
this result it was found that both chorismic acid and
prephenic acid were efficiently incorporated into ketomycin.
This result in turn raised the question whether prephenic
acid, a symmetrical intermediate, is processed asymmetrically to generate the ring of ketomycin.  This question was
answered by degrading a sample of ketomycin produced from

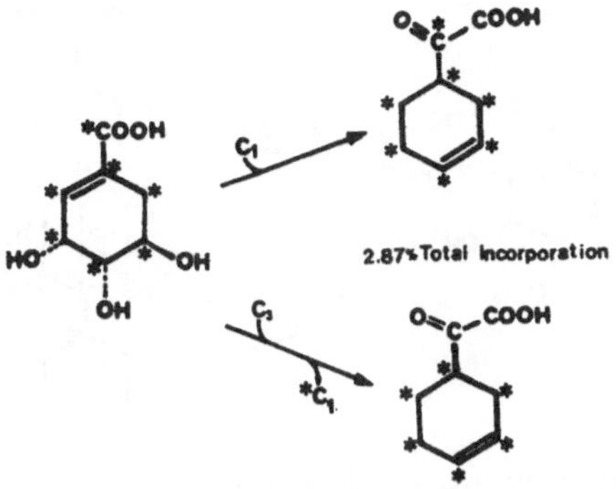

Fig. 43. Mode of incorporation of shikimic acid into keto-mycin.

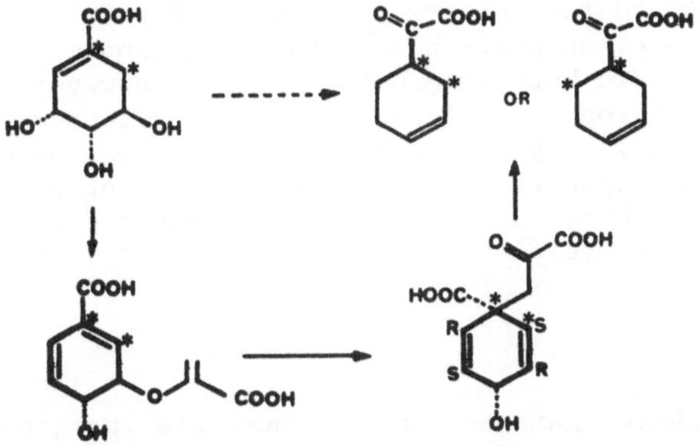

Fig. 44. Stereochemical alternatives in the conversion of shikimate into ketomycin via chorismate and prephenate.

[1,6-$^{14}$C]shikimic acid. As shown in Figure 44 the label may reside either in positions 1 and 2 or in positions 1 and 6 or it may be scrambled between all three positions. The degradation produced succinic acid from carbon atoms 4, 5, 6 and 1, which could contain either 50%, 100% or 75% of the radioactivity of ketomycin depending on the three possible distributions of label. The experimentally observed values of 51% and 46% in two independent degradations clearly indicate that prephenic acid is processed asymmetrically

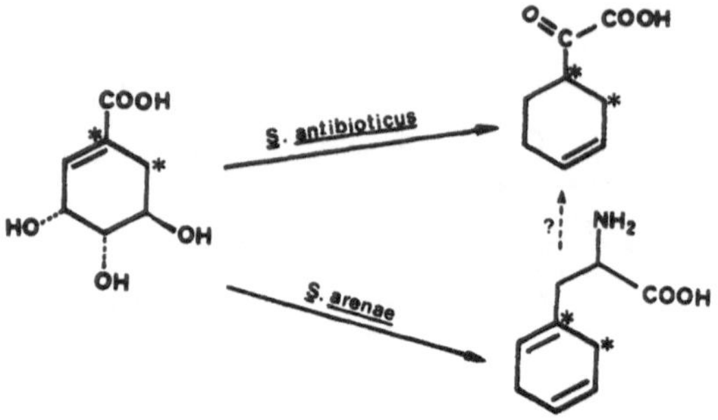

Fig. 45.   Steric course of the conversion of $[1,6-^{14}C]$ shikimate into ketomycin and 2,5-dihydrophenylalanine.

and that carbon 6 of shikimate gives rise to carbon 2 of the ring of ketomycin (Fig. 45).  This labeling pattern corresponds to that observed earlier[66] for 2,5-dihydro-phenylalanine, suggesting that this compound may be an intermediate in ketomycin biosynthesis.  However, when labeled dihydrophenylalanine was fed to Streptomyces anti-bioticus, no incorporation into ketomycin was observed. It therefore appears that yet another variant of the shikimate pathway must exist which accounts for the elaboration of ketomycin.

CONCLUSION

     The above examples amply demonstrate the great variety of products formed from the shikimate pathway and the multitude of variants of the pathway which lead to their formation.  Much has been learned about the pathway in recent years, but much more even remains to be unraveled.  It seems safe to predict that the shikimate pathway of aromatic biosynthesis will continue to constitute a fruitful topic for future symposia of the Phytochemical Society.

ACKNOWLEDGMENTS

     The author acknowledges with gratitude the enthusiastic collaboration of a number of able coworkers, whose names appear in the pertinent references.  I am equally pleased to acknowledge collaboration with a number of laboratories around the world on topics of mutual interest relating to the shikimate pathway. Our work was supported by grants from the National Institutes of Health.

REFERENCES

1. SWAIN, T., J.B. HARBORNE, C.F. VAN SUMERE. 1979. Biochemistry of plant phenolics. In Recent Advances in Phytochemistry. Vol. 12, Plenum Press, New York.
2. STEINRÜCKEN, H.C., N. AMRHEIN. 1980. The herbicide glyphosate is a potent inhibitor of 5-enolpyruvyl-shikimic acid 3-phosphate synthase. Biochem. Biophys. Res. Commun. 94: 1207-1212.
3. HASLAM, E. 1974. The Shikimate Pathway. Butterworths, London, 316 pp.
4. WEISS, U., J.M. EDWARDS. 1980. The Biosynthesis of Aromatic Compounds. Wiley, New York, 728 pp.
5. DANISHEFSKY, S., J. MORRIS, C.A. CLIZBE. 1981. Total synthesis of pretyrosine (arogenate). J. Am. Chem. Soc. 103: 1602-1604.
6. DANISHEFSKY, S., M. HIRAMA, N. FRITSCH, J. CLARDY. 1979. Synthesis of disodium prephenate and disodium epiprephenate. Stereochemistry of prephenic acid and an observation on the base-catalyzed rearrangements of prephenic acid to p-hydroxyphenyllactic acid. J. Am. Chem. Soc. 101: 7013-7018.
7. FROST, J.W., J.R. KNOWLES. 1984. 3-Deoxy-D-arabino-heptulosonic acid 7-phosphate: Chemical synthesis and isolation from Escherichia coli auxotrophs. Biochemistry 23: 4465-4469.
8. KOREEDA, M., M.A. CIUFOLINI. 1982. Natural product synthesis via allylsilanes. 1. Synthesis and reactions of (1E,3E)-4-acetoxy-1-(trimethylsilyl)-1,3-butadiene and its use in the total synthesis of (-)-shikimic acid. J. Am. Chem. Soc. 103: 2308-2310.
9. COBLENS, K.E., V.B. MURALIDHARAN, B. GANEM. 1982. Shikimate-derived metabolites. 12. Stereocontrolled total synthesis of shikimic acid and 6-β-deuterioshikimate. J. Org. Chem. 47: 5041-5042.
10. FLEET, G.W., T.K.M. SHING. 1983. An entry to chiral cyclohexanes from carbohydrates: A short, efficient and enantiospecific synthesis of (-)-shikimic acid from D-mannose. J. Chem. Soc., Chem. Commun. 849-850.

11. CAMPBELL, M.M., A.D. KAYE, M. SAINSBURY, R. YAVARZADEH. 1984. Brief synthesis of (±)-methyl shikimate, (±)-methyl epishikimate and structural variants. Tetrahedron 40: 2461-2470.

12. RAJAPAKSA, D., B.A. KEAY, R. RODRIGO. 1984. Shikimic acids from furan: Methods of stereocontrolled access to 3,4,5-trioxygenated cyclohexenes. Can. J. Chem. 62: 826-827.

13. BARTLETT, P.A., L.A. McQUAID. 1984. Total synthesis of (±)-3-phosphoshikimic acid. J. Am. Chem. Soc. 106: 7854-7860.

14. TENG, C.-Y.P., Y. YUKIMOTO, B. GANEM. 1985. Shikimate-derived metabolites. 14. Chiral synthesis of 5-enolpyruvyl-shikimate 3-phosphate. Tetrahedron Lett. 21-24.

15. McGOWAN, C.A., G.A. BERCHTOLD. 1982. Total synthesis of racemic chorismic acid and (-)-5-enolpyruvylshikimic acid ("Compound Z"). J. Am. Chem. Soc. 104: 7036-7041.

16. HOARE, J.H., P.P. POLICASTRO, G.A. BERCHTOLD. 1983. Improved synthesis of racemic chorismic acid. Claisen rearrangement of 4-epi-chorismic acid and dimethyl 4-epi-chorismate. J. Am. Chem. Soc. 105: 6264-6267.

17. GANEM, B., N. IKOTA, V.B. MURALIDHARAN, W.S. WADE, S.D. YOUNG, Y. YUKIMOTO. 1982. Total synthesis of (±)-chorismic acid. J. Am. Chem. Soc. 104: 6787-6788.

18. BUSCH. F.R., G.A. BERCHTOLD. 1983. Total synthesis of racemic isochorismic acid. J. Am. Chem. Soc. 105: 3346-3347.

19. TENG, C.-Y.P., B. GANEM. 1984. Shikimate-derived metabolites. 13. A key intermediate in the biosynthesis of anthranilate from chorismate. J. Am. Chem. Soc. 106: 2463-2464.

20. POLICASTRO, P.P., K.G. AU, C.T. WALSH, G.A. BERCHTOLD. 1984. Trans-6-amino-5-[(1-carboxyethyl)oxy]-1,3-cyclohexadiene-1-carboxylic acid: An intermediate in the biosynthesis of anthranilate from chorismate. J. Am. Chem. Soc. 106: 2443-2444.

21. TENG, C.-Y.P., B. GANEM, S.Z. DOKTOR, B.P. NICHOLS, R.K. BHATNAGAR, L.C. VINING. 1985. Total synthesis of (±)-4-amino-4-deoxychorismic acid: A key intermediate in the biosynthesis of para-aminobenzoic acid and L-para-aminophenylalanine. J. Am. Chem. Soc. 107: 5008-5009.

22. CHRISTOPHERSON, R.E., J.F. MORRISON. 1983. Synthesis and separation of tritium-labeled intermediates of the shikimate pathway. Arch. Biochem. Biophys. 220: 444-450.

23. ZAMIR, L.O., C. LUTHE. 1984. Chemistry of shikimic acid derivatives. Synthesis of specifically labeled shikimic acid at C-3 or C-4. Can. J. Chem. 62: 1169-1175.

24. HOARE, J.H., G.A. BERCHTOLD. 1984. Chemical synthesis of stereoselectively labeled [9-$^2$H,$^3$H]chorismate. J. Am. Chem. Soc. 106: 2700-2701.

25. SOGO, S.G., T.S. WIDLANSKI, J.H. HOARE, C.E. GRIMSHAW, G.A. BERCHTOLD, J.R. KNOWLES. 1984. Stereochemistry of the rearrangement of chorismate to prephenate: Chorismate mutase involves a chair transition state. J. Am. Chem. Soc. 106: 2701-2703.

26. McCANDLISS, R.J., K.M. HERRMANN. 1979. Immunological studies on 3-deoxy-D-arabino-heptulosonate 7-phosphate synthase isoenzymes. J. Biol. Chem. 254: 3761-3764.

27. SHULTZ, J., M.A. HERMODSON, K.M. HERRMANN. 1981. A comparison of the amino-terminal sequences of 3-deoxy-D-arabino-heptulosonate 7-phosphate synthase isoenzymes from Escherichia coli. FEBS Lett. 131: 108-110

28. SHULTZ, J., M.A. HERMODSON, C.C. GARNER, K.M. HERRMANN. 1984. The nucleotide sequence of the aroF gene of Escherichia coli and the amino acid sequence of the encoded protein, the tyrosine-sensitive 3-deoxy-D-arabino-heptulosonate 7-phosphate synthase. J. Biol. Chem. 259: 9655-9661.

29. HERRMANN, K.M., J. SHULTZ, M.A. HERMODSON. 1980. Sequence homology between the tyrosine-sensitive 3-deoxy-D-arabino-heptulosonate 7-phosphate synthase from Escherichia coli and hemerythrin from Sipunculida. J. Biol. Chem. 255: 7079-7081.

30. STEINRÜCKEN, H.C., N. AMRHEIN. 1984. 5-Enolpyruvyl-shikimate-3-phosphate synthase of Klebsiella pneumoniae. 2. Inhibition by glyphosphate [N-(phosphonomethyl)glycine]. Eur. J. Biochem. 143: 351-357.

31. BONDINELL, W.E., J. VNEK, P.F. KNOWLES, M. SPRECHER, D.B. SPRINSON. 1971. On the mechanism of 5-enol-pyruvylshikimate 3-phosphate synthetase. J. Biol. Chem. 246: 6191-6196.

32.  IFE, R.J., L.F. BALL, P. LOWE, E. HASLAM. 1976. The shikimate pathway. Part V. Chorismic acid and chorismate mutase. J. Chem. Soc., Perkin Trans. I. 1776-1783.

33.  ANTON, D.L., L. HEDSTROM, S.M. FISH, R.H. ABELES. 1983. Mechanism of enolpyruvylshikimate-3-phosphate synthase. Exchange of phosphoenolpyruvate with solvent protons. Biochemistry 22: 5903-5908.

34.  GRIMSHAW, C.E., S.G. SOGO, S.D. COPLEY, J.R. KNOWLES. 1984. Synthesis of stereoselectively labeled [$^2$H,$^3$H]chorismate and the stereochemical course of enolpyruvoylshikimate-3-phosphate synthetase. J. Am. Chem. Soc. 106: 2699-2700.

35.  FLOSS, H.G. 1979. The shikimate pathway. In T. Swain, J.B. Harborne, C.F. Van Sumere, eds., op. cit. Reference 1, pp. 59-89.

36.  LEE, J.J., Y. ASANO, T.-L. SHIEH, F. SPREAFICO, K. LEE, H.G. FLOSS. 1984. Steric course of the 5-enolpyruvylshikimate-3-phosphate synthetase and anthranilate synthetase reactions. J. Am. Chem. Soc. 106: 3367-3368.

37.  ASANO, Y., J.J. LEE, T.-L. SHIEH, F. SPREAFICO, C. KOWAL, H.G. FLOSS. 1985. Steric course of the reactions catalyzed by 5-enolpyruvylshikimate-3-phosphate synthase, chorismate mutase and anthranilate synthase. J. Amer. Chem. Soc. 107, 4314-4320.

38.  ANDREWS, P.R., E.N. CAIN, E. RIZZARDO, G.D. SMITH. 1977. Rearrangement of chorismate to prephenate. Use of chorismate mutase inhibitors to define the transition state structure. Biochemistry 16: 4848-4852.

39.  ANDREWS, P.R., R.C. HADDON. 1979. Molecular orbital studies on enzyme catalyzed reactions. Rearrangements of chorismate to prephenate. Aust. J. Chem. 32: 1921-1929.

40.  MILES, E.W. 1979. Tryptophan synthase: Structure, function, and subunit interaction. Adv. Enzymol. 49: 127-186.

41.  SNELL, E.E. 1975. Tryptophanase: Structure, catalytic activities, and mechanism of action. Adv. Enzymol. 42: 287-333.

42.  MILES, E.W., D.R. HOUCK, H.G. FLOSS. 1982. Stereochemistry of sodium borohydride reduction of tryptophan synthase from Escherichia coli and its amino acid Schiff's bases. J. Biol. Chem. 257: 14203-14210.

43. PHILLIPS, R.S., E.W. MILES, L.A. COHEN. 1984. Interaction of tryptophan synthase, tryptophanase and
pyridoxal phosphate with oxindolyl-L-alanine and
2,3-dihydro-L-tryptophan: Support for an indolenine
intermediate in tryptophan metabolism. Biochemistry
23: 6228-6234.

44. PHILLIPS, R.S., E.W. MILES, L.A. COHEN. 1985. Differential inhibition of tryptophan synthase and of
tryptophanase by the two diastereoisomers of
2,3-dihydro-L-tryptophan: Implications for the
stereochemistry of the reaction intermediates. J.
Biol. Chem. 260: 14665-14670.

45. SCHÖPPNER, A., H. KINDL. 1984. Purification and
properties of a stilbene synthase from induced cell
suspension cultures of peanut. J. Biol. Chem. 259:
6806-6811.

46. ROLFS, C.-H., H. KINDL. 1984. Stilbene synthase and
chalcone synthase. Two different constitutive
enzymes in cultured cells of Picea excelsa. Plant
Physiol. 75: 489-492.

47. HAHLBROCK, K., H. GRISEBACH. 1979. Enzymatic controls
in the biosynthesis of lignin and flavonoids. Annu.
Rev. Plant Physiol. 30: 105-130.

48. KREUZALER, F., H. RAGG, E. FAUTZ, D.N. KUHN, K.
HAHLBROCK. 1983. UV-Induction of chalcone
synthase mRNA in cell suspension cultures of
Petroselinum hortense. Proc. Natl. Acad. Sci.
USA 80: 2591-2593.

49. EBEL, J., W.E. SCHMIDT, R. LOYAL. 1984. Phytoalexin
synthesis in soybean cells: Elicitor induction of
phenylalanine ammonia-lyase and chalcone synthase
mRNAs and correlation with phytoalexin accumulation.
Arch. Biochem. Biophys. 232: 240-248.

50. BENTLEY, R., R. MEGANATHAN. 1982. Biosynthesis of
vitamin K (menaquinone) in bacteria. Microbiol.
Revs. 46: 241-280.

51. BENTLEY, R., R. MEGANATHAN. 1983. Vitamin K biosynthesis in bacteria - Precursors, intermediates,
enzymes and genes. J. Nat. Prod. 46: 44-59.

52. MEGANATHAN, R., R. BENTLEY. 1983. Thiamine pyrophosphate requirement for o-succinylbenzoic acid
synthesis in Escherichia coli and evidence for an
intermediate. J. Bacteriol. 153: 739-746.

53. WEISCHE, A., E. LEISTNER. 1985. Cell free synthesis
of o-succinylbenzoic acid from iso-chorismic acid.
The key reaction in vitamin $K_2$ (menaquinone)

biosynthesis. Tetrahedron Lett. 1487-1490.

54. KOLKMANN, R., G. KNAUEL, S. ARENDT, E. LEISTNER.
    1982. Site of activation of o-succinylbenzoic
    acid during its conversion to menaquinones (vitamin
    $K_2$). FEBS Lett. 137: 53-56.

55. KOLKMANN, R., E. LEISTNER. 1985. Synthesis and
    revised structure of the o-succinylbenzoic acid
    coenzyme A ester, an intermediate in menaquinone
    biosynthesis. Tetrahedron Lett. 1703-1704.

56. RINEHART, K.L. Jr., M. POTGIETER, W.-Z. JIN, C.J.
    PEARCE, D.A. WRIGHT, J.L.C. WRIGHT, J.A. WALTER,
    A.G. McINNES. 1982. Biosynthetic studies on
    antibiotics employing stable isotopes. In Trends
    in Antibiotic Research. Genetics, Biosyntheses,
    Actions and New Substances. (H. Umezawa, A.L.
    Demain, T. Hata, C.R. Hutchinson, eds.), Japan
    Antibiotics Res. Assoc., Tokyo, pp. 353-389.

57. HORNEMANN, U., J.H. EGGERT, D.P. HONOR. 1980. Role
    of D-[4-$^{14}$C]erythrose and [3-$^{14}$C]pyruvate in the
    biosynthesis of the meta-C-$C_6$-N unit of the mito-
    mycin antibiotics in Streptomyces verticillatus.
    J. Chem. Soc., Chem. Commun. 11-13.

58. GHISALBA, O., J. NÜESCH. 1981. A genetic approach to
    the biosynthesis of the rifamycin chromophore in
    Nocardia mediterrani. IV. Identification of 3-
    amino-5-hydroxybenzoic acid as a direct precursor
    of the seven-carbon amino starter unit. J.
    Antibiot. 34: 64-71.

59. KIRBY, J.J., I.A. McDONALD, R.W. RICKARDS. 1980.
    3-Amino-5-hydroxybenzoic acid as a key intermediate
    in ansamycin and maytansinoid biosynthesis. J.
    Chem. Soc., Chem. Commun. 768-769.

60 ANDERSON, M.G., J.J. KIRBY, R.W. RICKARDS, J.M.
    ROTHSCHILD. 1980. Biosynthesis of the mitomycin
    antibiotics from 3-amino-5-hydroxybenzoic acid.
    J. Chem. Soc., Chem. Commun. 1277-1278.

61. TSAO, S.-W. 1983. Biosynthesis of microbial metabo-
    lites. Part I: Studies on a red pigment from Strep-
    tomyces. Part II: Tracer studies on ansamycin type
    antibiotics. Ph.D. thesis, Purdue University.

62. RINEHART, K.L., M. POTGIETER, D.A. WRIGHT. 1982.
    Use of D-[$^{13}C_6$]glucose together with $^{13}$C-depleted
    glucose and homonuclear $^{13}$C decoupling to identify
    the labeling pattern by this precursor of the "m-
    $C_7$N" unit of geldanamycin. J. Am. Chem. Soc. 104:
    2649-2652.

63. RINEHART, K.L., M. POTGIETER, D.L. DELAWARE, H. SETO. 1981. Direct evidence from multiple $^{13}$C labeling and homonuclear decoupling for the labeling pattern by glucose of the m-aminobenzoyl ($C_7N$) unit of pactamycin. J. Am. Chem. Soc. 103: 2099-2101.

64. Unpublished results.

65. TAKEDA, Y., V. MAK, C.-C. CHANG, H.G. FLOSS. 1984. Biosynthesis of ketomycin. J. Antibiot. 37: 868-875.

66. SHIMADA, K., D.J. HOOK, G.F. WARNER, H.G. FLOSS. 1978. Biosynthesis of the antibiotic 2,5-dihydrophenylalanine by Streptomyces arenae. Biochemistry 17: 3054-3058.

67. BALDWIN, G.S., B.E. DAVIDSON. 1983. Kinetic studies on the mechanism of chorismate mutase/prephenate dehydratase from Escherichia coli. Biochim. Biophys. Acta 742: 374-383.

68. GYGAX, D., M. CHRIST, O. GHISALBA, J. NÜESCH. 1982. Regulation of 3-deoxy-D-arabino-heptulosonic acid 7-phosphate synthetase in Nocardia mediterrani. FEMS Microbiol. Lett. 15: 169-173.

Chapter Three

# TYROSINE AND PHENYLALANINE BIOSYNTHESIS: RELATIONSHIP BETWEEN ALTERNATIVE PATHWAYS, REGULATION AND SUBCELLULAR LOCATION

ROY A. JENSEN

Center for Somatic-cell Genetics and Biochemistry
State University of New York at Binghamton
Binghamton, New York 13901

## INTRODUCTION

The metabolic pathway responsible for biosynthesis of aromatic amino acids and for vitamin-like derivatives such as folic acid and ubiquinones is a major enzyme network in nature.[1] In higher plants this pathway plays an even larger role since it is the source of precursors for numerous phenylpropanoid compounds, lignins, auxins, tannins, cyanogenic glycosides and an enormous variety of other secondary metabolites.[2] Such secondary metabolites may originate from the amino acid end products or from intermediates in the pathway (Fig. 1). The aromatic pathway interfaces with carbohydrate metabolism at the reaction catalyzed by 3-deoxy-D-arabino-heptulosonate 7-phosphate (DAHP) synthase, the condensation of erythrose-4-phosphate and PEP to form

57

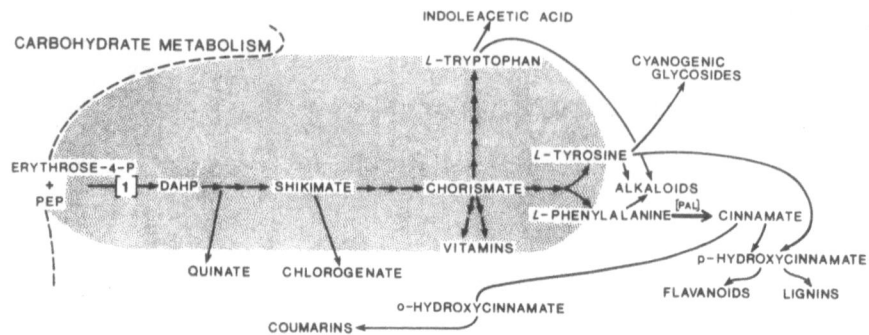

Fig. 1.   Interface of the pathway for aromatic biosynthesis
with carbohydrate metabolism and secondary metabolism.
Enzyme [1] is DAHP synthase, the initial enzyme of aromatic
biosynthesis.   [PAL] is phenylalanine ammonia-lyase, the
major enzymic gateway to phenylpropanoid synthesis.

the 7-carbon sugar, DAHP.   Evidence is accumulating to show
that separate biochemical pathways of aromatic biosynthesis
may exist in the spatially separated microenvironments of
the plastid and cytosolic compartments.[3]   Separate enzyme
systems of carbohydrate metabolism exist in the cytosol and
in plastids[4] that would be able to generate erythrose-4-
phosphate and PEP for entry into aromatic biosynthesis.
Secondary metabolites appear to be synthesized by enzymes
located primarily in the cytosol.

This review deals with the following questions.   (i)
What enzyme steps are utilized for aromatic amino acid
biosynthesis in higher plants?   (ii) What is the subcellular location of these enzymes?   (iii) What regulatory
mechanisms govern the output of these enzyme networks?

ENZYMIC CONSTRUCTION OF HIGHER-PLANT PATHWAYS TO PHENYLALA-
NINE AND TYROSINE

Figure 2 illustrates the recent observation[5] that
tyrosine may be formed from either L-arogenate or from
4-hydroxyphenylpyruvate, while phenylalanine may be formed
from either L-arogenate or from phenylpyruvate.   Some
organisms fail to use L-arogenate for aromatic biosynthesis

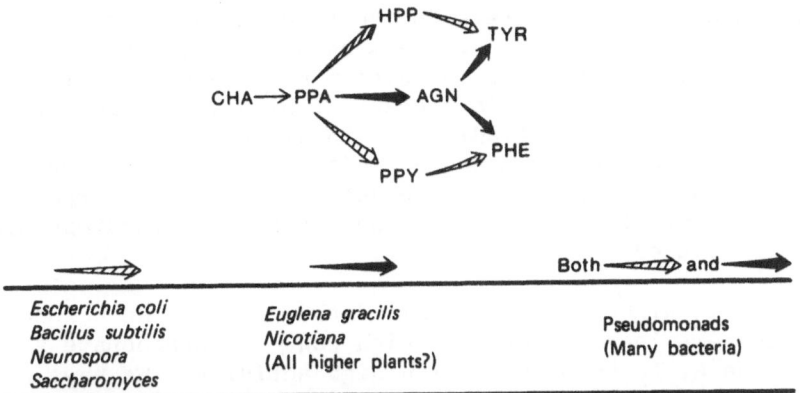

Fig. 2. Flow routes to phenylalanine and tyrosine in nature. Abbreviations: CHA, chorismate; PPA, prephenate; HPP, 4-hydroxyphenylpyruvate; AGN, L-arogenate; PPY, phenylpyruvate; TYR, L-tyrosine; PHE, L-phenylalanine.

at all while others rely exclusively upon L-arogenate as a precursor of both L-phenylalanine and L-tyrosine. It has been perhaps surprising to find that many microorganisms possess both sets of enzymes. Other pathway combinations not shown in Figure 2 have been found in microorganisms. For example, the arogenate pathway-to-tyrosine and the phenylpyruvate pathway-to-phenylalanine is an exceedingly common biochemical arrangement in prokaryotes.

In higher plants prephenate dehydratase, which converts prephenate to phenylpyruvate, has never been demonstrated. Since we have recently demonstrated the presence of arogenate dehydratase, which converts L-arogenate to L-phenylalanine, in several higher plants (this paper), the arogenate route to phenylalanine may be characteristic of higher plants.

Arogenate dehydrogenase, which converts L-arogenate to L-tyrosine, has been identified in mung bean,[6] tobacco,[7] corn[8] and sorghum.[9] We have also noted the presence of arogenate dehydrogenase activity in duckweed, spinach and buckwheat (unpublished data). In all cases prephenate dehydrogenase was absent except in the case of mung bean,[6,10] and other bean species.[10] The mung bean dehydrogenase has broad specificity for substrate, being reactive with either

prephenate or L-arogenate.  Unlike specific arogenate dehy-
drogenases in other plants, the mung bean dehydrogenase is
relatively insensitive to inhibition by L-tyrosine.  The
mung bean dehydrogenase is present at high levels in coty-
ledon tissue of young seedlings.  In all of the species
mentioned, the dehydrogenase activity is linked to $NADP^+$.
Although the Zea mays dehydrogenase was reported to require
$NAD^+$,[8] recent studies have now shown that the dehydrogenase
is $NADP^+$-linked.

In all higher plants examined thus far, we have found
prephenate aminotransferase, which converts prephenate to
L-arogenate, to be present with high activity.  We have
consistently found that crude extracts prepared from a
variety of higher plant organisms are able to transaminate
prephenate much better than either phenylpyruvate or 4-
hydroxyphenylpyruvate.  Thus, L-arogenate is emerging as a
major if not exclusive precursor of both L-tyrosine and
L-phenylalanine in higher plants, although aspects of
diversity that have yet to be fully evaluated have been
observed.

It should be noted that evidence in the older literature that is sometimes cited as proof for existence of the
phenylpyruvate and 4-hydroxyphenylpyruvate pathway routes
in vivo is not valid.  Thus, the demonstration by Gamborg
and Simpson[11] that prephenate plus L-glutamate formed
L-phenylalanine in mung bean does not distinguish between
phenylpyruvate or L-arogenate as an intermediate.  Labeling
studies such as those of Widholm[12] showed that phenylpyruvate or 4-hydroxyphenylpyruvate can be converted to their
amino acid counterparts when supplied in the growth medium.
However, non-specific aminotransferases are commonly able
to transaminate phenylpyruvate and 4-hydroxyphenylpyruvate,
even though this potential may ordinarily not be realized
in vivo.  This is exemplified by results obtained with
Euglena gracilis which utilizes only L-arogenate in vivo.[13]
The essential point is that higher plants cannot transaminate phenylpyruvate and 4-hydroxyphenylpyruvate if these
keto acids are not formed in vivo for lack of prephenate
dehydratase and prephenate dehydrogenase, respectively.  In
Figure 1 of their paper on biosynthesis of prenylquinones,
Fiedler et al.[14] assumed 4-hydroxyphenylpyruvate to be a
branchpoint leading to L-tyrosine and to the prenylquinones.
However, in spinach extracts we obtained the following
specific activities for prephenate dehydrogenase, 4-

hydroxyphenylpyruvate aminotransferase, prephenate
aminotransferase and arogenate dehydrogenase:  0, 24.3,
28.0 and 2.1 nmoles/min/mg, respectively.  Thus, it
appears that the 4-hydroxyphenylpyruvate required for
prenylquinone biosynthesis is formed by transamination
of L-tyrosine following its formation from L-arogenate.


ISOZYME PAIRS OF AROMATIC-PATHWAY ENZYMES

    Most of our studies have been carried out in Nicotiana
silvestris, chosen for the following advantageous characteristics.  It is a true diploid (2n=24).  A background
of basic classical genetics exists.  It is self-fertile,
flowers rapidly and produces numerous seeds.  It has a
relatively short life cycle.  Haploids are readily obtained.
It grows rapidly in suspension culture and is totipotent.
Protoplast techniques have been developed and somatic-cell
hybridization is feasible.

## Chorismate Mutase Isozymes

    Two distinct isozymes of chorismate mutase have been
demonstrated in N. silvestris.[15]  Form CM-1 is subject to
allosteric control by phenylalanine, tyrosine and
tryptophan.[16]  Form CM-2 is not affected by biosynthetic-
pathway compounds, although the secondary metabolite,
caffeic acid, was inhibitory.  Figure 3 shows that the
ratio of these isozymes varies considerably in mesophyll
cells from leaf tissue and in suspension cell populations
harvested after about two generations of exponential growth.
A similar pair of isozymes from sorghum exhibits immunological differences[17] that suggest separate evolutionary
origins.

## DAHP Synthase Isozymes

    Distinct isozymes of DAHP synthase have been resolved
from both mung bean[18] and N. silvestris (Ganson and Jensen,
submitted).  The fact that the properties of the two
isozymes of DAHP synthase are so similar in mung bean and
tobacco (as, indeed, is also the case with the chorismate-
mutase isozyme pair) suggests that the results may
generally apply to higher plants.  DAHP synthase-Mn and
DAHP synthase-Co can be separated by DEAE-cellulose
chromatography.  Their considerable differential properties

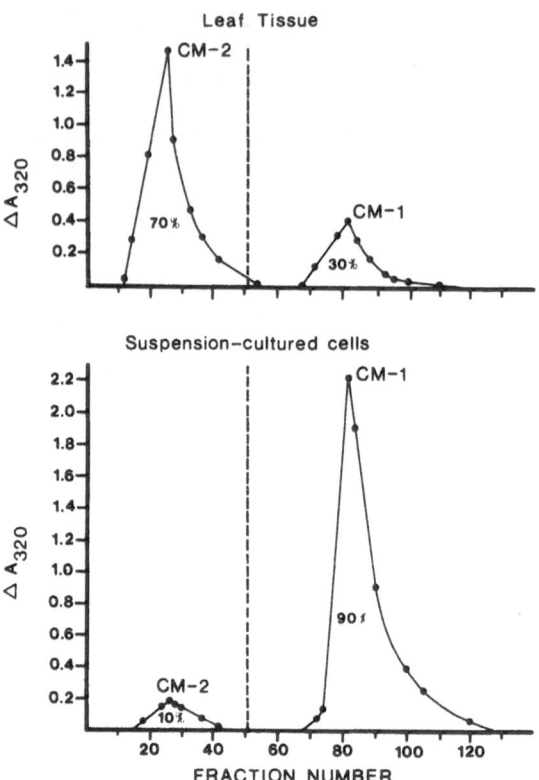

Fig. 3. Chromatographic resolution of isozymes of chorismate mutase from N. silvestris. The isozymes were separated from organismal tissue (top) and from cells in culture (bottom) as described by Goers and Jensen[15] using DEAE-cellulose chromatography. Prephenate was converted to phenylpyruvate at acidic pH, and the phenylpyruvate was quantified at basic pH by measuring absorbance at 320 nm. The vertical, dashed line indicates the beginning of the linear gradient of KCl used to elute isozyme CM-1.

are listed in Table 1. It is noteworthy that optimal conditions for assay of either isozyme are poor conditions for the other, thus enhancing the possibility that one isozyme present in a mixture could go undetected under any given assay conditions. DAHP synthase-Co can substitute $Mg^{++}$ for $Co^{++}$, although much higher concentrations of $Mg^{++}$ are required. DAHP synthase-Mn is activated by dithio-

Table 1.  Differential properties of isozymes of DAHP
synthase.

| Property | DAHP Synthase-Mn | DAHP Synthase-Co |
|---|---|---|
| Erythrose-4-P | Saturates at 0.6 m$\underline{M}$ | Saturates at 6.0 m$\underline{M}$ |
| Dithiothreitol | Needed for activity | Inhibits activity |
| Metal | Activated by 0.5 m$\underline{M}$ Mn$^{++}$ | Requires 0.5 m$\underline{M}$ Co$^{++}$ |
| pH optimum | pH 6.9 | pH 8.8 |
| Glyphosate | Does not inhibit | Inhibits |

threitol in a relatively slow process that yields a hyster-
etic progress curve when dithiothreitol is added at zero
time.  Complete activation requires at least 7 min at 37°C
when dithiothreitol is added in the absence of substrate.

DAHP synthase-Co was insensitive to allosteric effects
of aromatic-pathway compounds, although caffeic acid was
inhibitory.  On the other hand, DAHP synthase-Mn was found
to be sensitive to feedback inhibition by L-arogenate in
both mung bean and in N. silvestris.  In mung bean,[18] where
the most detailed studies have been done thus far, a number
of other pathway intermediates produced allosteric effects.

The complexity of DAHP synthase-Mn is further under-
scored by results obtained in an experiment where phenylalanine ammonia-lyase was activated by treatment with white
light.  N. silvestris plants growing in a growth chamber
providing 7,500 lux of white light were covered with a
black cloth for 68 h before exposure to 15,000 lux for 22 h.
Control plants were maintained in the dark during the time
of light treatment.  The specific activities of PAL and
DAHP synthase-Mn increased 14-fold and 11-fold respectively
when extracts prepared from dark-treated and light-treated
plants were compared.  On the other hand, DAHP synthase-Co
activity from light-treated tissue was unchanged (± 5%)
from that of dark-treated tissue.

Other Isozyme Pairs?

The differential allosteric properties of the isozymes
of chorismate mutase and DAHP synthase have provided obvious

advantages for recognition of separate isozymes. Evidence
for unregulated (allosteric) isozymes of anthranilate
synthase, arogenate dehydrogenase and arogenate dehydratase
is currently being sought in N. silvestris. It is sugges-
tive that a minor isozyme of anthranilate synthase, insensitive to feedback inhibition by L-tryptophan, has been
detected in potato.[19] We have resolved a minor species of
arogenate dehydrogenase from N. silvestris; its allosteric
properties have not yet been examined. Chromatographically
separable isozymes of 5-enolpyruvylshikimate-3-phosphate
synthase have been obtained from crude extracts of N.
silvestris. Mousdale and Coggins[20] recently demonstrated
the existence of major and minor isozymes of a number of
pre-chorismate enzymes from pea. These major and minor
isozyme species appeared to have plastidic and cytosolic
subcellular locations, respectively.


POST-PREPHENATE ENZYMES

     A preliminary description of prephenate aminotrans-
ferase from N. silvestris has been published.[21] It
possesses a strong preference for prephenate as keto-acid
substrate and uses either L-glutamate or L-aspartate as
amino-donor substrate. Substrate inhibition occurs at
concentrations of prephenate in excess of 0.8 $mM$ in the
presence of 20 $mM$ L-glutamate. Prephenate aminotransferase
possesses remarkable thermal tolerance, being stable during
reaction up to 70°C in the presence of a combination of
three protectants: pyridoxal-5'-phosphate, glycerol and
EDTA. It can be detected as a single, highly mobile band
following gel electrophoresis. Prephenate aminotransferase
can also be visualized as a faint band among nine bands
resolved with an activity stain for oxaloacetate following
reaction with 2-ketoglutarate and L-aspartate. After
thermal treatment of extracts for 10 min at 70°C, only a
single band corresponding to the mobility position of
prephenate aminotransferase could be visualized.

     Arogenate dehydrogenase can be assayed by continuously
monitoring the formation of NADPH. Controls are essential
in crude extracts to ensure that an unidentified substrate
that sometimes contaminates L-arogenate preparations is not
being utilized, or that an endogenous cofactor-reducing
activity is not present. A second assay procedure employs
measurement of tyrosine formation by high pressure liquid

chromatography (HPLC). L-Tyrosine has been shown to be an effective inhibitor of the major form of arogenate dehydrogenase present in N. silvestris[7] and in sorghum,[9] but not in the enzyme obtained from mung bean.[6]

Until recently no dehydratase reaction (arogenate or prephenate) leading to the formation of phenylalanine has been demonstrated in any higher-plant system. We report here the successful assay of arogenate dehydratase from both N. silvestris cell cultures and spinach leaf tissue. Figure 4 illustrates a reaction mixture assayed by HPLC which before reaction (left) contains L-arogenate (substrate), L-tyrosine (activator), and L-phenylalanine, an inevitable minor contaminant of L-arogenate preparations.[22] Following reaction for 30 min at 33°C in the presence of N. silvestris enzyme, the appearance of reaction product at the retention time of L-phenylalanine is readily apparent, and this is matched by the diminution in the amount of L-arogenate. Table 2 summarizes some properties of arogenate dehydratase thus far determined in N. silvestris. The enzyme has also been successfully assayed by coupling it with commercial PAL from yeast. Since yeast PAL reacts with L-tyrosine, it is necessary to omit the tyrosine activator (about 2-fold activation of arogenate dehydratase at 1 mM substrate) when using the coupled assay. Arogenate dehydratase from N. silvestris is sensitive to feedback inhibition by L-phenylalanine. This inhibition has been technically difficult to assess quantitatively since one must measure product (phenylalanine) formation against a large background of inhibitor (phenylalanine). The alternative of HPLC measurement of substrate disappearance also has the disadvantage of measuring a small change against a large background. A rigorous quantitative characterization of the kinetics of inhibition of arogenate dehydratase by phenylalanine awaits the preparation of radioactive L-arogenate so that labelled product molecules of phenylalanine can be distinguished from unlabelled phenylalanine molecules present as inhibitor. Crude extracts that contain high protease activities may accumulate free phenylalanine and therefore can tend to obscure the low activities of arogenate dehydratase that are generally present. Arogenate dehydratase is precipitated by 40% ammonium sulfate, yielding an apparent purification of about 3-fold and eliminating a significant fraction of protease activity.

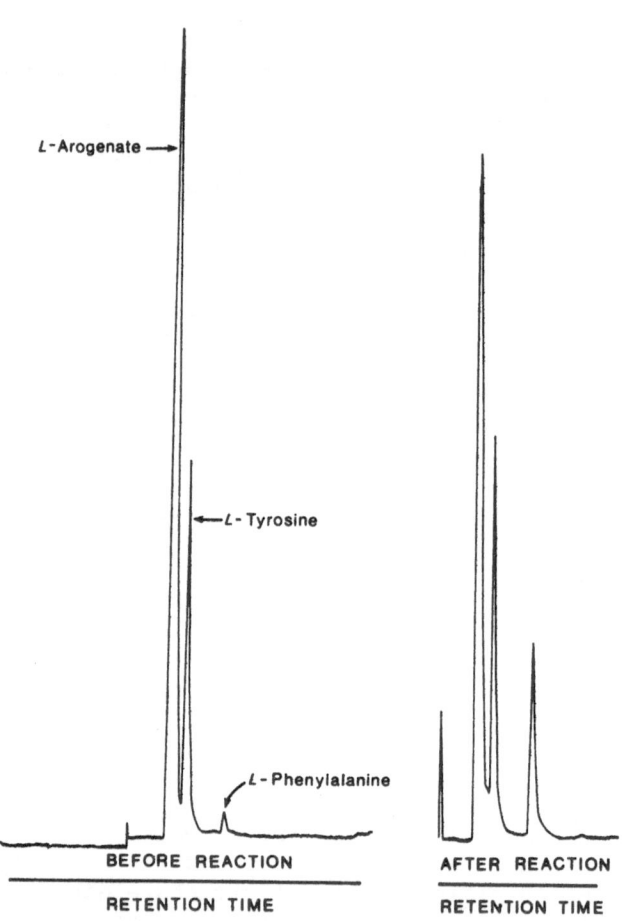

Fig. 4. HPLC assay for L-phenylalanine produced as the result of the reaction catalyzed by arogenate dehydratase. The reaction mixture contained 1 mM L-arogenate, 0.5 mM L-tyrosine and enzyme purified about 3-fold by precipitation with NH₄SO₄ at 40% of saturation. Reaction time was 30 min at 33°C.

Table 2. Arogenate dehydratase from N. silvestris.

| | |
|---|---|
| Temperature optimum | 34°C |
| pH optimum | 9.0 |
| $Km_{app}$ | 0.33 mM |
| Feedback inhibitor | L-Phenylalanine |
| Allosteric activator | L-Tyrosine |
| Specific activity (crude extract) | 1 nmole/min/mg |
| Assay procedure | (i) HPLC detection of L-phenylalanine |
| | (ii) Coupled assay with PAL |

SUBCELLULAR LOCATION OF AROMATIC-PATHWAY ENZYMES

The CM-1 and CM-2 isozymes of chorismate mutase in N. silvestris have been shown to exist in plastidial and cytosolic compartments.[23] Similar evidence also exists in support of a parallel compartmentation of DAHP synthase isozymes, DAHP synthase-Mn being plastid-localized and DAHP synthase-Co being cytosolic (d'Amato and Ganson, unpublished data). Prephenate aminotransferase activity of N. silvestris is also largely or entirely localized within plastids (d'Amato and Bonner, unpublished data).

In one experiment washed chloroplasts were isolated and assayed for nitrite reductase, DAHP synthase-Mn and chorismate mutase-1 activities (Table 3). Since enzymes may fractionate with organelles by non-specific (or specific) association with the organelle surface, latency determinations were made. With this approach, activity determinations are made before and after rupture of the washed chloroplasts. If activities are located within the organelle, they are expected to increase dramatically following organelle disruption. Thus, nitrite reductase (chloroplast marker enzyme) gave a latency value of 16, a value similar to those obtained for DAHP synthase-Mn and chorismate mutase-1. The identity of chorismate mutase as the CM-1 isozyme was confirmed by its sensitivity to inhibition by L-tyrosine.

Spinach chloroplasts were isolated by a procedure modified from the method of Mills and Joy.[24] Chloroplasts

Table 3.  Latency determinations of enzymes localized in chloroplasts from leaf tissue of N. silvestris.

| Enzyme | Chloroplast Enzyme Activities* | | Latency Ratio |
|--------|------------|------------|---------|
|        | Intact | Disrupted | |
| Nitrite reductase | 0.42 | 6.75 | 16 |
| DAHP synthase-Mn | 0.03 | 0.46 | 15 |
| Chorismate mutase-1 | 0.04 | 0.89 | 22 |
| (+0.5 mM L-tyrosine) | (0.04) | (0.23) | |

* Enzyme activities are expressed as nmol/min/µg of chlorophyll, except for nitrite reductase:  µmoles/min/µg of chlorophyll.  See Reference 23 for experimental details.

were sedimented through a 40% (v/v) Percoll medium.  Washed chloroplasts were broken by osmotic rupture, and the clarified extract obtained after low-speed centrifugation was desalted.  Table 4 shows that 7 aromatic-pathway enzymes were detected in these preparations.  DAHP synthase-Mn, having properties similar to the N. silvestris isozyme, was found in chloroplasts.  DAHP synthase-Co activity was not found in the chloroplast preparation, even though this isozyme was very active in the total homogenate.  The CM-1 isozyme of chorismate mutase was not detected in spinach and may be unstable.  It is noteworthy that arogenate dehydratase activity was found in the chloroplasts.  The activities of both anthranilate synthase and arogenate dehydrogenase in chloroplast preparations were highly sensitive to feedback inhibition by L-tryptophan and L-tyrosine, respectively.  This experiment indicates that an intact pathway is present in the chloroplast compartment.  Furthermore, this pathway appears to form both phenylalanine and tyrosine from L-arogenate precursors since neither prephenate dehydratase nor prephenate dehydrogenase activities were found.


PHYSIOLOGICAL VARIATION OF ENZYME LEVELS DURING GROWTH

    Suspension cultures of N. silvestris are routinely transferred with a subculture routine in which exponential

Table 4. Comparison of key enzymes of the aromatic pathway in extracts prepared from homogenates and from chloroplasts of spinach leaves.

| Enzyme | Specific Activity | |
| --- | --- | --- |
| | Crude Extract | Chloroplast Extract |
| DAHP synthase-Co | 60 nmols/min/mg | -0- nmols/min/mg |
| DAHP synthase-Mn | 9.8 | 7.5 |
| Shikimate dehydrogenase | 22.0 | 22.9 |
| 5-Enolpyruvylshikimate 3-phosphate | 10.1 | 17.1 |
| Anthranilate synthase | 0.2 | 0.4 |
| Chorismate mutase | 3.1 | -0- |
| Prephenate aminotransferase | 14.3 | 28.0 |
| Arogenate dehydratase | 0.1 | 0.9 |
| Arogenate/NADP dehydrogenase | 0.5 | 2.0 |

growth is only maintained for about two generations.  The
doubling time during exponential growth is 36-40 hours
(Fig. 5).  Cell populations can be maintained continuously
in exponential phase by dilution at appropriate cell densi-
ties.  The righthand portion of Figure 5 depicts the growth
of cells between 30 and 32 generations of continuous
exponential growth.  Since DAHP synthase-Mn and DAHP
synthase-Co can be readily discriminated in mixtures,
possible differences in their levels during the various
physiological phases of growth were examined.  DAHP synthase-
Mn and DAHP synthase-Co reached different dramatic extremes

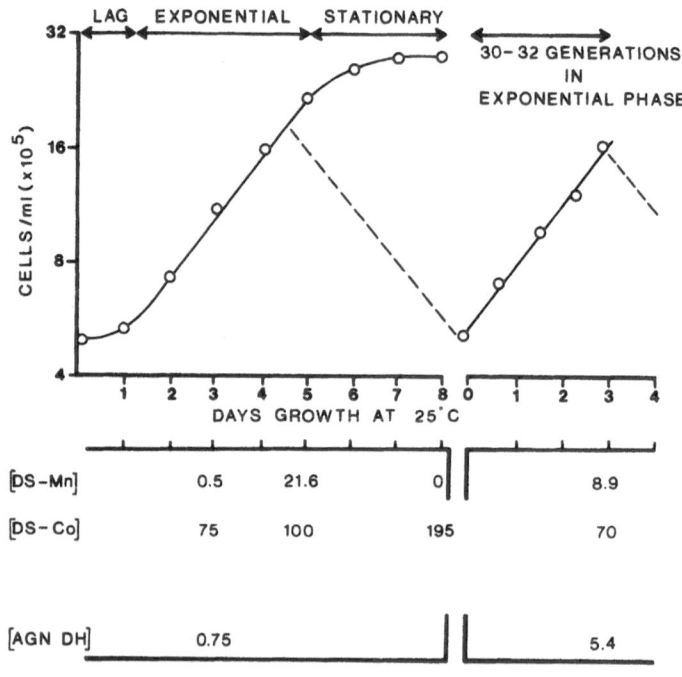

Fig. 5.  Variation of levels of enzyme activity during growth
of cultured cells of N. silvestris.  Dashed line indicates
transfer of cells in late-exponential growth to fresh medium
(16-fold dilution).  Data shown at the upper right is a
portion of a growth curve obtained between 30 and 32 generations of continuous exponential growth.  The bottom section
shows specific activities (nmoles/min/mg) obtained for DAHP
synthase-Mn [DS-Mn], DAHP synthase-Co [DS-Co], and arogenate
dehydrogenase [AGN DH] in samples taken from cultures at
times (in days) aligned with the values of specific activity.

in the stationary phase of growth. DAHP synthase-Co was at
its highest level during stationary-phase physiology. Upon
subculture, it reached its lowest level during exponential
growth, and cell populations 32 generations removed from
previous stationary-phase physiology had about one-third
the levels seen in stationary-phase cells. DAHP synthase-Mn
showed the opposite behavior, rising sharply during the
transition from stationary phase to exponential phase,
apparently overshooting the level found to be characteristic
of cells maintained continuously in exponential phase.
Limited data with arogenate dehydrogenase suggest a pattern
that may be similar to that of DAHP synthase-Mn.

If stationary-phase physiology of cultured-cell populations is equated with the slowed growth of mature organismal
tissues, then the results obtained with the chorismate
mutase isozyme pair (Fig. 3) seem to be consistent with the
results shown in Figure 5 for the isozymes of DAHP synthase.
Thus, cytosolic CM-2 is elevated in organismal tissue and
cytosolic DAHP synthase-Co (DS-Co) is elevated during
stationary-phase physiology of tissue culture.

Figure 6 shows results of a similar experiment in which
shikimate dehydrogenase activity was followed throughout a
single subculture that was sampled for 11 days. The activity level rose to its highest level 4-5 days after the
onset of stationary-phase physiology and declined there-
after. Upon subculture the enzyme level diminished
progressively until the early stages of stationary phase.
The inset shows that if cells were maintained continuously
in exponential phase, then shikimate dehydrogenase activity
leveled off after eight generations to a lower activity than
was ever reached in ordinary subcultures. It will be interesting to see whether stationary-phase cultures might yield
a cytosolic form of shikimate dehydrogenase that has not
been seen in extracts thus far studied from exponential-
phase cells.

Results similar to those obtained with shikimate dehydrogenase were obtained with prephenate aminotransferase
(Fig. 7). High levels of enzyme activity that were present
in stationary phase declined upon dilution into fresh
medium, reaching a low point during exponential phase.
When cells were maintained in exponential-phase growth, a
progressive decline in specific activity was obtained
throughout perhaps six generations. Activity stabilized

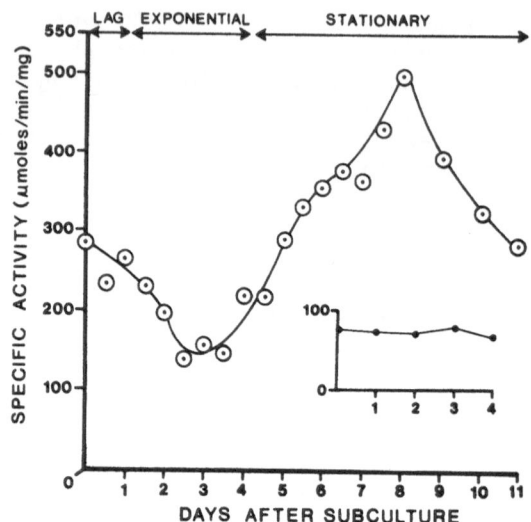

Fig. 6. Variation of the levels of shikimate dehydrogenase
in N. silvestris during the various growth phases following
subculture from cells in stationary phase. The inset shows
the constant and reduced levels of activity obtained from
cultures maintained continuously in exponential phase (data
shown for the 8-to-11 generation interval).

at a level at least twice as low as ever obtained during
exponential growth of a routine subculture.

Declining specific activities of enzymes such as
shikimate dehydrogenase and prephenate aminotransferase
during transition to exponential-phase growth could reflect
the new synthesis of one or more abundant proteins. It is
clear that an adequate number of generations must be
sustained in exponential growth to realize an enzyme balance
that truly reflects the levels characteristic of exponential
growth. In some cases the results indicate that in addition
to allosteric control, the levels of aromatic-pathway
enzymes vary in response to the physiological conditions
of growth. It will be interesting to see whether pathway
enzymes that share compartmented space also share common
trends in the regulation of their expression.

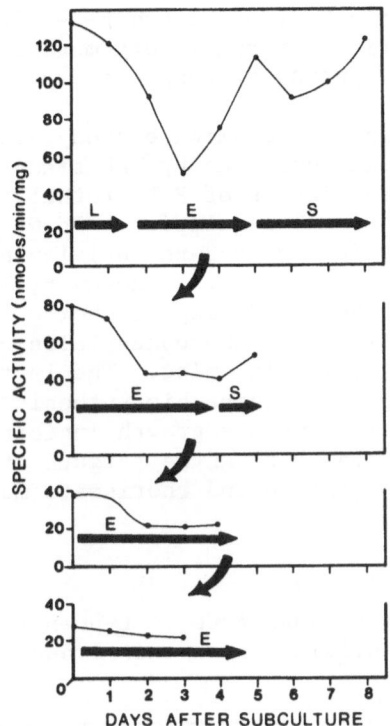

Fig. 7. Variation of the levels of prephenate aminotransferase as a function of growth phase. After 3.5 days
(indicated by arrow) a portion of the cell culture monitored in the top panel was diluted 5-fold into fresh
medium to initiate the culture shown in the 2nd panel.
Subsequent transfers were carried out at 3 days and 4 days,
respectively, to yield the cultures shown in panels 3 and
4 (bottom). L, E, and S denote lag, exponential and
stationary phases of growth.

## STRATEGIES OF MUTANT ISOLATION

Regulatory mutants are highly desirable in order to
probe the physiological and regulatory roles of pathway
enzymes that may function in different compartments in the
form of different isozymes. Analog selection is a powerful
approach because mutants can be selected directly since
mutations are expected to be dominant or semi-dominant. In

$\underline{N}$. silvestris $\underline{m}$-fluorotyrosine and β-2-thienylalanine ought
to provide selection for regulatory mutants of tyrosine and
phenylalanine biosynthesis, respectively.

It is desirable to eliminate phenylalanine analogs that
are capable of being degraded by PAL since resistant mutants
might have elevated levels of PAL, rather than being altered
in biosynthetic-pathway regulation. In $\underline{N}$. silvestris both
$\underline{p}$-fluorophenylalanine and β-2-thienylalanine are exceedingly
effective inhibitors of growth. However, Table 5 shows β-2-
thienylalanine to be the analog of choice since the PAL of
$\underline{N}$. silvestris utilizes $\underline{p}$-fluorophenylalanine quite well in
comparison to β-2-thienylalanine. The latter analog specif-
ically inhibits phenylalanine biosynthesis since $\underline{L}$-phenyl-
alanine effectively reverses growth inhibition provoked by
the analog. Two feasible enzyme targets of analog action
are arogenate dehydratase and chorismate mutase-1. The

Table 5. Ability of analogues of $\underline{L}$-phenylalanine to act as
substrates of phenylalanine ammonia-lyase (PAL) in $\underline{N}$.
silvestris.

| PAL Substrate | Apparent $K_m$* |
|---|---|
| $\underline{L}$-Phenylalanine | 0.026 m$\underline{M}$ |
| $\underline{p}$-Fluorophenylalanine | 0.099 m$\underline{M}$ |
| $\underline{o}$-Fluorophenylalanine | 0.250 m$\underline{M}$ |
| β-2-Thienylalanine | 1.000 m$\underline{M}$ |
| $\underline{L}$-Tyrosine | -0- |
| $\underline{m}$-Fluorotyrosine | -0- |
| $\underline{m}$-Tyrosine | -0- |

* Apparent $K_m$ values were estimated from double reciprocal
  plots of substrate saturation data obtained at 37°C using
  unfractionated extracts from $\underline{N}$. silvestris prepared from
  a light-induced population of suspension-cultured cells.
  Cinnamate was separated from $\underline{L}$-phenylalanine by isocratic,
  reverse-phase high pressure liquid chromatography on an
  ultrasphere-ODS C18 column; elution solution contained
  0.1% acetic acid in 50% aqueous methanol. The specific
  activity of PAL with $\underline{L}$-phenylalanine was 42 nanomoles/hr/
  mg of extract protein.

latter isozyme is not severely inhibited by the analog in
vitro (Fig. 8, right side).  The sensitivity of arogenate
dehydratase has not yet been tested.

m-Fluorotyrosine has been shown to be an extremely
effective analog of L-tyrosine as an inhibitor of arogenate
dehydrogenase activity.[7]  Since m-fluorotyrosine inhibits
arogenate dehydrogenase more effectively than even L-
tyrosine, (the analog is less effective than L-tyrosine
against chorismate mutase-1 (Fig. 8, left side), m-
fluorotyrosine-resistant mutants are expected as a consequence of the genetic deregulation of arogenate dehydrogenase.  m-Fluorotyrosine is inert as a substrate for PAL
(Table 5).  If only the plastid contains a regulated
arogenate dehydrogenase, then regulatory mutations will
impact directly upon plastid physiology.  Such mutants
could provide insight into the biochemical interplay between
intracellular compartments.  Other scenarios for an enzyme
like arogenate dehydrogenase can be envisioned that would be
revealed by the nature of a defined mutation.

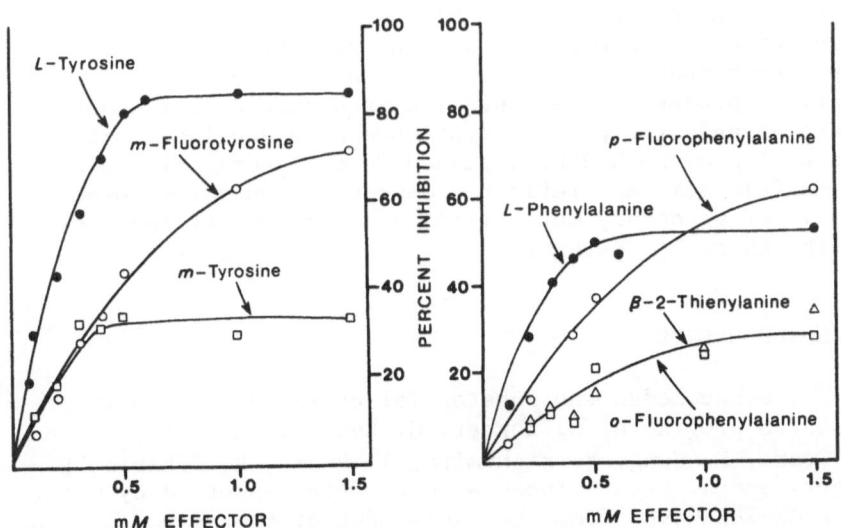

Fig. 8.  Ability of analogs of L-tyrosine (left chart) and
of L-phenylalanine (right chart) to inhibit chorismate
mutase-1 isolated from N. silvestris.[15]

CONCLUSION

Plastids of higher plants must contain an intact assemblage of enzymes for biosynthesis of aromatic amino acids. Virtually all aromatic-pathway enzymes have been detected in plastids in studies dealing with a number of higher-plant species. This pathway probably employs L-arogenate as an exclusive precursor of phenylalanine and tyrosine and is subject to tight allosteric control at all branchpoint junctions. The presence of a complete aromatic pathway within plastid organelles is consistent with results showing that spinach chloroplasts can assimilate either $CO_2$ or shikimate into aromatic amino acids.[25,26] The construction and regulation of this plastidic pathway is diagramed in Figure 9.

A separate complete pathway probably exists in the cytosol. Although Figure 10 shows L-arogenate to be the exclusive precursor of both phenylalanine and tyrosine in the cytosol, little evidence bearing on this point actually exists. The cytosolic branchpoint enzymes, DAHP synthase-Co and chorismate mutase-2, are not subject to allosteric control by compounds of the biosynthetic pathway. Other branchpoint enzymes may exist in the cytosol that are insensitive to feedback inhibition. I suggest that the allosterically unregulated pathway present in the cytosol supplies connecting networks of secondary metabolism by an overflow process whose operation is enhanced in tissues where demands for protein synthesis have declined. The inhibition of both DAHP synthase-Co and chorismate mutase-2 by caffeic acid may reflect a degree of control between levels of secondary metabolites and activity of the enzymes of the cytosolic pathway.

ACKNOWLEDGMENTS

I acknowledge the substantial contributions to this research program of C. Bonner, G. Byng, T. d'Amato, R. Ganson, S. Goers, E. Jung, B. Rightmire, J. Rubin, R. Tiberio, S. Unger, and J. Vrba. These studies were supported by Grant DE-ACO2-78ER04967 from the Department of Energy and by Grant PCM-8309070 from the National Science Foundation.

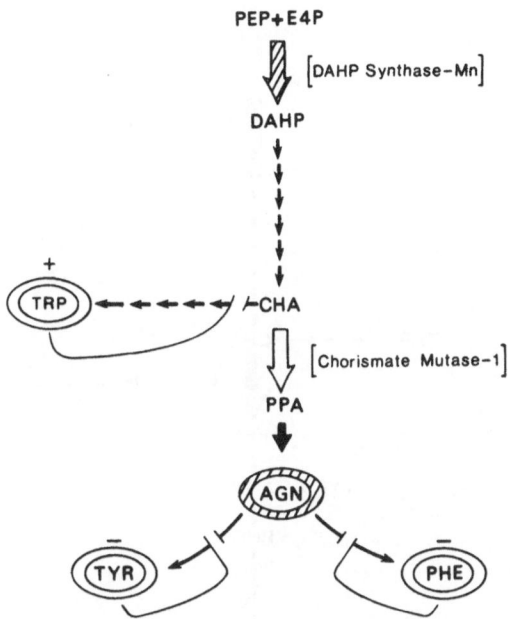

Fig. 9. Sequential pattern of allosteric control over biosynthesis of aromatic amino acids in the plastid compartment. In the presence of excess aromatic amino acids, L-tyrosine (TYR) inhibits arogenate dehydrogenase, L-phenylalanine (PHE) inhibits arogenate dehydratase and L-tryptophan (TRP) inhibits anthranilate synthase. The three aromatic amino acids exert allosteric inhibition (-) or activation (+) effects upon chorismate mutase-1 as symbolized. However, activation dominates over inhibition. The outcome of these events is to trap L-arogenate (AGN) between the various foci of control in the pathway. As shown symbolically, L-arogenate (AGN) then acts to feedback inhibit DAHP synthase-Mn.

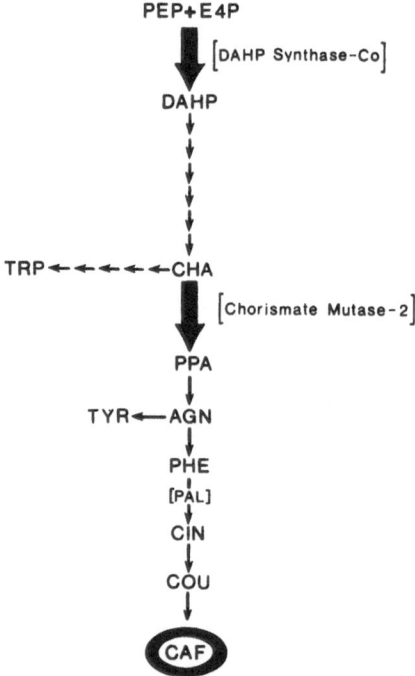

Fig. 10.  The pathway of aromatic biosynthesis in the cytosol and its point of interface with phenylpropanoid biosynthesis at the reaction catalyzed by phenylalanine ammonia-lyase (PAL).  Enzymes sensitive to inhibition by caffeic acid (CAF) are indicated by dark shading.  Abbreviations as in Figure 9; additionally, CIN, cinnamic acid; COU, coumaric acid.

REFERENCES

1. WEISS, U., J.M. EDWARDS. 1980. The biosynthesis of aromatic compounds. John Wiley and Sons, New York, 728 pp.
2. STAFFORD, H.A. 1974. The metabolism of aromatic compounds. Annu. Rev. Plant Physiol. 25: 459-486.
3. JENSEN, R.A. 1986. The shikimate/arogenate pathway: link between carbohydrate metabolism and secondary metabolism. Physiol. Plant., in press.
4. GOTTLIEB, L.D. 1982. Conservation and duplication of isozymes in plants. Science 216: 373-380.
5. BYNG, G.S., J.F. KANE, R.A. JENSEN. 1982. Diversity in the routing and regulation of complex biochemical pathways as indicators of microbial relatedness. Crit. Rev. Microbiol. 9: 227-252.
6. RUBIN, J.L., R.A. JENSEN. 1979. The enzymology of L-tyrosine biosynthesis in mung bean (Vigna radiata [L.] Wilczek). Plant Physiol. 64: 727-734.
7. GAINES, C.G., G.S. BYNG, R.J. WHITAKER, R.A. JENSEN. 1982. L-Tyrosine regulation and biosynthesis via arogenate dehydrogenase in suspension-cultured cells of Nicotiana silvestris (Speg. et Comes). Planta 156: 233-240.
8. BYNG, G.S., R.J. WHITAKER, C. FLICK, R.A. JENSEN. 1981. Enzymology of L-tyrosine biosynthesis in corn (Zea mays). Phytochemistry 20: 1289-1292.
9. CONNELLY, J.A., E.E. CONN. 1986. Tyrosine biosynthesis in Sorghum bicolor: isolation and regulatory properties of arogenate dehydrogenase. Z. Naturforsch., in press.
10. GAMBORG, O.L. 1966. Aromatic metabolism in plants. II. Enzymes of the shikimate pathway in suspension cultures of plant cells. Can. J. Biochem. 44: 791-799.
11. GAMBORG, O.L., F.J. SIMPSON. 1964. Preparation of prephenic acid and its conversion to phenylalanine and tyrosine by plant enzymes. Can. J. Biochem. 42: 583-591.
12. WIDHOLM, J.M. 1974. Control of aromatic amino acid biosynthesis in cultured plant tissues: effect of intermediates and aromatic amino acids on free levels. Physiol. Plant. 30: 13-18.
13. BYNG, G.S., R.J. WHITAKER, C.L. SHAPIRO, R.A. JENSEN. 1981. The aromatic amino acid pathway branches at L-arogenate in Euglena gracilis. Mol. Cell Biol. 1: 426-438.

14. FIEDLER, E., J. SOLL, G. SCHULTZ. 1982. The forma-
    tion of homogentisate in the biosynthesis of
    tocopherol and plastoquinone in spinach chloro-
    plasts. Planta 155: 511-515.
15. GOERS, S.K., R.A. JENSEN. 1984. Separation and
    characterization of two chorismate-mutase isoenzymes
    from Nicotiana silvestris. Planta 162: 109-116.
16. GOERS, S.K., R.A. JENSEN. 1984. The differential
    allosteric regulation of two chorismate-mutase
    isoenzymes of Nicotiana silvestris. Planta 162:
    117-124.
17. SINGH, B.K., E.E. CONN. 1986. Immunological
    characterization of chorismate mutase from
    Sorghum bicolor. Arch. Biochem. Biophys.,
    in press.
18. RUBIN, J.L., R.A. JENSEN. 1985. Differentially
    regulated isozymes of 3-deoxy-D-arabino-heptulo-
    sonate 7-phosphate synthase from seedlings of
    Vigna radiata [L.] Wilczek. Plant Physiol. 79:
    711-718.
19. CARLSON, J.E., J.M. WIDHOLM. 1978. Separation of two
    forms of anthranilate synthetase from 5-methyltryp-
    tophan-susceptible and -resistant cultured Solanum
    tuberosum cells. Physiol. Plant. 44: 251-255.
20. MOUSDALE, D.M., J.R. COGGINS. 1985. Subcellular
    localization of the common shikimate-pathway
    enzymes in Pisum sativum L. Planta 163: 241-249.
21. BONNER, C.A., R.A. JENSEN. 1985. Novel features of
    prephenate aminotransferase from cell cultures of
    Nicotiana silvestris. Arch. Biochem. Biophys. 238:
    237-246.
22. ZAMIR, L.O., R.A. JENSEN, B. ARISON, A. DOUGLAS, G.
    ALBERS-SCHONBERG, J.R. BOWEN. 1980. Structure of
    arogenate (pretyrosine), an amino acid intermediate
    of aromatic biosynthesis. J. Am. Chem. Soc. USA
    102: 4499-4504.
23. d'AMATO, T.A., R.J. GANSON, C.G. GAINES, R.A. JENSEN.
    1984. Subcellular localization of chorismate-mutase
    isoenzymes in protoplasts from mesophyll and
    suspension-cultured cells of Nicotiana silvestris.
    Planta 162: 104-108.
24. MILLS, W.R., K.W. JOY. 1980. A rapid method for
    isolation of purified, physiologically active
    chloroplasts used to study the intracellular distri-
    bution of amino acids in pea leaves. Planta 148:
    75-83.

25.  BICKEL, H., L. PALME, G. SCHULTZ.  1978.  Incorporation
     of shikimate and other precursors into aromatic
     amino acids and prenylquinones of isolated spinach
     chloroplasts.  Phytochemistry 17: 119-124.
26.  BUCHHOLZ, B., B. REUPKE, H. BICKEL, G. SCHULTZ.  1979.
     Reconstruction of amino acid synthesis by combining
     spinach chloroplasts with other leaf organelles.
     Phytochemistry 18: 1109-1111.

Chapter Four

SPECIFIC INHIBITORS AS PROBES INTO THE BIOSYNTHESIS AND
METABOLISM OF AROMATIC AMINO ACIDS

NIKOLAUS AMRHEIN

Lehrstuhl für Pflanzenphysiologie
Ruhr-Universität Bochum
D-4630 Bochum
Federal Republic of Germany

INTRODUCTION

Inhibitors of enzymic and metabolic processes are
invaluable tools in biochemical and physiological research,
and their application as drugs or pesticides ranges from
medicine to agriculture. The information one can extract
from their judicious use depends, on the one hand, on the
complexity of the system to which they are applied and, on
the other hand, on their selectivity for a given target,
as well as on their access to this target. Accessibility
in this context is meant to include the arrival of the
inhibitor at its target site in a state in which it is
capable of exerting its inhibitory action. It is obvious
that the chances for selectivity of a given inhibitor
decrease with the increasing complexity of a system as
measured, for example, by the number of enzymic reactions
involved and the degree of their interaction and inter-

83

dependence in the metabolic network of a cell.  To illustrate this point, α-aminooxy acetic acid (AOA) is a fairly
potent inhibitor of the biosynthesis of phenylpropanoid
compounds and has been used in complementation experiments
to study the biosynthesis of cyanidin in buckwheat.[1]
However, AOA can hardly be considered a selective inhibitor
since it exhibits a general reactivity towards pyridoxal
phosphate-dependent enzymes.[2]  In fact, as AOA is also a
potent inhibitor of ethylene biosynthesis,[3] it has been
employed to delay the senescence in carnation flowers.[4]

The problems of determining the selectivity of a
metabolic inhibitor and of identifying the site(s) of its
action are intimately related to each other.  This will
become clear in the discussion of the approaches to the
investigation of inhibitors of the shikimate pathway in
the author's laboratory presented below.  Fedtke[5] has classified the systems used in studies on the mechanism of
herbicide (i.e. in this context:  inhibitor) action as
"supercomplex" (growth related), "complex" (metabolism
related), and "defined systems" (site directed).  Systems
of all three levels of complexity have been used in studies
on inhibitors of the biosynthesis and metabolism of aromatic
amino acids.  The first indication that the action of the
herbicide glyphosate was related to an interference with
the shikimate pathway was obtained in a "supercomplex
system", i.e. growth experiments with Rhizobium japonicum
and Lemna gibba,[6] while the action of the inhibitor of
phenylalanine ammonia-lyase (PAL), L-α-aminooxy-β-
phenylpropionic acid (L-AOPP), was first investigated in
a "defined system".[7]  Eventually, of course, from whichever
end one starts to unravel the coil, the ends must meet!

If one considers, in a "complex system", a branched
metabolic pathway with the intermediate metabolites $M_{1 \to n}$
and the end products $P_{1 \to n}$, in which an inhibitor is assumed
to interfere with the conversion of $M_3$ to $M_4$ (Fig. 1), one
should be able to verify the following predictions:

1.    In the presence of the inhibitor the end products $P_{5 \to 8}$
      will fail to accumulate, and their level may even
      decline, if their turnover is sufficiently rapid
      within the observation period.  The accumulation of
      $P_2$ will not be affected, or it may actually increase
      if there is an "overflow" from $M_3$ accumulating in the
      presence of the inhibitor.

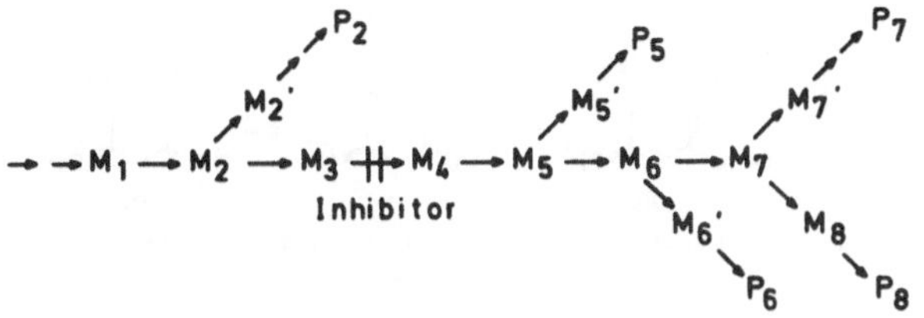

Fig. 1.  Scheme of a branched metabolic pathway with the
intermediate metabolites $M_{1\to n}$ and the end products $P_{1\to n}$.
The enzymatic conversion of $M_3$ to $M_4$ is blocked by the
inhibitor.

2.   Upon complementation of the inhibited system with $M_4$
     or $M_5$, the accumulation of $P_{5\to 8}$ will resume, while
     complementation with $M_7$ will restore only the capacity
     to accumulate $P_7$ and $P_8$.

3.   $M_3$ will accumulate in the presence of the inhibitor and
     its buildup may lead to an "overflow" into other path-
     ways (see 1).  This possibility can be checked by
     following the metabolic fate of radioactively labelled
     $M_1$, $M_2$ or $M_3$ in the absence and presence of the
     inhibitor.

4.   Lastly, at the level of the "defined system", it must
     be shown that the inhibitor actually interferes with
     the cell-free formation of $M_4$ from $M_3$.  Once this step
     is known, one can try to correlate the relative
     potency of inhibitor analogs in the "complex system"
     with their potency in the "defined system".  Further-
     more, one can check the selectivity of the inhibitor
     in closely related "defined systems".  Additional
     evidence for the identification of the "correct"
     target site of the inhibitor can be obtained by the
     investigation of target-related resistance (see below)
     or by "tagging" the target irreversibly (e.g. by
     photoaffinity labelling).

While one can envision numerous complications in this
scheme, it provides the investigator with some basic guide-
lines for studying the action of inhibitors as one
progresses from one level of complexity to the next, the
direction of this progression depending on the starting

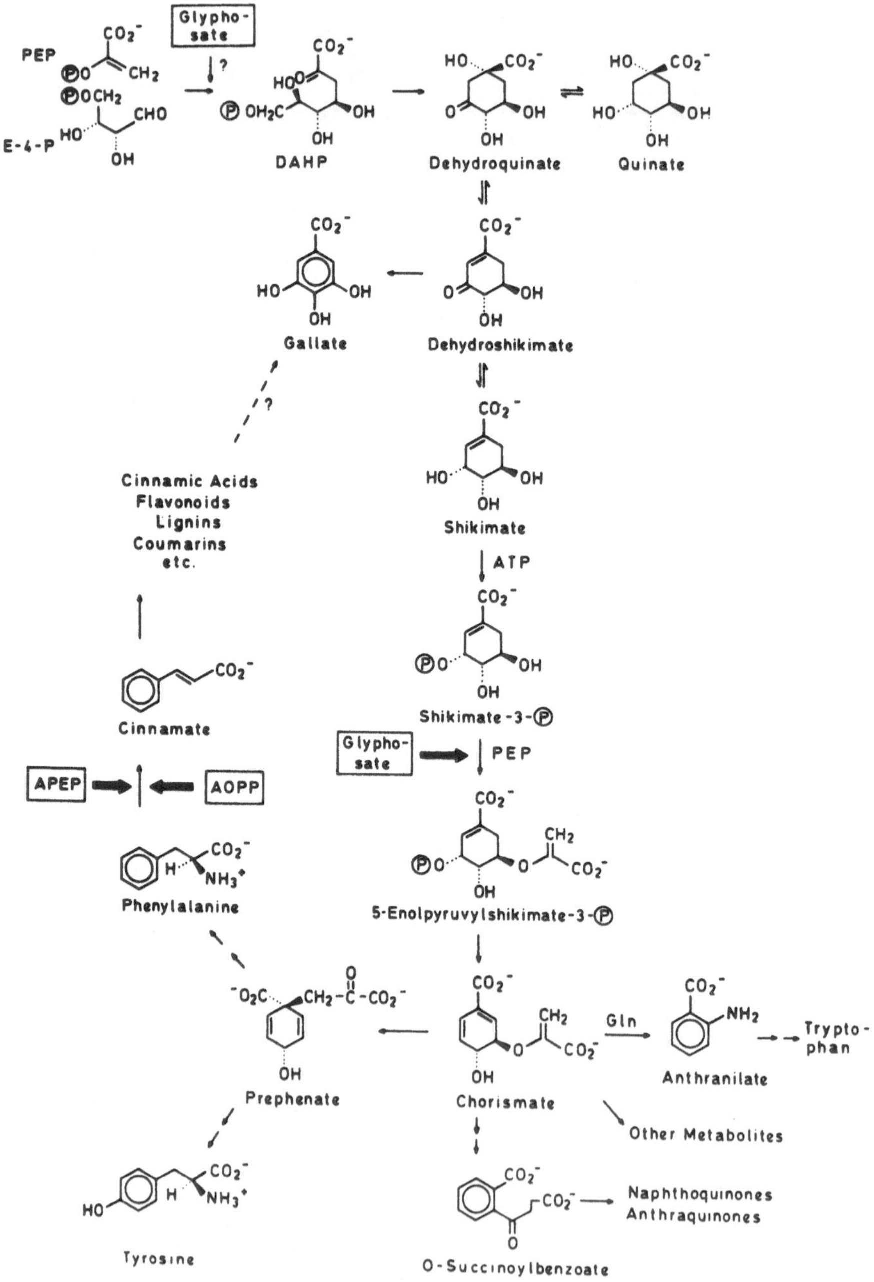

Fig. 2.   The shikimate pathway with the targets of inhibitor action.   For abbreviations, see text.

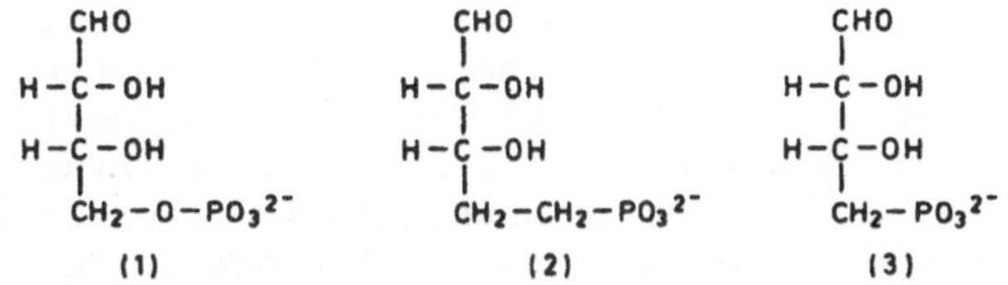

```
     CHO              CHO                CHO
      |                |                  |
  H–C–OH           H–C–OH             H–C–OH
      |                |                  |
  H–C–OH           H–C–OH             H–C–OH
      |                |                  |
  CH₂–O–PO₃²⁻      CH₂–CH₂–PO₃²⁻       CH₂–PO₃²⁻
     (1)              (2)                (3)
```

Fig. 3. Structures of erythrose 4-phosphate ($\underline{1}$) and its homophosphonate ($\underline{2}$) and phosphonate ($\underline{3}$) analogs.


point. For further discussions of these strategies, in relation to the design and the mode of action of pesticides, see the valuable monographs by Fedtke[5] and Corbett et al.[8]

In this chapter, the discussion will concentrate on two inhibitors with a reasonable claim to selective action on enzymes related to the shikimate pathway: glyphosate, which inhibits 5-enolpyruvylshikimate 3-phosphate (EPSP) synthase; and L-α-aminooxy-β-phenylpropionic acid (L-AOPP), an inhibitor of phenylalanine ammonia-lyase (PAL) (Fig. 2). In addition to introducing a novel inhibitor of PAL, ($\underline{R}$)-(1-amino-2-phenylethyl)phosphonic acid (APEP), previous and current efforts to design inhibitors of other shikimate pathway enzymes will be described. The treatment presented here will show that the deductions and predictions made on the basis of the abstract scheme in Figure 1 can be, and have been, tested on the basis of the real pathway presented in Figure 2.


INHIBITORS OF EARLY REACTIONS IN THE SHIKIMATE PATHWAY

The first enzyme which is specific for aromatic amino acid biosynthesis is 3-deoxy-D-arabino-heptulosonic 7-phosphate (DAHP) synthase (EC 4.1.2.1.5), which utilizes D-erythrose 4-phosphate (E4P) and phosphoenolpyruvate (PEP) as substrates (Fig. 2).[9–11] Le Maréchal et al.[12] prepared the isosteric homophosphonate of E4P, in which the C–O–P group is replaced by the C–CH₂–P group as well as the non-isosteric phosphonate analogue, in which the C–O–P group is replaced by the C–P group (Fig. 3). Both analogues served as substrates for the tyrosine-sensitive DAHP synthase of Escherichia coli, with $V_{max}$ values in the ratio 20:3.5:1 for E-4-P, the phosphonate, and the homophosphonate,

respectively.[12]   Inhibition of isozymes of DAHP synthase of
certain microorganisms and higher plants by glyphosate[13-16]
will be discussed in another section.   In the next reaction
in the pathway, catalyzed by 3-dehydroquinate synthase (EC
4.6.1.3) (Fig. 2), inorganic phosphate is released from the
substrate, DAHP.   Because of the resistance of the C-P bond
to hydrolytic cleavage,[17] both the phosphonate and homophosphonate of DAHP resist enzyme catalysis and act as
competitive inhibitors of dehydroquinate synthase.   Specifically, for the enzyme in E. coli, the $K_m$ for DAHP was
50 µM while the $K_i$ for the phosphonate and homophosphonate
analogs were 2.5 µM and 260 µM, respectively.[18]   Activities
of these compounds in vivo have not been reported.   The
phosphoenolpyruvate analogue, (Z)-phosphoenol-3-fluoropy-
ruvate, served as a substrate, as well as an inhibitor of
the phenylalanine-sensitive DAHP synthase of E. coli, and
3-fluoro-DAHP, in turn, served as a substrate for dehydroquinate synthase.[19]   These compounds appear to be interesting mainly for mechanistic studies.

While no studies on inhibitors of dehydroquinate
dehydratase (EC 4.2.1.10) are found in the literature, the
following enzyme in the pathway, shikimate dehydrogenase
(EC 1.1.1.25) was chosen by Baillie et al.[20] as the target
for a biochemically designed[8] herbicide.   To quote these
authors,

> "the shikimic acid biosynthetic pathway,
> which operates in microorganisms and higher
> plants but not in mammals, leads to several
> essential amino acids and also to a variety
> of other metabolites.   It was hoped, there-
> fore, that blockage of the pathway might
> have a damaging effect on plant metabolism,
> giving a useful commercial herbicide with
> selective toxicity."[20]

Derivatives of the dehydroshikimate analogue 1,6-dihydroxy-
2-oxoisonicotinic acid (Fig. 4) were fairly powerful
inhibitors of shikimate dehydrogenase from pea, but they
failed as herbicides in vivo.[20]   There is a certain irony
in the fact that in the same year in which Baillie et al.[20]
announced the inconclusive outcome of their efforts,
Jaworski[6] proposed inhibition of aromatic amino acid biosynthesis as the mode of action of glyphosate, which was just
emerging as a highly successful herbicide.

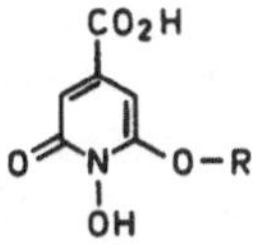

| R | $K_i$ ($\mu$M) |
|---|---|
| H | 120 |
| $CH_3$ | 800 |
| $C_2H_5$ | 100 |
| $n\,C_3H_7$ | 80 |
| $iso\,C_3H_7$ | 70 |

Fig. 4.  $K_i$ values for inhibition of pea shikimate dehydrogenase by 1,6-dihydroxy-2-oxoisonicotinic acid derivatives (see Reference 20).

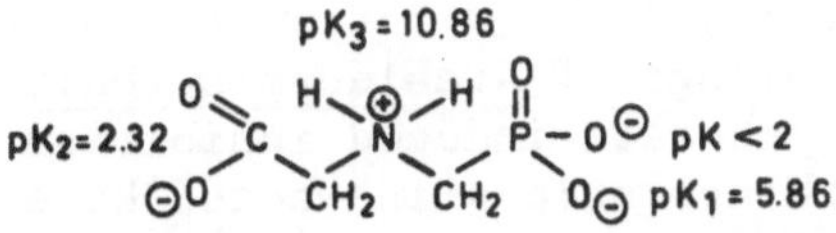

Fig. 5.  Structure and pK values of glyphosate.  Ionic form shown is prevalent at pH 8 ± 1 (see Reference 119).

GLYPHOSATE:  AN INHIBITOR OF AROMATIC AMINO ACID BIOSYNTHESIS

Studies in "Supercomplex Systems"

Since its introduction in 1971,[21] glyphosate [N-(phosphonomethyl)glycine, Fig. 5] has attracted much attention because of its novel chemistry,[22] its unique herbicidal properties,[23] and, last but not least, its commercial success.  It is a non-selective, broad spectrum, post-emergence herbicide that is phloem-mobile and is therefore readily translocated in plants following a typical source-sink relationship.[24]  As higher plants generally do not appear to be able to degrade, or otherwise inactivate glyphosate to any significant extent, the herbicide accumulates in the meristematic areas of the plant (i.e. areas of high metabolic activity[25-27]) where it produces its phytotoxic effects.  In the earliest study on the biochemical mode of action of glyphosate, Jaworski[6] observed that the inhibition of growth of the aquatic plant L. gibba (duckweed) by glyphosate was alleviated by phenylalanine.  Similarly, reversal of the inhibition of growth of R. japonicum by glyphosate required the presence of phenylalanine and tyrosine in the nutrient medium.  As chorismate is the last common precursor of these two

aromatic amino acids (Fig. 2), Jaworski suggested that
glyphosate may inhibit or repress chorismate mutase and/or
prephenate dehydratase. However, when Roisch and
Lingens[13,14] tested this hypothesis using a "defined (i.e.
cell-free) system" from E. coli, the activities of the
indicated enzymes were not affected by glyphosate. Inhibition of DAHP synthase and dehydroquinate synthase by mM
concentrations of glyphosate was reversed by $Co^{2+}$ ions and
may thus have been the result of the metal chelating properties of glyphosate, which it shares with other phosphonates.[28,29] Reversal of glyphosate toxicity by aromatic
amino acids was found in a variety of microorganisms and
cultured plant cells,[14,30,31] but, with the exception of
duckweed[6] and seedlings of Arabidopsis thaliana,[31] attempts
to reverse the glyphosate-induced inhibition of growth of
intact plants with aromatic amino acids have not met with
success. The anticipated depletion of pool sizes of the
aromatic amino acids when cells or plants were treated with
glyphosate, has been observed in some, but not all cases,[23]
and this has cast further doubt on the direct interference of
glyphosate with the biosynthesis of the aromatic amino
acids. In our hands, however, when the concentration of
phenylalanine in buckwheat hypocotyls was raised by blocking
its utilization in phenylpropanoid synthesis, glyphosate
clearly reduced the increase in this amino acid[32] (see also
Table 3).

## Studies in "Complex Systems"

While the results of the growth reversal experiments in
the "supercomplex systems" described above had stimulated
investigations at the level of "defined systems" (i.e. the
examination of certain metabolic reactions in cell-free
systems), no definitive answers as to the biochemical mode
of action of glyphosate had been obtained. Based on our
experience with inhibitors of phenylalanine ammonia-lyase
(see below), we included the "complex system" level (i.e.
the examination of a metabolic pathway in vivo) in our
strategy in order to define the limits more closely.
Hypocotyls from etiolated buckwheat seedlings provided a
system in which the rapid synthesis of phenylalanine-
derived products, such as anthocyanin and other phenylpropanoid compounds, can be very simply induced by
illumination.[33] Anthocyanins, in particular, can be
conveniently extracted and quantified and, at least in
buckwheat, are not subject to measurable turnover within

a period of days (Amrhein, unpublished). Their accumulation therefore reflects the net synthesis of aromatic compounds derived from an endproduct of the shikimate pathway, phenyl-alanine (Fig. 2). An inhibitor of a shikimate pathway enzyme might thus be expected to reduce the light-induced accumulation of the pigment. Glyphosate was found to inhibit anthocyanin, as well as general phenylpropanoid, synthesis in buckwheat hypocotyls, and the inhibition could be reversed by phenylalanine.[32] Compounds structurally related to glyphosate had little, or no effect. Inhibition of phenylalanine accumulation by glyphosate could be clearly demonstrated when the further metabolism of endogenous phenylalanine by deamination was inhibited by the simul-taneous application of an inhibitor of phenylalanine ammonia-lyase (see also Table 3).

The following experiments enabled us to reduce considerably the number of putative enzyme target sites of the herbicide: glyphosate inhibited the incorporation of radioactively labelled shikimate into all three aromatic amino acids, and the labelled precursor was metabolized to only a minor extent.[32] Following this lead, we found that buckwheat hypocotyls, as well as intact plants and cultured plant cells, accumulate large amounts of shikimic acid in the presence of the herbicide.[34,35] In buckwheat hypo-cotyls, there was an excellent correlation between the accumulation of shikimic acid and the depletion of antho-cyanin,[36] and a correlation between the accumulation of shikimic acid and the reduced growth of cultured plant cells (in the absence of amino acids in the nutrient medium) was also established.[35] In cultured cells of Galium mollugo, shikimic acid constituted up to 10% of the dry weight of the cells after 10 days of growth in the presence of 0.5 mM glyphosate.[34] Production of anthraquinones, for which chorismate and O-succinoylbenzoic acid (OSB) serve as precursors (Fig. 2; see also Chapter 9 by E. Leistner in this volume), was inhibited by glyphosate, and the inhibi-tion could be partially alleviated by the addition of chorismate or OSB to the nutrient medium. Inhibition of the synthesis of phenylpropanoid compounds by glyphosate and glyphosate-induced accumulation of shikimic acid were subsequently also observed in cultured cells of other plant species.[37,38] Although the level of total soluble hydroxy-phenolic compounds has been used to measure the effect of glyphosate on the metabolism of phenolic compounds,[39] this approach is not recommended because the concentrations of some phenolic compounds, such as protocatechuic acid and gallic acid, actually increase in glyphosate-treated tissues.[40] These phenolic compounds are presumably directly derived from dehydroshikimic acid,[41] and their increased formation in the presence of glyphosate probably reflects an "overflow" from accumulated shikimate (see Fig. 1; $P_2$). By allowing leaves of appropriate plant species to photoassimilate $^{14}CO_2$ in the presence of glyphosate, we have developed simple and inexpensive methods to produce uniformly $^{14}C$-labelled shikimic and gallic acids of high specific radioactivity in good yield[35] (Amrhein, Joop, and Topp, unpublished). While no accumulation of quinic acid (Fig. 2) was observed in glyphosate-treated buckwheat hypocotyls, quinic acid might be expected to accumulate in some tissues under the influence of glyphosate as another "overflow metabolite". The rate of shikimate accumulation in a tissue in the presence of glyphosate can be used as an indicator of the flow of carbon through the pathway under various physiological conditions.[37] Light stimulated the flux through the pathway in buckwheat hypocotyls, but the light-controlled, rate-limiting step was not identified.[42] Neither aromatic amino acids nor cinnamic acid added exogenously affected the flux in buckwheat hypocotyls, while these metabolites caused some depression of the flux in cauliflower florets (Reich, Holländer, and Amrhein, unpublished). In vivo experiments of this type may thus help to solve the question if the early section of the shikimate pathway is subject to feedback control by the end products.[43]

The information obtained from the application of glyphosate to "complex systems" strongly pointed to one of the following three enzymes as the target of the inhibitor in the shikimate pathway: shikimate kinase (EC 2.7.1.71), 5-enolpyruvylshikimate 3-phosphate (EPSP) synthase (EC 2.5.1.19), and chorismate synthase (EC 4.6.1.4). Jointly, these three enzymes convert shikimic acid to chorismic acid in a series of interesting reactions[9],[10] (Fig. 2). A "defined system" had therefore to be found in which the conversion of shikimic acid to chorismic acid could be conveniently studied.

Studies in "Defined Systems"

The paucity of information on shikimate pathway enzymes from higher plants, in particular shikimate

kinase,[44-48] EPSP-synthase,[47,48] and chorismate synthase
(still no report available at the time of this writing), as
well as the initial unavailability, with the exception of
shikimic acid, of the required substrates for these enzymes
called for the adoption of a cell-free bacterial system
capable of carrying out the desired sequence of reactions.[34,49] The conversion of shikimic acid to anthranilic
acid catalyzed by a preparation from the Aerobacter
aerogenes (=Klebsiella pneumoniae) strain 62-1 in the
presence of the appropriate cofactors, was inhibited by
low concentrations of glyphosate,[34] and EPSP synthase was
identified as the glyphosate-sensitive enzyme.[49] This
finding generated much interest in the enzyme and its
mechanism of action. As a result EPSP-synthase (alternative
name: 3-phosphoshikimate 1-carboxyvinyltransferase) is now
one of the better understood shikimate pathway enzymes from
microorganisms as well as plants. It has been purified,
partially or to homogeneity, from E. coli,[50,51] K. pneumoniae,[52-54] Neurospora crassa,[55] and seedlings of Pisum
sativum,[56] as well as from cultured cells of Corydalis
sempervirens[57] (Smart, Johänning, Müller and Amrhein,
submitted), Nicotiana silvestris,[58] and Petunia hybrida
(Steinrücken, Schulz, Amrhein, Porter, and Fraley,
submitted). The aroA gene which codes for EPSP synthase
has been cloned, and strains of E. coli which overproduce
EPSP synthase have been obtained by transformation with a
multicopy plasmid carrying this gene.[51,59,60] Mechanistic
and structural studies of the EPSP synthase of E. coli are
thus greatly facilitated by the availability of milligram,
and even gram amounts of purified enzyme. Furthermore,
the fact that these genetically engineered overproducing
strains are highly tolerant to glyphosate[51,59] provides
in vivo genetic evidence that EPSP synthase is the target
of glyphosate in these bacteria (see also section on
glyphosate tolerance below). EPSP synthases from plant
and bacterial sources are monofunctional, i.e. separable
from other enzymes of the shikimate pathway, while in
N. crassa EPSP synthase is part of a multifunctional enzyme
with four other shikimate pathway activities.[61] The monofunctional EPSP synthases are generally monomeric, and their
molecular weights are around 45,000 D. The amino acid
sequences for EPSP synthases from E. coli[62] and Salmonella
typhimurium[63] have been deduced from the nucleotide
sequences of the respective aroA genes. In both species,
the complete polypeptide chain of EPSP synthase consists of
427 amino acids and has an $M_r$ of 46, 100. The known

sequences will greatly aid efforts to define the active site
of the enzyme, as well as its catalytic mechanism and the
mechanism of its inhibition by glyphosate.

EPSP synthase catalyzes a rare type of reaction in
which the intact enolpyruvyl (carboxyvinyl) group is transferred from phosphoenolpyruvate (PEP) to the 5-hydroxyl
group of shikimate 3-phosphate (Fig. 6). This type of
reaction is encountered in only one other known enzymatic
process: the transfer of the enolpyruvyl group to UDP-N-
acetylglucosamine during the biosynthesis of the bacterial
cell wall peptidoglycan.[64]

A reversible addition/elimination mechanism was
proposed for EPSP synthase[65] and supported by the results
of isotope-exchange studies[66,67] (see also Chapter 2 by
H.G. Floss in this volume). The 5-hydroxyl group of S-3-P
is thought to add to C-2 of PEP, and C-3 thus transiently
becomes a methyl group. Elimination of $P_i$ from the intermediate then generates the product (Fig. 6). However,
using 4,5-dideoxyshikimate, an analogue of S-3-P which
lacks the 5-OH group, Anton et al.[53] showed that this group
of S-3-P is not required to form the complex in which C-3
of PEP becomes a methyl group. These authors therefore
proposed the intermediate formation of an enzyme-PEP
complex.

All published reports on the inhibition of EPSP
synthase by glyphosate agree that the inhibition is competitive with respect to PEP. Table 1 surveys the published
$K_i$ values. The values must be compared with caution,
however, because the conditions under which they were

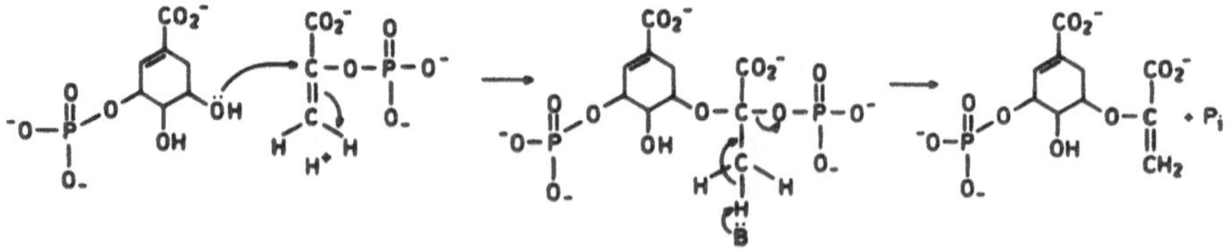

Fig. 6. Mechanism proposed for the reaction catalyzed by
5-enolpyruvylshikimate 3-phosphate synthase (see Reference
66).

Table 1. $K_i$ values for competitive inhibition, with respect to PEP, of EPSP synthases by glyphosate.

| Sources of Enzyme | $K_i$ (µM) | Reference |
|---|---|---|
| Escherichia coli | 0.9 | 62 |
| Escherichia coli | ca. 40 | 70 |
| Escherichia coli | 0.8 | a |
| Klebsiella pneumoniae | 1 | 68 |
| Klebsiella pneumoniae | 10 | 53 |
| Neurospora crassa | 1.1 | 55 |
| Candida maltosa | 12 | 69 |
| Corydalis sempervirens | 10 | 57 |
| Corydalis sempervirens | 0.32 | b |
| Nicotiana silvestris | 1.25 | 58 |
| Petunia hybrida | 0.17 | c |
| Pisum sativum | 0.08 | 56 |

a   Krüper and Amrhein, unpublished.
b   Smart, Johänning, Müller, and Amrhein, submitted.
c   Steinrücken, Schulz, Amrhein, Porter, and Fraley, submitted.

obtained may not be comparable. In particular, the pH of the incubation mixture strongly affects the sensitivity of EPSP synthase to glyphosate; between pH 6 and 8 there is a striking increase in the sensitivity. That is, glyphosate inhibits the enzyme more strongly with increasing pH.[58,68] The ionic form inhibiting EPSP synthase must therefore be

$$^-O_2CCH_2\overset{+}{N}H_2CH_2\ PO_3{}^{2-}$$

which is evident from glyphosate's pK values (Fig. 5), but the pH effect may as well indicate the presence of a weak acid in the glyphosate-binding region of the enzyme. On the basis of inactivation experiments with group specific reagents we have suggested that the weak acid may be the sulfhydryl group of a cysteinyl residue of EPSP synthase.[68] It has been shown for EPSP synthase from E. coli,[50] N. crassa[55] and N. silvestris[58] that glyphosate displays uncompetitive inhibition with respect to S-3-P. This type of inhibition would indicate a sequential binding mechanism for EPSP synthase in which glyphosate competes with PEP for

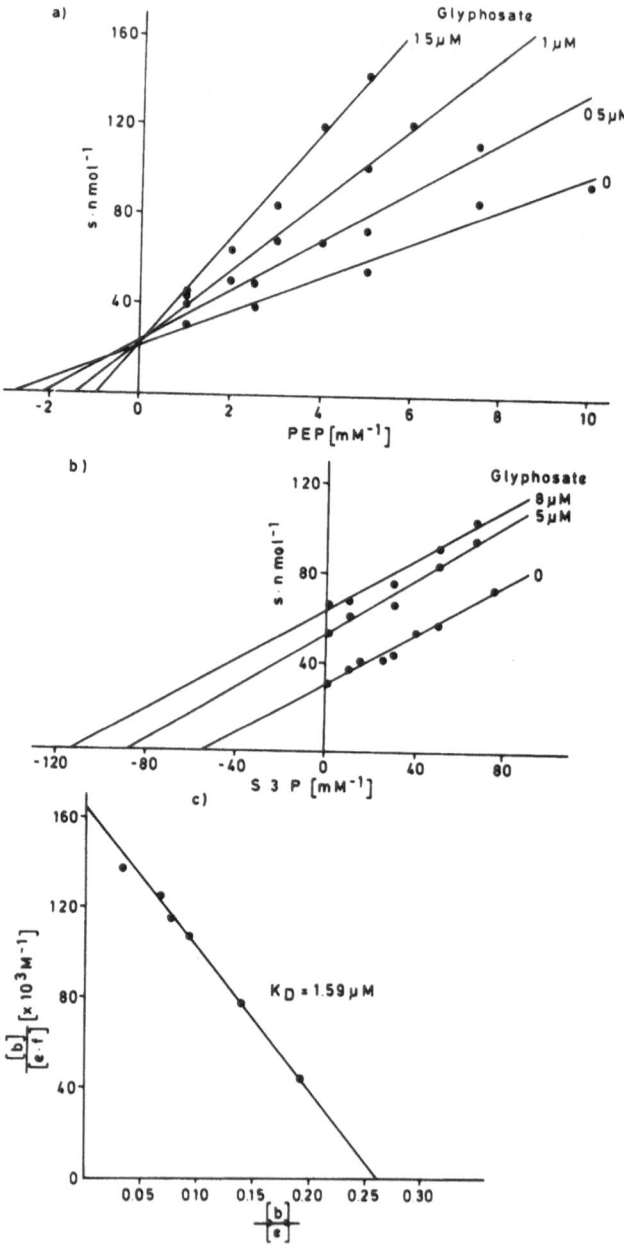

Fig. 7. Lineweaver–Burk plots of inhibition of E. coli EPSP synthase by glyphosate with (a) PEP or (b) S-3-P as variable substrates. (c) Scatchard plot of binding of glyphosate to E. coli EPSP synthase in the absence of substrates.

binding to an [EPSP synthase · S-3-P] complex.  This
mechanism predicts that neither PEP nor glyphosate can
bind to the enzyme in the absence of S-3-P.  We have
pointed out, however, that glyphosate must be able to bind
to the EPSP synthase of K. pneumoniae, because it effectively protects the enzyme against inactivation by group
specific reagents.[68]  Likewise, glyphosate protected the
EPSP synthase of E. coli against inactivation by N-
ethylmaleimide and against tryptic digestion, and,
furthermore, it provided protection against inactivation
of the Petunia enzyme by $O_2$ (Schulz and Amrhein,
unpublished).

Unequivocal evidence for the binding of glyphosate
to EPSP synthase was provided by equilibrium dialysis of
the E. coli enzyme with $^{14}C$-labelled glyphosate.  The
Scatchard plot (Fig. 7) revealed a single binding site for
glyphosate with $K_D$ = 1.59 µM, which is somewhat higher than
the $K_i$ value for competitive inhibition (0.8 µM using the
coupled assay,[55,56] 0.65 µM when $P_i$ release[68] was measured)
obtained by kinetic analysis.  In the presence of
increasing concentrations of S-3-P the $K_D$ decreased until
it came close to the $K_i$ value (Krüper and Amrhein, unpublished).  This phenomenon indicates that S-3-P enhances
the affinity of EPSP synthase for glyphosate, but that it
is not required per se for the binding of the inhibitor.
Using the sensitive coupled assay[55,56] for determination
of EPSP synthase activity at low substrate concentrations,
we have now also found that glyphosate inhibition is
uncompetitive with respect to S-3-P for the enzymes of
E. coli (Fig. 7), C. sempervirens, and P. hybrida.  We
have difficulties in reconciling the results of the kinetic
analyses with the results of the inactivation and binding
studies.  Bode et al.[69] have reported that inhibition of
the EPSP synthase of Candida maltosa by glyphosate is
uncompetitive with respect to S-3-P at substrate concentrations lower than 40 µM, but non-competitive at higher
concentrations.  A satisfactory answer will, however,
depend on the availability of more information on the
structure of the active center of EPSP synthase.

We[68] and others[53] have proposed that glyphosate may
act as a transition-state analogue of PEP in which the
positive charge of the imino group of glyphosate is equivalent to the charge of the carbonium ion (C-2) of the
presumed transition state of PEP (Fig. 8).  Thus,

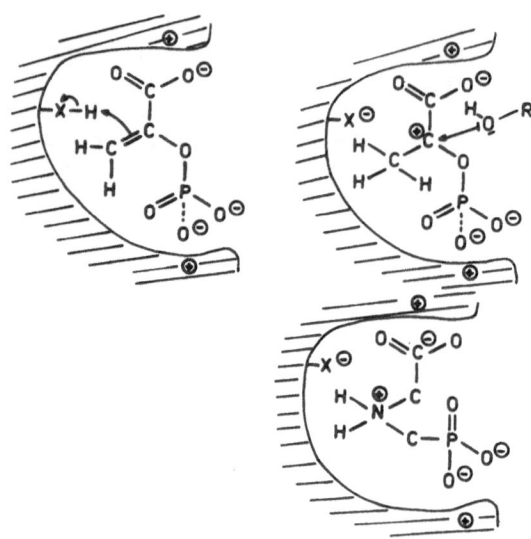

Fig. 8.  Proposed binding of PEP and glyphosate to the
active center of EPSP synthase (reprinted with permission
from Reference 68).

glyphosate cannot be considered simply an analog of PEP,
and the lack of an effect of glyphosate on nine other
enzymes which utilize PEP as substrate[68] is in agreement
with this conclusion.

It must be mentioned here that the cobalt-dependent
isozyme of mungbean DAHP synthase,[15] as well as the tyrosine-
sensitive isozyme of the DAHP synthase of C. maltosa[16] are
inhibited by glyphosate in millimolar concentrations.
Surprisingly, with the mungbean enzyme the inhibition was
found to be competitive with respect to erythrose 4-
phosphate, while it was competitive with PEP in the case
of the C. maltosa enzyme.  Rubin et al.[15,58] have suggested
that the inhibition of the $Co^{2+}$-dependent DAHP synthase
isozyme may have physiological significance.  According to
the interpretation of their data for mungbeans and N.
silvestris, which have not yet fully appeared in print,
the $Mn^{2+}$-dependent, glyphosate-insensitive, DAHP synthase
isozyme is localized in the chloroplast, while the $Co^{2+}$-
dependent, glyphosate-sensitive isozyme is localized in the
cytoplasm.  In the presence of glyphosate, EPSP synthase

(thought to be present in the cytoplasm and the chloroplast) as well as $Co^{2+}$-dependent DAHP synthase activities would be blocked, and S-3-P would originate through the action of the $Mn^{2+}$-dependent DAHP synthase in the chloroplast. As the latter isozyme appears to be subject to feedback inhibition by L-arogenate, depletion of this metabolite (due to inhibition of EPSP synthase) would lead to the continuous flow of substrate (PEP and E-4-P) into the shikimate pathway and to the unrestrained accumulation of S-3-P, or shikimate, respectively, which would ultimately result in an energy-drain.

While the scope of our contribution does not permit a full discussion of the intricate question of the compartmentation of the shikimate pathway in the cells of higher plants,[71-76] (see also Chapter 3 by R.A. Jensen in this volume) the following observations are relevant to the mode of action of glyphosate: using aqueous cell fractionation procedures, it has been shown that EPSP synthase in pea seedling shoots,[75] sunflower cotyledons, and young spinach leaves (Joop and Amrhein, unpublished) is predominantly chloroplastic. Fractionation of freeze-stopped spinach leaves in nonaqueous media[76] resulted in the recovery of $\geq$ 90% of EPSP synthase in the chloroplast fraction, and $\leq$ 10% in the cytosol fraction. After application of $^{14}C$-glyphosate to an assimilate exporting leaf of spinach, and transport of glyphosate into the young leaves for three days, the concentrations of glyphosate in the chloroplast stroma, the cytosol, and the vacuole were 65, 1140, and 13 μM respectively. S-3-P was recovered mainly from the chloroplast and cytosol, while shikimate was predominantly in the cytosol and vacuole (Joop and Amrhein, unpublished). The cytosolic concentration of glyphosate ($\geq$ 1 mM) would be high enough to significantly block the (cytoplasmic) $Co^{2+}$-dependent isozyme of DAHP synthase described by Rubin et al.[58] Inhibition of this isozyme therefore appears possible under physiological conditions.

The much lower concentration of glyphosate in the chloroplast stroma, on the other hand, would still be sufficiently high to inhibit the EPSP synthase which is highly sensitive to glyphosate. Chloroplasts in young leaves of tomato plants into which glyphosate had been imported by phloem transport from an older leaf appeared enormously swollen in electron micrographs after S-3-P and shikimate had begun to accumulate in the leaves, but long before the leaves exhibited toxic symptoms such as chlorosis (Mollenhauer, Smart, and Amrhein, unpublished). This observation indicates that the action of glyphosate might, indeed, be chloroplast-related.

Tissues and cultured cells accumulate predominantly shikimate, rather than S-3-P, in the presence of glyphosate,[57] and in cultured buckwheat cells shikimate was shown to accumulate in the vacuole.[77] As neither S-3-P nor shikimate compete with glyphosate for binding sites on EPSP synthase, and as PEP does not accumulate in glyphosate-treated cells (Amrhein, unpublished), the inhibition of EPSP synthase is not overcome. This example illustrates that in order to effectively inhibit an enzyme in vivo the inhibitor need not necessarily be of the tight-binding or suicide ($K_{cat}$; active-site-directed irreversible) type.

MECHANISMS OF GLYPHOSATE TOLERANCE

Tolerance to a metabolic inhibitor can be acquired by cells and multicellular organisms by a number of mechanisms. These are:

1. Reduced uptake of the inhibitor
2. Detoxification (by degradation, conjugation, etc.)
3. Increase in the concentration of the metabolite which competes with the inhibitor for a binding site
4. Reduced affinity of the inhibitor for the target (binding) site
5. Increased number of target sites as a consequence of their overproduction.

The latter two mechanisms are target-related and are of particular interest here, because discovery of a target (i.e. EPSP synthase)-related mechanism of glyphosate tolerance would lend further credibility to the postulated mode of action of glyphosate. It has already been mentioned that the amplification of the aroA gene from E. coli results in overproduction of EPSP synthase and a concomitantly increased tolerance to glyphosate.[51,59] During repeated transfer into nutrient media containing glyphosate A. aerogenes developed a tolerance to glyphosate which was paralleled by an increased specific activity of EPSP synthase.[78] The same phenomenon (i.e. a relationship

between the adaptation to glyphosate and large increases in
the extractable activity of EPSP synthase) has been observed
in cultured cells of the higher plants C. sempervirens[78,79]
(Smart, Johänning, Müller, and Amrhein, submitted), Daucus
carota,[80] Fagopyrum esculentum (Holländer-Czytko,
unpublished), and P. hybrida (Steinrücken, Schulz, Amrhein,
Porter, and Fraley, submitted).  For C. sempervirens and
P. hybrida it was conclusively established that the
glyphosate-tolerant cells produced the same molecular
(glyphosate-sensitive) species of EPSP synthase as the
glyphosate-sensitive cells.  Quantitation of EPSP synthase
by immunoassay in C. sempervirens cells showed that the
EPSP synthase protein constituted ca. 0.06% of the total
soluble protein in glyphosate-sensitive cells, while its
fraction in the glyphosate-tolerant cells was nearly 2.6%.
The extractable activities of five other shikimate pathway
enzymes were not increased in the glyphosate-tolerant
cells, which suggests a selective overproduction of EPSP
synthase.  It remains to be seen, if increased EPSP synthase
gene copy number,[81] or an alteration of transcriptional
regulation is responsible for the overproduction of EPSP
synthase.

EPSP synthases resistant to glyphosate have been
identified and partially characterized in glyphosate-
resistant mutants of S. typhimurium[63,82] and K.
pneumoniae.[83,84]  Furthermore, EPSP synthases with a
sensitivity to glyphosate 50 to 100-fold lower than the
sensitivity of other bacterial and plant EPSP synthases
were identified in a number of species of the genus
Pseudomonas.[85]  PEP, but not glyphosate, protected the
mutant EPSP synthase of K. pneumoniae against inactivation
by group-specific reagents, which indicates that the mutant
enzyme no longer efficiently binds glyphosate.[84]  The
mutation in the S. typhimurium enzyme has been traced back
to a single base pair change resulting in a substitution
of proline for serine at the 101st codon of the protein.[63]
When introduced into E. coli, the mutation conferred high
resistance to glyphosate.[82]

Taken together, the investigation of glyphosate
resistance, both in cultured plant cells and in bacteria,
supports the conclusion that EPSP synthase is a major
target of glyphosate.  While no glyphosate-resistant plant
EPSP synthase has been identified, current efforts are
directed towards the transformation of crop plants with

the gene coding for the glyphosate-resistant EPSP
synthase.[63,82,86,87]

Of the enzymes subsequent to EPSP synthase, only
chorismate mutase has been the subject of inhibitor
studies (see Chapter 5 by P. Bartlett in this volume).

INHIBITORS OF PHENYLALANINE AMMONIA-LYASE

Phenylalanine ammonia-lyase (PAL, EC 4.3.1.5) is not
considered an enzyme of the shikimate pathway proper, but
since in higher plants it channels a large flux of fixed
carbon into phenolic compounds, in particular lignin, the
shikimate pathway must furnish its substrate, phenylalanine,
in sufficient amounts. Literature on PAL has periodically
been reviewed (literature in References 88 and 89) and I
will concentrate here on a selective and brief discussion
of our work on PAL inhibitors in the Bochum laboratory.

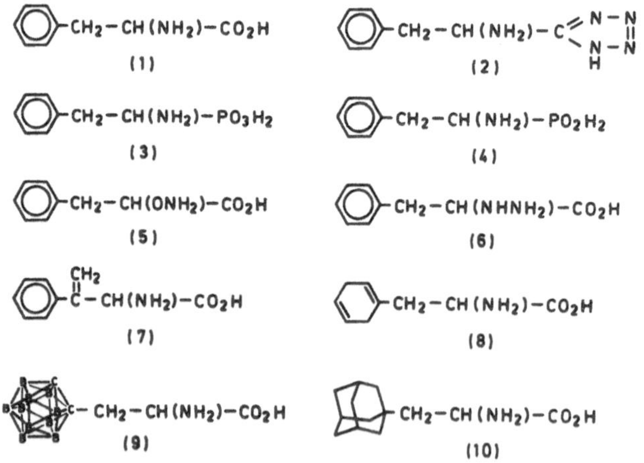

Fig. 9. Structures of phenylalanine (1) and of phenylalanine
analogs discussed in the text. (2), 5-(1-amino-2-
phenylethyl)tetrazole; (3), (1-amino-2-phenylethyl)
phosphonic acid; (4), (1-amino-2-phenylethyl)phosphonous
acid; (5), α-aminooxy-β-phenylpropionic acid; (6),
α-hydrazino-β-phenylpropionic acid; (7), β-methylene-
phenylalanine; (8), (1,4-cyclohexadienyl)-alanine; (9),
o-carboranylalanine; (10), adamantylalanine.

When we were developing an assay for the estimation
of PAL activity in intact cells,[90] we employed $\alpha$-aminooxy-
acetate[2] (AOA) as a transaminase inhibitor.  When it became
clear that AOA was a weak inhibitor of buckwheat PAL ($K_i$ =
120 $\mu$M)[91] we rationalized that the aminooxy analog of
L-phenylalanine, L-$\alpha$-aminooxy-$\beta$-phenylpropionic acid (AOPP,
5 in Fig. 9) should have an increased inhibitory ability.
This was confirmed beyond our expectations, as L-AOPP
inhibited buckwheat PAL with a $K_i$ of 1.4 nM[7] (Table 2) and
also inhibited the biosynthesis of phenylpropanoid compounds
in vivo without deleterious effects on normal plant development.[7,92]  Selective accumulation of phenylalanine in
AOPP-treated tissues provided evidence for inhibition of
PAL in vivo.[93]  As the result of a detailed study[94] it was
subsequently proposed by Hanson that the enantiomers of
AOPP pack into the active site of PAL in a mirror image
relationship and act as transition state analogs in the
elimination reaction.  Soybean PAL appeared to be irrever-
sibly inhibited by L-AOPP,[95] which can presumably be
explained by the slow dissociation of the enzyme-ligand
complex.[96]

L-AOPP has been used with advantage in studies on the
regulation of the level of PAL in plant tissues,[97-100] on
the turnover of isoflavones,[101,102] on phytoalexin function
in pathogen resistance in soybean,[103] on the involvement of
products of PAL activity in the regulation of cell elonga-
tion,[104] and on the function of lignin deposition in the
secondary cell wall of xylem vessels.[105,106]  In the latter
investigation, the acknowledged primary role of xylem
lignification in the ability of the dead, water-conducting
vessel to withstand the tensile forces generated during
transpiration[107] could for the first time be experimentally
examined.  (A promising approach to reduce lignification by
inhibition of cinnamyl alcohol dehydrogenase activity has
recently been reported.[108])  However, considering its low
$K_i$ value for PAL, L-AOPP is not a very efficient inhibitor
of phenylpropanoid synthesis in vivo.  In addition it
cannot be considered an absolutely specific inhibitor for
PAL.[3,91,109,110]  Over the years therefore, we have
continued our search for other suitable inhibitors of PAL.
In Figure 9, the structures of some of the more interesting
phenylalanine analogs which have been tested for their
effect on PAL are shown and their $K_i$ values for buckwheat
PAL are summarized in Table 2.  The data for the $\alpha$-aminooxy
and $\alpha$-hydrazino analogs (5 and 6) have previously been

Table 2.  Inhibition constants ($K_i$) of competitive inhibitors of buckwheat phenylalanine ammonia-lyase

| Inhibitors | $K_i$ ($\mu$M) |
|---|---|
| L-$\alpha$-aminooxy-$\beta$-phenylpropionic acid | 0.0014 |
| D-$\alpha$-aminooxy-$\beta$-phenylpropionic acid | 0.025 |
| L-$\alpha$-hydrazino-$\beta$-phenylpropionic acid | 0.15 |
| D-$\alpha$-hydrazino-$\beta$-phenylpropionic acid | 18 |
| (R)-(1-amino-2-phenylethyl)phosphonic acid | 1.5 |
| (S)-(1-amino-2-phenylethyl)phosphonic acid | 11.6 |
| (R)-(1-amino-2-phenylethyl)phosphonous acid | 35 |
| (S)-(1-amino-2-phenylethyl)phosphonous acid | 205 |
| D,L-$\beta$-methylene-phenylalanine | 35 |
| L-carboranylalanine | 29 |
| L-adamantylalanine | no inhibition |
| L-5-(1-amino-2-phenylethyl)tetrazole | weak inhibition |

Data from Reference 93 and unpublished data from the author's laboratory.  For structures of the inhibitors, see Figure 9.

reported,[93] and 3-(1,4-cyclohexadienyl)-L-alanine (8), which we have not tested ourselves, is included because of the very thorough study of Hanson and his colleagues[111] on the interaction of this compound with PAL ($K_i$ for PAL from maize = 2.3 mM).  Compound (3), the phosphonic analog of phenylalanine, (1-amino-2-phenylethyl)phosphonic acid (APEP) is a promising novel competitive inhibitor of PAL (Laber, Kiltz, and Amrhein, submitted for publication).  The $K_i$ value of (R)-APEP (which has the same configuration as L-phenylalanine) is 1.5 $\mu$M, which is $10^3$ times higher than the $K_i$ value of L-AOPP, but the potencies of the two inhibitors in vivo compare quite favourably.  APEP caused a dramatic and specific increase in the concentration of free phenylalanine in buckwheat hypocotyls, which could effectively be reduced by the simultaneous application of glyphosate (Table 3).  The increase must therefore have been due to the synthesis of phenylalanine de novo in the tissue.  The potency of the corresponding phosphonous analog (4), an isostere of phenylalanine,[112] is about 20 times less than the phosphonic analog (Table 2).  The additional charge carried by the phosphonate at pH 8.8 (pH of PAL assay) is

Table 3.  Effect of (R,S) (1-amino-2-phenylethyl)phosphonic
acid (APEP) and glyphosate on the phenylalanine content of
buckwheat hypocotyls.

| | Phenylalanine (nmol $\cdot$ g$^{-1}$ fresh weight) |
| --- | --- |
| Zero time control | 59 |
| 24 h, Control | 31 |
| 24 h, 1 mM APEP | 1321 |
| 24 h, 1 mM Glyphosate | 19 |
| 24 h, APEP + Glyphosate | 105 |

presumably responsible for its higher affinity for PAL in
comparison with the corresponding phosphonous and carboxylic
acids, because the active site of PAL must contain a counter
ion to the carboxyl group of phenylalanine.  As might be
expected, the methylphosphinic analog of phenylalanine,
which carries the - C - PH(CH$_3$)0 group, does not inhibit
PAL.  Likewise, the tetrazole analog of phenylalanine (2)
is only a weak inhibitor of PAL.  In this instance, the
steric fit and the degree of ionization may not be significantly different, but there is a greater area available
within the tetrazole ring for the distribution of the
negative charge.[113]  β-Methylene-phenylalanine (7), which
was synthesized as a potential irreversible inhibitor of
pyridoxal dependent enzymes such as aromatic amino acid
decarboxylases or aminotransferases,[114] was a fairly potent
inhibitor considering that the racemate was tested (Table 2).
L-o-carboranylalanine (9) containing 10-boron atoms is a
very interesting analog of phenylalanine in that the dimen-
sions of the carborane cage correspond to the steric bulk of
a benzene ring rotating about its 1,4-axis.[115]  L-adamantyl-
alanine (10) has space-filling properties similar to those
of carboranylalanine,[116] but it lacks the pseudoaromaticity
of the latter compound.  Of these two "fat" amino acids,
only the pseudoaromatic carboranylalanine exhibits affinity
for PAL as indicated by a K$_i$ value of 29 μM (Table 2), which
is even lower than the K$_m$ value of buckwheat for its
substrate, phenylalanine (45 μM).  Thus, the active center
of PAL must be sufficiently spacious to accommodate the
steric bulk of the carborane cage.

CONCLUSION

When knowledge was needed of the biochemistry of the shikimic acid pathway in higher plants, studies on the mode of action of glyphosate showed that only scattered information was available. This situation is changing, and while the specific inhibitory interaction of glyphosate with EPSP synthase is at the center of much current interest and many efforts, one can expect that the entire pathway, its biochemistry as well as its physiology, will be closely scrutinized. The recent discovery that another class of herbicides, the sulfonylureas, also attack an enzyme of an amino acid biosynthetic pathway (acetolactate synthase, the first enzyme in the biosynthesis of the branched chain amino acids valine and isoleucine)[86,117,118] emphasizes the need for a better understanding of the biosynthesis of the essential amino acids in higher plants. The strategies used in elucidating the respective modes of action of glyphosate and the sulfonylureas, the rational progression from higher to lower levels of complexity, and the exploitation of the potential of microorganisms as well as of plant tissue cultures, were remarkably similar. While the results of these investigations will stimulate efforts to rationally design new herbicides, it is hoped that, as the "shavings" of these designs, new metabolic inhibitors will become available which will help us to learn more about the mechanisms of plant enzymes, the regulation of plant metabolism, and the function of plant products.

ACKNOWLEDGMENTS

The author's investigations were supported by grants from the Deutsche Forschungsgemeinschaft, Bonn, the Fonds der Chemischen Industrie, Frankfurt, and Monsanto Agricultural Products Co., St. Louis. The willingness of the following chemists to generously supply us with samples of phenylalanine analogs is much appreciated: L. Maier (Ciba Geigy AG); J.S. Morley (ICI Pharmaceuticals Division); R. Schwyzer (ETH Zürich); and J. Wemple (University of Detroit).

REFERENCES

1. AMRHEIN, N. 1979. Biosynthesis of cyanidin in buck-wheat hypocotyls. Phytochemistry 18: 585-589.
2. JOHN, R.A., A. CHARTERS, L.J. FOWLER. 1978. The reaction of amino-oxyacetate with pyridoxal phosphate-dependent enzymes. Biochem. J. 171: 771-779.
3. AMRHEIN, N., D. WENKER. 1979. Novel inhibitors of ethylene production in higher plants. Plant Cell Physiol. 20: 1635-1642.
4. FUJINO, D.W., M.S. REID, S.F. YANG. 1980. Effects of amino-oxyacetic acid on postharvest characteristics of carnation. Acta Hortic. 113: 59-64.
5. FEDTKE, C. 1982. Biochemistry and physiology of herbicide action. Springer, Berlin - Heidelberg - New York, 202 pp.
6. JAWORSKI, E.G. 1972. The mode of action of N-phosphonomethylglycine. Inhibition of aromatic amino acid biosynthesis. J. Agric. Food Chem. 20: 1195-1198.
7. AMRHEIN, N., K.-H. GÖDEKE. 1977. α-Aminooxy-β-phenylpropionic acid, a potent inhibitor of L-phenylalanine ammonia-lyase in vitro and in vivo. Plant Sci. Lett. 8: 313-317.
8. CORBETT, J.R., K. WRIGHT, A.C. BAILLIE. 1984. The biochemical mode of action of pesticides. 2nd Edition, Academic Press, London, 382 pp.
9. HASLAM, E. 1974. The shikimate pathway. Butterworths, London, 316 pp.
10. WEISS, U., J.M. EDWARDS. 1980. The biosynthesis of aromatic compounds. John Wiley and Sons, New York, 728 pp.
11. HERRMANN, K.M. 1983. The common aromatic biosynthetic pathway. In Amino Acids: Biosynthesis and Genetic Regulation. (K.M. Herrmann, R.L. Somerville, eds.), Addison-Wesley, Reading, Massachusetts, pp. 301-322.
12. LE MARÉCHAL, P., C. FROUSSIOS, M. LEVEL, R. AZERAD. 1980. Enzyme properties of phosphonic analogues of D-erythrose 4-phosphate. Biochem. Biophys. Res. Commun. 92: 1097-1103.
13. ROISCH, U., F. LINGENS. 1974. Wirkung des Herbizids N-(Phosphonomethyl)glycin auf die Biosynthese der aromatischen Aminosäuren. Angew. Chem. 13: 400.
14. ROISCH, U., F. LINGENS. 1980. Zur Wirkungsweise des Herbizids N-(Phosphonomethyl)Glycin. Einfluß von

N-(Phosphonomethyl)Glycin auf das Wachstum und auf die Enzyme der Aromatenbiosynthese von Escherichia coli. Hoppe-Seyler's Z. Physiol. Chem. 361: 1049-1058.

15. RUBIN, J.L., C.G. GAINES, R.A. JENSEN. 1982. Enzymological basis for herbicidal action of glyphosate. Plant Physiol. 70: 833-839.

16. BODE, R., C. MELO RAMOS, D. BIRNBAUM. 1984. Inhibition of tyrosine-sensitive 3-deoxy-D-arabino-heptulosonate 7-phosphate synthase by glyphosate in Candida maltosa. FEMS Microbiol. Lett. 23: 7-10.

17. ENGEL, R. 1977. Phosphonates as analogues of natural phosphates. Chem. Rev. 77: 349-367.

18. LE MARÉCHAL, P., C. FROUSSIOS, N. LEVEL, R. AZERAD. 1980. The interaction of phosphonate and homophosphonate analogues of 3-deoxy-D-arabino heptulosonate 7-phosphate with 3-dehydroquinate synthetase from Escherichia coli.

19. PILCH, P.F., R.L. SOMERVILLE. 1976. Fluorine-containing analogues of intermediates in the shikimate pathway. Biochemistry 15: 5315-5320.

20. BAILLIE, A.C., J.R. CORBETT, J.R. DOWSETT, P. McCLOSKEY. 1972. Inhibitors of shikimate dehydrogenase as potential herbicides. Pestic. Sci. 3: 113-120.

21. BAIRD, D.D., R.P. UPCHURCH, W.B. HOMESLEY, J.E. FRANZ. 1971. Introduction of a new broad spectrum post emergence herbicide class with utility for herbaceous perennial weed control. Proc. North Centr. Weed Control Conf. 26: 64-68.

22. FRANZ, J.E. 1979. Glyphosate and related chemistry. In Advances in Pesticide Science, Part 2. (H. Geissbühler, ed.), Pergamon Press, Oxford and New York, pp. 139-147.

23. HOAGLAND, R.E., S.O. DUKE. 1982. Biochemical effects of glyphosate [N-(phosphonomethyl)glycine]. In Biochemical responses induced by herbicides. (D.E. Moreland, J.B. St. John, F.D. Hess, eds.), ACS Symposium Series 181: 175-205.

24. GOUGLER, J.A., D.R. GEIGER. 1981. Uptake and distribution of N-(phosphonomethyl)glycine in sugar beet plants. Plant Physiol. 68: 668-672.

25. SPRANKLE, P., W.F. MEGGITT, D. PENNER. 1975. Absorption, action, and translocation of glyphosate. Weed Sci. 23: 235-240.

26. HADERLIE, L.C., F.W. SLIFE, H.S. BUTLER. 1978. $^{14}$C-
    glyphosate absorption and translocation in
    germinating maize (Zea mays) and soybean (Glycine
    max) seeds and in soybean plants. Weed Res. 18:
    269-273.

27. ASHTON, F.M., A.S. CRAFTS. 1981. Mode of Action of
    Herbicides. 2nd Edition, John Wiley and Sons,
    New York, pp. 236-253.

28. KABACHNIK, M.I., T.Y. MEDVED, N.M. DYATLOVA, M.V.
    RUDOMINO. 1974. Organophosphorus complexones.
    Russ. Chem. Rev. (Engl. Transl.) 43: 733-744.

29. GLASS, R.L. 1984. Metal complex formation by
    glyphosate. J. Agric. Food Chem. 32: 1249-1253.

30. HADERLIE, L.C., J. WIDHOLM, F.W. SLIFE. 1977. Effect
    of glyphosate on carrot and tobacco cells. Plant
    Physiol. 60: 40-43.

31. GRESSHOFF, P. 1979. Growth inhibition of glyphosate
    and reversal of its action by phenylalanine and
    tyrosine. Aust. J. Plant Physiol. 6: 177-185.

32. HOLLÄNDER, H., N. AMRHEIN. 1980. The site of the
    inhibition of the shikimate pathway by glyphosate.
    I. Inhibition by glyphosate of phenylpropanoid
    synthesis in buckwheat (Fagopyrum esculentum
    Moench). Plant Physiol. 66: 823-829.

33. SCHERF, H., M.H. ZENK. 1967. Der Einfluß des Lichtes
    auf die Flavonoidsynthese und die Enzyminduktion
    bei Fagopyrum esculentum Moench. Z. Pflanzenphy-
    siol. 57: 401-418.

34. AMRHEIN, N., B. DEUS, P. GEHRKE, H.C. STEINRÜCKEN.
    1980. The site of the inhibition of the shikimate
    pathway by glyphosate. II. Interference of
    glyphosate with chorismate formation in vivo and
    in vitro. Plant Physiol. 66: 830-834.

35. AMRHEIN, N., B. DEUS, P. GEHRKE, H. HOLLÄNDER, J.
    SCHAB, A. SCHULZ, H.C. STEINRÜCKEN. 1981. Inter-
    ference of glyphosate with the shikimate pathway.
    Proc. Plant Growth Regul. Soc. Am. 8: 99-106.

36. AMRHEIN, N., J. SCHAB, H.C. STEINRÜCKEN. 1981. The
    mode of action of the herbicide glyphosate.
    Naturwissenschaften 67: 356-357.

37. BERLIN, J., L. WITTE. 1980. Effect of glyphosate on
    shikimic acid accumulation in tobacco cell cultures
    with low and high yields of cinnamoyl putrescines.
    Z. Naturforsch. 36c: 210-214.

38. ISHIKURA, N., Y. TAKESHIMA. 1984. Effects of
    glyphosate on caffeic acid metabolism in Perilla

cell suspension cultures. Plant Cell Physiol. 25: 185–189.

39. DUKE, S.O., R.E. HOAGLAND, C.D. ELMORE. 1979. Effects of glyphosate on metabolism of phenolic compounds. IV. Phenylalanine ammonia-lyase activity, free amino acids, and soluble hydroxyphenolic compounds in axes of light-grown soybeans. Physiol. Plant 46: 307–317.

40. AMRHEIN, N., H. TOPP, O. JOOP. 1984. The pathway of gallic acid biosynthesis in higher plants. Plant Physiol. 75: S96.

41. SAIJO, R. 1983. Pathway of gallic acid biosynthesis and its esterification with catechins in young tea shoots. Agric. Biol. Chem. 47: 455–460.

42. AMRHEIN, N., H. HOLLÄNDER. 1981. Light promotes the production of shikimic acid in buckwheat. Naturwissenschaften 68: 43.

43. SUZICH, J.A., R. RANJEVA, P.M. HASEGAWA, K.M. HERRMANN. 1984. Regulation of the shikimate pathway of carrot cells in suspension culture. Plant Physiol. 75: 369–371.

44. BOWEN, J.R., T. KOSUGE. 1977. The formation of shikimate-3-phosphate in cell-free preparations of Sorghum. Phytochemistry 16: 881–884.

45. BOWEN, J.R., T. KOSUGE. 1979. In vivo activity, purification, and characterization of shikimate kinase from Sorghum. Plant Physiol. 64: 382–386.

46. KOSHIBA, T. 1979. Alicyclic acid metabolism in plants 12. Partial purification and some properties of shikimate kinase from Phaseolus mungo seedlings. Plant Cell Physiol. 20: 803–809.

47. KOSHIBA, T. 1979. Shikimate kinase and 5-enolpyruvylshikimate-3-phosphate synthase in Phaseolus mungo seedlings. Z. Pflanzenphysiol. 88: 353–355.

48. KOSHIBA, T. 1979. Organization of enzymes in the shikimate pathway of Phaseolus mungo seedlings. Plant Cell Physiol. 20: 667–670.

49. STEINRÜCKEN, H.C., N. AMRHEIN. 1980. The herbicide glyphosate is a potent inhibitor of 5-enolpyruvylshikimic acid-3-phosphate synthase. Biochem. Biophys. Res. Commun. 94: 1207–1212.

50. LEWENDON, A., J.R. COGGINS. 1983. Purification of 5-enolpyruvylshikimate 3-phosphate synthase from Escherichia coli. Biochem. J. 213: 187–191.

51. DUNCAN, K., L. LEWENDON, J.R. COGGINS. 1984. The purification of 5-enolpyruvylshikimate 3-phosphate

synthase from an overproducing strain of Escherichia coli. FEBS Lett. 165: 121–127.

52. STEINRÜCKEN, H.C. 1982. Zur Wirkung des Herbizids Glyphosat: Einfluss auf die 5-Enolpyruvoylshikimisäure-3-phosphat Synthase aus Aerobacter aerogenes 62-1. Doctoral Dissertation. Ruhr-Universität Bochum, 153 pp.

53. ANTON, D.L., L. HEDSTROM, S.M. FISH, R.H. ABELES. 1983. Mechanism of enolpyruvyl shikimate 3-phosphate synthase exchange of phosphoenolpyruvate with solvent protons. Biochemistry 22: 5903–5908.

54. STEINRÜCKEN, H.C., N. AMRHEIN. 1984. 5-Enolpyruvylshikimate 3-phosphate synthase of Klebsiella pneumoniae. 1. Purification and properties. Eur. J. Biochem. 143: 341–349.

55. BOOCOCK, M.R., J.R. COGGINS. 1983. Kinetics of 5-enolpyruvylshikimate 3-phosphate synthase inhibition by glyphosate. FEBS Lett. 154: 127–133.

56. MOUSDALE, D.M., J.R. COGGINS. 1984. Purification and properties of 5-enolpyruvylshikimate 3-phosphate synthase from seedlings of Pisum sativum L. Planta 160: 78–83.

57. AMRHEIN, N., H. HOLLÄNDER-CZYTKO, J. LEIFELD, A. SCHULZ, H.C. STEINRÜCKEN, H. TOPP. 1982. Inhibition of the shikimate pathway by glyphosate. In Groupe Polyphenols. Journées Internationales d'Études et Assemblées Générales. (A.M. Boudet, R. Ranjeva, eds.), Bulletin d'Liaison 11: 21–30.

58. RUBIN, J.L., C.G. GAINES, R.A. JENSEN. 1984. Glyphosate inhibition of 5-enolpyruvylshikimate 3-phosphate synthase from suspension-cultured cells of Nicotiana silvestris. Plant Physiol. 75: 829–845.

59. ROGERS, S.G., L.A. BRAND, S.B. HOLDER, E.S. SHARPS, M.J. BRACKIN. 1983. Amplification of the aroA gene from Escherichia coli results in tolerance to the herbicide glyphosate. Appl. Environ. Microbiol. 46: 37–43.

60. JAWORSKI, E.G., T.J. MOZER, S.G. ROGERS, D. TIEMIER. 1983. Herbicide target sites, mode of action, and detoxification: Chloroacetanilides and glyphosate. In Biosynthesis of the Photosynthetic Apparatus: Molecular Biology, Development and Regulation. (J.P. Thornber, L.A. Staehelin, R.G. Hallick, eds.), UCLA Symp. Mol. Cell Biol., New Ser. Vol. 14, Alan R. Liss, Inc., New York, pp. 335–349.

61. LUMSDEN, J., J.R. COGGINS. 1977. The subunit structure of the arom multienzyme complex of Neurospora crassa. A possible pentafunctional polypeptide chain. Biochem. J. 161: 599-607.

62. DUNCAN, K., A. LEWENDON, J.R. COGGINS. 1984. The complete amino acid sequence of Escherichia coli 5-enolpyruvylshikimate 3-phosphate synthase. FEBS Lett. 170: 59-63.

63. STALKER, D.M., W.R. HIATT, L. COMAI. 1985. A single amino acid substitution in the enzyme 5-enolpyruvylshikimate 3-phosphate synthase confers resistance to the herbicide glyphosate. J. Biol. Chem. 260: 4724-4728.

64. CASSIDY, P.J., F.M. KAHAN. 1973. A stable enzyme-phosphoenolpyruvate intermediate in the synthesis of uridine-5-diphospho-N-acetyl-2-amino-2-deoxyglucose-3-0-enolpyruvylether. Biochemistry 12: 1363-1374.

65. LEVIN, J.G., D.B. SPRINSON. 1964. The enzymatic formation and isolation of 3-enolpyruvylshikimate-5-phosphate. J. Biol. Chem. 239: 1142-1150.

66. BONDINELL, W.E., J. VNEK, P.F. KNOWLES, M. SPRECHER, D.B. SPRINSON. 1971. On the mechanism of 5-enolpyruvylshikimate 3-phosphate synthetase. J. Biol. Chem. 246: 6191-6196.

67. GRIMSHAW, C.E., S.G. SOGO, J.R. KNOWLES. 1982. The fate of the hydrogens of phosphoenolpyruvate in the reaction catalyzed by 5-enolpyruvylshikimate 3-phosphate synthase. J. Biol. Chem. 257: 596-598.

68. STEINRÜCKEN, H.C., N. AMRHEIN. 1984. 5-Enolpyruvyl-shikimate 3-phosphate synthase of Klebsiella pneumoniae. 2. Inhibition by glyphosate [N-(phos-phonomethy)glycine]. Eur. J. Biochem. 143: 351-357.

69. BODE, R., C. MELO, D. BIRNBAUM. 1984. Mode of action of glyphosate in Candida maltosa. Arch. Microbiol. 140: 83-85.

70. SHARPS, E.S. 1984. A radiometric assay for 5-enolpyruvylshikimate-3-phosphate synthase. Anal. Biochem. 140: 183-189.

71. FEIERABEND, J., D. BRASSEL. 1977. Subcellular localization of shikimate dehydrogenase in higher plants. Z. Pflanzenphysiol. 82: 334-346.

72. BICKEL, H., L. PALME, G. SCHULTZ. 1978. Incorporation of shikimate and other precursors into aromatic amino acids and prenylquinones of isolated spinach chloroplasts. Phytochemistry 18: 498-499.

73. ROTHE, G.M., G. HENGST, I. MILDENBERGER, H. SCHARER, D. UTESCH. 1983. Evidence for an intra- and extraplastidic prechorismate pathway. Planta 157: 358-366.

74. D'AMATO, T.A., R.J. GANSON, C.G. GAINES, R.A. JENSEN. 1984. Subcellular localization of chorismate mutase isoenzymes in protoplasts from mesophyll and suspension-cultured cells of Nicotiana sylvestris. Planta 162: 104-108.

75. MOUSDALE, D.M., J.R. COGGINS. 1985. Subcellular localization of the common shikimate pathway enzymes in Pisum sativum L. Planta 163: 241-249.

76. GERHARDT, R., H.W. HELDT. 1984. Measurement of subcellular metabolite levels in leaves by fractionation of freeze-stopped material in nonaqueous media. Plant Physiol. 75: 542-547.

77. HOLLÄNDER-CZYTKO, H., N. AMRHEIN. 1983. Subcellular compartmentation of shikimic acid and phenylalanine in buckwheat cell suspension cultures grown in the presence of shikimate pathway inhibitors. Plant Sci. Lett. 29: 89-96.

78. AMRHEIN, N., D. JOHÄNNING, J. SCHAB, A. SCHULZ. 1983. Biochemical basis for glyphosate tolerance in a bacterium and a plant tissue culture. FEBS Lett. 157: 191-196.

79. AMRHEIN, N., D. JOHÄNNING, C.C. SMART. 1985. A glyphosate-tolerant plant tissue culture. In Primary and Secondary Metabolism of Plant Cell Cultures. (K.H. Neumann, ed.), Springer, Berlin-Heidelberg, New York, pp. 356-361.

80. NAFZIGER, E.D., J.M. WIDHOLM, H.C. STEINRÜCKEN, J.L. KILLMER. 1984. Selection and characterization of a carrot cell line tolerant to glyphosate. Plant Physiol. 76: 571-574.

81. STARK, G.R., F.M. WAHL. 1984. Gene amplification. Annu. Rev. Biochem. 53: 447-491.

82. COMAI, L., L.C. SEN, D.M. STALKER. 1983. An altered aroA gene product confers resistance to the herbicide glyphosate. Science 221: 370-371.

83. SCHULZ, A., D. SOST, N. AMRHEIN. 1984. Insensitivity of 5-enolpyruvylshikimic acid-3-phosphate synthase to glyphosate confers resistance to this herbicide in a strain of Aerobacter aerogenes. Arch. Microbiol. 137: 121-123.

84. SOST, D., A. SCHULZ, N. AMRHEIN. 1984. Characterization of a glyphosate insensitive 5-enolpyruvyl-

shikimic acid-3-phosphate synthase. FEBS Lett.
173: 238-241.

85. SCHULZ, A., A. KRÜPER, N. AMRHEIN. 1985. Differential
    sensitivity of bacterial 5-enolpyruvylshikimate-3-
    phosphate synthases to the herbicide glyphosate.
    FEMS Microbiol. Lett. 28: 297-301.

86. HARDY, R.W.F., R.T. GIAQUINTA. 1984. Molecular
    biology of herbicides. BioEssays 1: 152-156.

87. NETZER, W.F. 1984. Engineering herbicide tolerance:
    When is it worthwhile? Biotechnology 2: 939-944.

88. JONES, D.H. 1984. Phenylalanine ammonia-lyase:
    Regulation of its induction, and its role in plant
    development. Phytochemistry 23: 1349-1359.

89. HANSON, K.R., E.A. HAVIR. 1981. Phenylalanine
    ammonia-lyase. In The Biochemistry of Plants:
    A Comprehensive Treatise. (E.E. Conn, ed.), Vol.
    7, Academic Press, New York, pp. 577-625.

90. AMRHEIN, N., K.H. GÖDEKE, J. GERHARDT. 1976. The
    estimation of phenylalanine ammonia-lyase(PAL)-
    activity in intact cells of higher plant tissue.
    1. Parameters of the assay. Planta 131: 33-40.

91. AMRHEIN, N., K.H. GÖDEKE, V.I. KEFELI. 1976. The
    estimation of relative intracellular phenylalanine
    ammonia-lyase(PAL)-activities and the modulation
    in vivo and in vitro by competitive inhibitors.
    Ber. Deutsch. Bot. Ges. 89: 247-259.

92. AMRHEIN, N., H. HOLLÄNDER. 1979. Inhibition of
    anthocyanin formation in seedlings and flowers
    by the enantiomers of α-aminooxy-β-phenylpropionic
    acid and their N-benzyloxycarbonyl derivatives.
    Planta 144: 385-389.

93. HOLLÄNDER, H., H.H. KILTZ, N. AMRHEIN. 1979. Inter-
    ference of L-α-aminooxy-β-phenylpropionic acid with
    phenylalanine metabolism in buckwheat. Z.
    Naturforsch. 34c: 1162-1173.

94. HANSON, K.R. 1981. Phenylalanine ammonia-lyase:
    Mirror-image packing of D- and L-phenylalanine and
    D- and L-transition state analogs into the active
    site. Arch. Biochem. Biophys. 211: 575-588.

95. HAVIR, E.A. 1981. Modification of L-phenylalanine
    ammonia-lyase in soybean cell suspension cultures
    by 2-aminooxyacetate and L-2-aminooxy-3-phenylpro-
    pionate. Planta 152: 124-130.

96. JONES, H.D., D.H. NORTHCOTE. 1984. Stability of the
    complex formed between French bean (Phaseolus
    vulgaris) phenylalanine ammonia-lyase and its

transition-state analog. Arch. Biochem. Biophys. 235: 167-177.

97. AMRHEIN, N., J. GERHARDT. 1979. Superinduction of phenylalanine ammonia-lyase in gherkin hypocotyls caused by the inhibitor, L-α-aminooxy-β-phenylpropionic acid. Biochim. Biophys. Acta 583: 434-442.

98. NOE, W., C. LANGEBARTELS, H.U. SEITZ. 1980. Anthocyanin accumulation and PAL activity in a suspension culture of Daucus carota L. Inhibition by L-AOPP and t-cinnamic acid. Planta 149: 283-287.

99. NOE, W., H.U. SEITZ. 1982. Induction of mRNA activity for phenylalanine ammonia-lyase (PAL) by L-α-aminooxy-β-phenylpropionic acid, a substrate analogue of L-phenylalanine, in cell suspension cultures of Daucus carota L. FEBS Lett. 146: 52-54.

100. SHIELDS, S.E., V.P. WINGATE, C.J. LAMB. 1982. Dual control of phenylalanine ammonia-lyase production and removal by its product cinnamic acid. Eur. J. Biochem. 123: 389-395.

101. AMRHEIN, N., E. DIEDERICH. 1980. Turnover of isoflavones in Cicer arietinum L. Naturwissenschaften 67: 40.

102. JACQUES, U., J. KÖSTER, W. BARZ. 1985. Differential turnover of isoflavone 7-0-glucoside-6"-0-malonates in Cicer arietinum roots. Phytochemistry 24: 949-951.

103. MOESTA, P., H. GRISEBACH. 1982. L-2-Aminooxy-3-phenylpropionic acid inhibits phytoalexin accumulation in soybean with concomitant loss of resistance against Phytophthora megasperma f. sp. glycinea. Physiol. Plant Pathol. 21: 65-70.

104. BARNES, L., R.L. JONES. 1984. Regulation of phenylalanine ammonia-lyase activity and growth in lettuce by light and gibberellic acid. Plant Cell Environ. 7: 89-95.

105. AMRHEIN, N., G. FRANK, G. LEMM, H.B. LUHMANN. 1983. Inhibition of lignin formation by L-α-aminooxy-β-phenylpropionic acid, an inhibitor of phenylalanine ammonia-lyase. Eur. J. Cell Biol. 29: 139-144.

106. SMART, C.C., N. AMRHEIN. 1985. The influence of lignification on the development of vascular tissue in Vigna radiata L. Protoplasma 124: 87-95.

107. RAVEN, J.A. 1977. The evolution of vascular land plants in relation to supracellular transport processes. In Advances in Botanical Research.

(W.H. Woolhouse, ed.), Vol. 5, London, Academic
Press, pp. 153-219.

108. GRAND, C., F. SARNI, A.M. BOUDET. 1985. Inhibition
of cinnamyl alcohol dehydrogenase activity and
lignin synthesis in poplar (Populus X euramericana
Dode) tissues by two organic compounds. Planta
163: 232-237.

109. DE-EKNAMKUL, W., B.E. ELLIS. 1985. Characterization
of tyrosine aminotransferase, a key enzyme in
rosmarinic acid formation in Anchusa officinalis
cell cultures. Plant Physiol. 77: S112.

110. WILLIAMS, R., C. CHAPPLE, B.E. ELLIS. 1985. Characterization of tyrosine decarboxylase from Syringa
vulgaris cell cultures. Plant Physiol. 77: S112.

111. HANSON, K.R., E.A. HAVIR, C. RESSLER. 1979.
Phenylalanine ammonia-lyase: Enzymic conversion
of 3-(1,4-cyclohexadienyl)-L-alanine to trans-
3-(1,4-cyclohexadienyl)acrylic acid. Biochemistry
18: 1431-1438.

112. BAYLIS, E.K., C.D. CAMPUBELL, J.G. DINGWALL. 1984.
1-Amino-alkylphosphonous acids. Part 1.
Isosteres of the protein amino acids. J. Chem.
Soc., Perkin Trans. I: 2845-2853.

113. ELWOOD, J.K., R.M. HERBST, G.L. KILGOUR. 1965.
Tetrazole analogues of glutamic acid. I. Reaction
with glutamic dehydrogenase. J. Biol. Chem. 240:
2073-2076.

114. CHARI, R.V.J., J. WEMPLE. 1979. A simple, efficient
synthesis of β-methylene phenylalanine. A new
approach to the preparation of β,γ-unsaturated
α-amino acid enzyme substrate analogs. Tetrahedron
Lett. 111-114.

115. LEUKART, O., M. CAVIEZEL, A. EBERLE, E. ESCHER, A.
TUN-KYI, R. SCHWYZER. L-o-Carboranylalanine, a
boron analogue of phenylalanine. Helv. Chim. Acta
59: 2184-2187.

116. DO, K.Q., P. THANEI, M. CAVIEZEL, R. SCHWYZER. 1979.
98. The synthesis of (S)-(+)-2-amino-3-(1-adaman-
tyl)-propionic acid (L-(+)-adamantylalanine, Ada)
as a 'fat' or 'super' analogue of leucine and
phenylalanine. Helv. Chim. Acta 62: 956-964.

117. RAY, T.B. 1984. Site of action of chlorsulfuron.
Inhibition of valine and isoleucine biosynthesis
in plants. Plant Physiol. 75: 827-831.

118. LAROSSA, R.A., J.V. SCHLOSS. 1984. The sulfonylurea
herbicide sulfometuron methyl is an extremely

potent and selective inhibitor of acetolactate
synthase in <u>Salmonella</u> <u>typhimurium</u>.  J. Biol.
Chem. 259: 8753-8757.
119.  WAUCHOPE, D.  1976.  Acid dissociation constants of
arsenic acid, methylarsonic acid (MAA), dimethyl-
arsinic acid (cacodylic acid), and N-(phosphonomethyl)glycine (glyphosate).  J. Agric. Food Chem.
24: 717-721.

Chapter Five

SYNTHETIC ORGANIC CHEMISTRY AND THE SHIKIMATE PATHWAY:
INHIBITORS AND INTERMEDIATES

PAUL A. BARTLETT

Department of Chemistry
University of California
Berkeley, California 94720

INTRODUCTION

Scientists have been interested in the shikimate pathway (Fig. 1) for decades, yet it shows no sign of losing
this popularity.[1,2] Part of the reason for this is that it
offers something for everyone. Until quite recently a
number of the key intermediates in the sequence had never
been synthesized non-enzymatically. The unusual mechanisms
of several of the enzymes involved have still not been fully
elucidated. And the search for inhibitors of the various
steps continues to be spurred by the success of glyphosate
as an herbicide.

SYNTHESIS OF INTERMEDIATES

When we began our synthetic program, shikimic acid had
been prepared by a number of routes,[3] and the groups of
Danishefsky[6] and Plieninger[7] had reported syntheses of
prephenic acid. The intervening intermediates, shikimate-3-
phosphate (S-3-P), 5-enolpyruvylshikimate-3-phosphate (EPSP),
and chorismate, were only available from natural sources or
by enzymatic transformations.

119

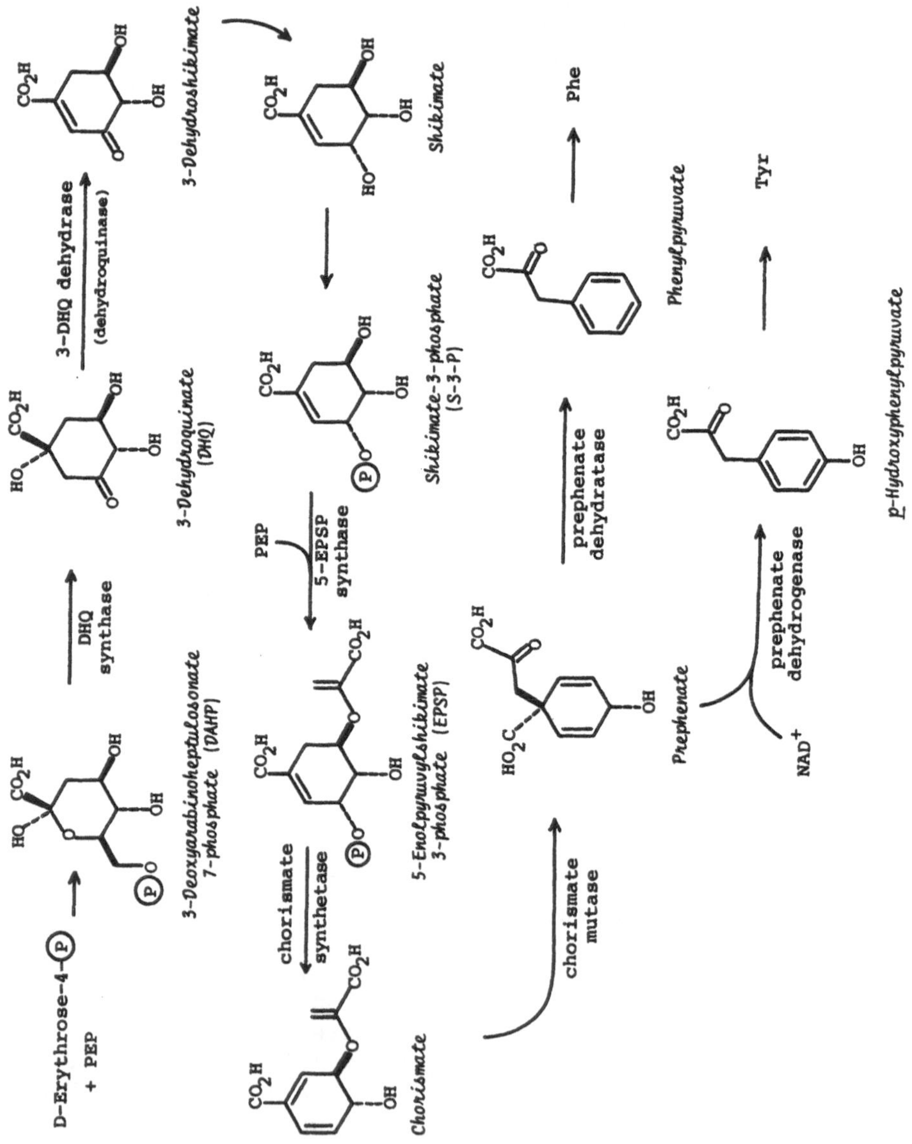

Fig. 1. The shikimate pathway.

## Shikimate

A key element in any contemplated route to S-3-P or EPSP is a method for distinguishing between the various hydroxyl groups of the shikimate nucleus. We envisaged accomplishing this by the sequence depicted below, in which the three ring oxygens of the epoxy lactone 1 would be sequentially unmasked and protected or derivatized.

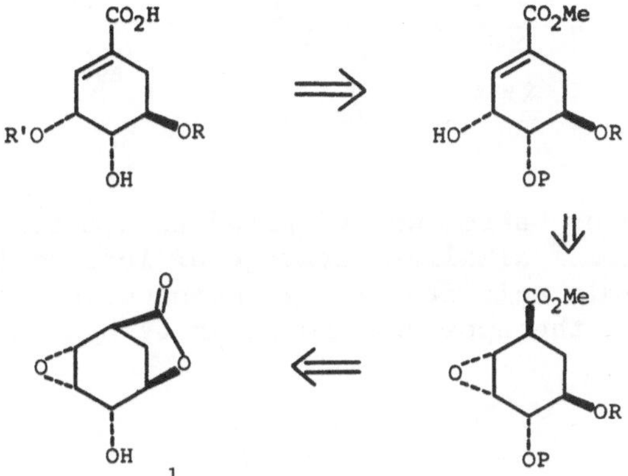

The key lactone 1 was synthesized in high yield starting with 3-cyclohexenecarboxylic acid 2. Iodolactonization and elimination afford the allylic lactone 4 and epoxidation with 3,5-dinitroperbenzoic acid[8] provides exo epoxide 5 with high stereoselectivity. This epoxide

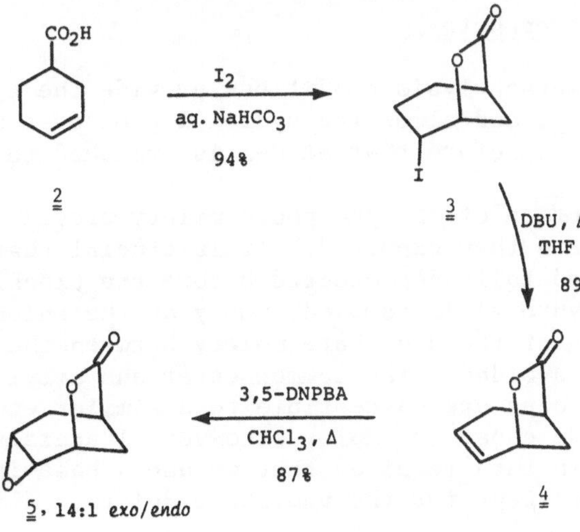

5, 14:1 exo/endo

can be rearranged to the allylic alcohol 6 on treatment
with trimethylsilyl bromide (TMSiBr) and 1,8-diazabicyclo
[5.4.0]undec-7-ene (DBU).  The trimethylsilyl bromohydrin
is an intermediate in this transformation.  Peracid treatment again effects exo epoxidation and gives the desired
lactone intermediate 1.[4]

As shown by Ganem, who prepared this material by a
different route,[9] alkaline cleavage of lactone 1 in methanol
affords shikimic acid itself, via methanolysis of the lactone,
elimination of the epoxide, and ultimate hydrolysis of the
ester.

*Shikimic Acid*

49% overall yield
from 3-cyclohexenecarboxylic acid

## Shikimate-3-phosphate

Milder methanolysis conditions provide the epoxy ester
intermediate 7, and allow the hydroxyl groups at C-4 and C-5
to be protected before that at C-3 is unmasked (Scheme 1).

Introduction of the phosphate moiety proved less
straightforward than expected.  It is crucial that it be
introduced and fully deprotected before the blocking group
at the C-4 hydroxyl is removed.  Only as the anion can
equilibration of the phosphate moiety between the vicinal
hydroxyls be avoided.  The common ester and silyl hydroxyl
protecting groups are susceptible to a similar equilibration, hence we chose to employ ethoxyethyl acetals as shown.
This choice in turn required that we use a base-labile
protection strategy for the phosphate moiety.  Although

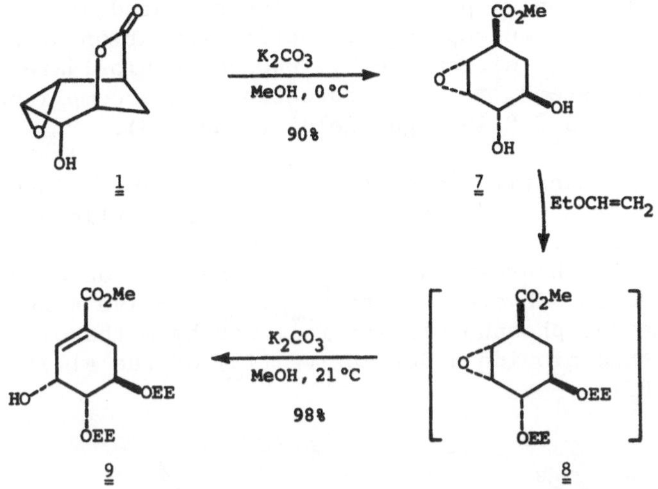

Scheme 1

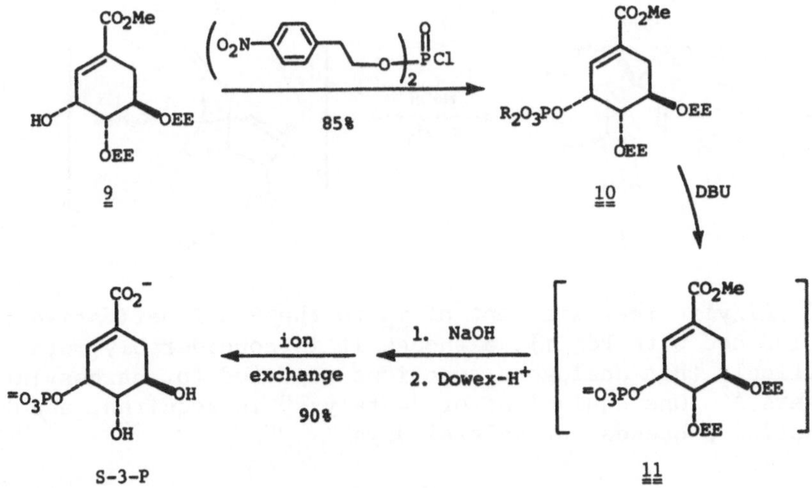

Scheme 2

phosphorylation using dimethyl phosphorochloridate followed by deprotection with TMSiBr was investigated at length,[10] we were unable to suppress competing formation of allylic bromide. Bis(p-nitrophenethyl)phosphorochloridate was employed instead,[11] and deprotection was accomplished with DBU to give S-3-P in high yield[4] (Scheme 2).

An alternative route to S-3-P was briefly investigated, starting from 1-hydroxy-3-cyclohexenecarboxylic acid (12). In a manner similar to that described above for transformation of 2 to hydroxy lactone 6, the hydroxy acid 12 can be converted to dihydroxy lactone 13. Selective silylation and dimethyl phosphorylation followed by methanolysis of the lactone provide a derivative (15) of the allylic isomer of S-3-P.

Allylic rearrangement of 15 to the S-3-P derivative can be induced with Pd(II), although it is considerably more difficult than analogous reactions reported for carboxylate esters.[12] One equivalent of "catalyst" is required, and the reaction proceeds for several days.

## 5-Enolpyruvylshikimate-3-phosphate

After silylation of epoxy lactone 1 and opening to the epoxy ester 18, we expected that application of Ganem's procedure [13] for the introduction of the enolpyruvyl side chain would afford an intermediate appropriate for synthesis of EPSP.  Unfortunately, the epoxide functionality is incompatible with the rhodium-catalyzed diazomalonate insertion reaction and the desired intermediate 19 is not obtained.  The major product appears to result from deoxygenation of the epoxide.

As a more practical route to EPSP, we elected to pursue a partial synthesis, starting with methyl shikimate acetonide (20) which is readily available from shikimic acid itself.[14] While our work on this transformation was in progress, Ganem et al. reported essentially the same sequence of reactions.[15]  Diazomalonyl insertion proceeds quite efficiently in this instance, as do the subsequent steps for introduction of the enolpyruvyl side chain.[13]

After removal of the acetonide protecting group with
mild acid, alkaline hydrolysis affords "Compound $Z_1$" (25),
a branch metabolite along the pathway.[16]  Alternatively,
treatment of the dimethyl ester 24 with $K_2CO_3$ in acetoni-
trile leads to the <u>trans</u> lactone 26, in which the C-3 and
C-4 hydroxyl groups have been conveniently distinguished
(Scheme 3).

Although faced with a similar problem with respect to
choice of phosphorylation and deprotection strategies, we
selected a strategy different from that described above for
the synthesis of S-3-P.  We investigated further the TMSiBr-
induced dealkylation of dimethyl phosphate ester 27a, and
found that the formation of allylic bromide 28 could be
moderated by the inclusion of catalytic amounts of triphenyl-
phosphine in the reaction mixture, and by neutralization or
strict exclusion of HBr.  Although in some instances we
obtained nearly complete demethylation without formation of
the byproduct, these results were not reproducible.  A much
simpler solution is the use of benzyl esters as the protecting group.  Phosphorylation of the lithium salt of 26 with
tetrabenzyl pyrophosphate[17] affords dibenzyl phosphate 27b
in high yield, and dealkylation can be accomplished with
TMSiBr at 0°C without competing production of the allylic
bromide 28 (Scheme 4).  Simple alkaline hydrolysis of the
product and ion exchange afford EPSP.

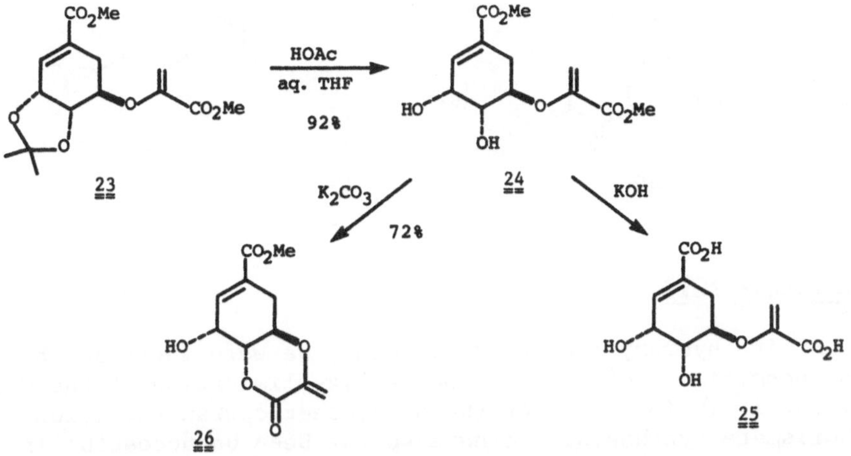

Scheme 3

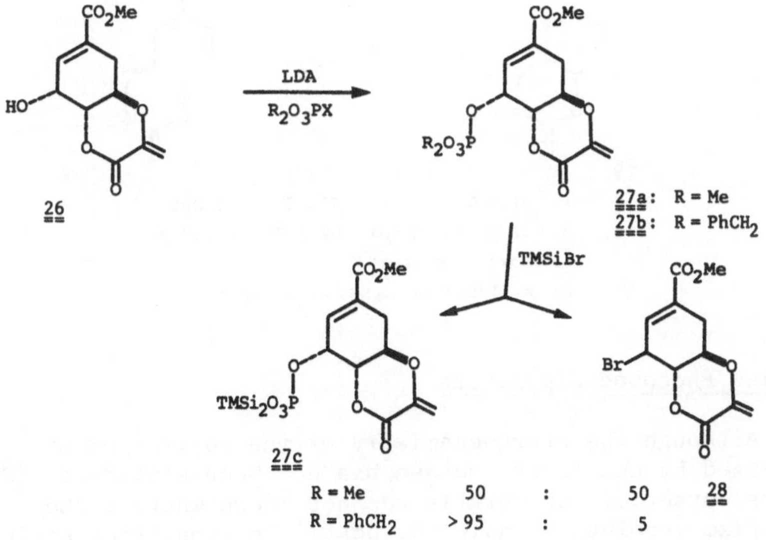

Scheme 4

## Chorismic Acid

With hydroxy lactone 26 in hand, we were intrigued by the possibility of carrying out a 1,4-elimination of the C-3 hydroxyl, in imitation of the biosynthetic transformation in chorismate synthesis. We have so far been unsuccessful in this endeavor, however, obtaining (not too surprisingly) primarily aromatic materials under a variety of conditions. Both Berchtold[18] and Ganem[13] have reported total syntheses of chorismic acid.

1: a) MsCl or 2,6-Cl$_2$PhCOCl    b) DBU
2: a) MsCl or Ac$_2$O    b) Pd(Ph$_3$P)$_4$, Et$_3$N
3: Martin's salt
4: a) ArSeX or ArSX, Et$_3$N    b) Δ

## Labeled Phosphoenolpyruvate

Although the stereochemistry of the rearrangement catalyzed by chorismate mutase has now been elucidated,[19] at the outset of our work it was not known whether the mechanism involves a chair- or boat-like transition state conformation.

Since the solution to this problem required chorismate with a stereoisotopically labeled enolpyruvyl moiety, we addressed an intermediate question, that of devising a chemical route to such compounds, as exemplified by phosphoenolpyruvate (PEP) itself.[20] Although stereoisotopically labeled PEP is available through enzymatic methods,[21] a convenient chemical synthesis was not available. The starting material, ethyl (E)-2,3-dideuterioacrylate, is prepared from the ethyl propiolate-anthracene Diels-Alder adduct as described by Hill and Newkome.[22] Epoxidation with 3,5-dinitroperoxybenzoic acid affords the labeled glycidate with a 96:4 E/Z ratio.

Depending on the sequence selected, this epoxide can
be converted into either the (Z)- or (E)-isomer of
3-deuterio-PEP.  Opening of the epoxide with diisobutyl-
aluminum phenylselenide in hexane affords selectively
(3:1) the 2-hydroxy-3-phenylseleno ester.  Phosphorylation,
oxidation to the selenoxide, and syn-elimination then give
the Z-isomer.

Alternatively, opening of the epoxide with TMSiBr
and ZnBr$_2$ catalysis provides almost exclusively the desired
bromohydrin, which on phosphorylation and anti-elimination
leads to the E-product.  Each of these materials was
hydrolyzed in a two-step procedure (TMSiBr, followed by
KOH) with purification by ion exchange chromatography and
crystallization of the cyclohexylammonium salt.

SYNTHESIS OF INHIBITORS OF CHORISMATE MUTASE

A key strategy for the design of enzyme inhibitors is to devise molecules which mimic the transition state or a reactive intermediate along the reaction coordinate.[23,24] For an enzymatic reaction for which the "non-catalyzed" transformation would be unimolecular, one can demonstrate that the transition state form of the substrate should have a (hypothetical) dissociation constant that is as much lower than that of the ground state form of the substrate as the enzymatic reaction is faster than the non-catalyzed process (Fig. 2).

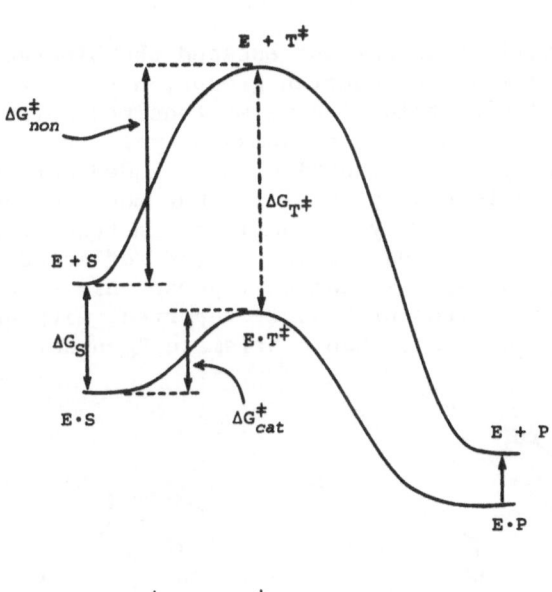

$$\Delta G_S + \Delta G_{cat}^{\ddagger} = \Delta G_{non}^{\ddagger} + \Delta G_{T^{\ddagger}}$$

$$- \Delta G_{T^{\ddagger}} = \Delta G_{non}^{\ddagger} - \Delta G_{cat}^{\ddagger} - \Delta G_S$$

$$K_{T^{\ddagger}} = \frac{k_{non-cat}}{k_{cat}} \cdot K_S$$

Fig. 2. Comparison of enzymatic and nonenzymatic transformations.

Although this comparison of reaction coordinate
diagrams provides an intuitive argument for the synthesis of
transition state analog enzyme inhibitors, in a quantitative
sense it is applicable only to unimolecular $S \rightarrow T^{\ddagger} \rightarrow P$
transformations.  When one recognizes that most enzymatic
mechanisms involve at least acid- or base-catalysis, nucleophilic attack or electron transfer, it is clear that transformations with unimolecular counterparts are exceedingly
rare.  The conversion of chorismate to prephenate is
exceptional in this respect, since the apparently unimolecular, uncatalyzed rearrangement does take place, at a rate
which is estimated to be $2 \times 10^{6}$-fold slower than the
enzymatic reaction.[25]  If the enzymatic rearrangement in
fact involves the same mechanism, then, a perfect transition
state analog should have a dissociation constant which is
two million-fold lower than that of chorismate, i.e., on the
order of $10^{-11}$ M.

This realization has not escaped the bioorganic commu-
nity, and a number of research groups, notably those of
Andrews[26] and Berchtold,[27] have synthesized molecules that
are designed to mimic the presumed bicyclic transition
state.  A truly potent inhibitor has eluded our grasp,
however, as reflected by the fact that most compounds investigated bind more weakly to chorismate mutase/prephenate
dehydrogenase than does chorismate (as reflected by the
substrate $K_m$ value).  1-Adamantyl phosphonic acid is the
most potent inhibitor previously reported, with an $I_{50}$
value twenty-fold less than chorismate $K_m$ under comparable
conditions.

In view of the non-linear kinetics frequently observed
with chorismate mutases and the resultant difficulty of
determining true $K_i$ values, most inhibitor binding
constants[26,27] are reported as $I_{50}$ values, equal to the
concentration of inhibitor which produces 50% inhibition
when the chorismate concentration equals $K_m$.  In Table 1,
the compounds are compared as the ratio of reported $I_{50}/K_m$
values, since they were evaluated against enzymes from
different sources and under different conditions.

We became interested as well in the possibility of
synthesizing a transition state analog inhibitor for the
chorismate mutases.  In our mind, the relationship between
the carboxyl groups is crucial, in particular the conformation of that representing the bridging enolpyruvyl moiety.
Whereas one can readily conceive of improving the transition
state mimicry of the bicyclic hydroxy diacids 28 and 32 by
incorporation of an ether oxygen and the allylic double
bond, the introduction of a trigonal carbon at the point
where the bridging carboxyl is attached is less straightforward.  A double bond in the bridge would position this
carboxyl perpendicular to the lower ring, for example.  Our
solution to this dilemma was to do away with the carboxyl
group entirely, replacing it with a nitronate anion, as
illustrated by structure 36.  Nitronate anions can be generated in mildly alkaline solution and have been employed
frequently in the past[28] to mimic carboxyl enolate and
enediolate intermediates.

36

Our intended synthesis of this compound began with the
Diels-Alder condensation between butadiene and dimethyl
itaconate.  The less congested carboxyl group of the diester
product can be hydrolyzed selectively and converted to the
aldehyde via reduction of the acid chloride with a cuprous
borohydride derivative.[29]

## Table 1

| Structure | $I_{50}/K_m$ (chorismate) |
|---|---|
| 28 (e) | > 100 |
| 29 (e) | 50 |
| (e) | 18 |
| 30 (a), 31 (e) | 6 |
| 32 (e) | 3.5 |
| 33 (a,e), 34 (a) | ~ 1 |
| 35 (a) | 0.05 |

(e): Escherichia coli enzyme, pH 7.5, $K_m$ (chorismate) = $1.1 \times 10^{-4}$ M (Andrews et al., Biochemistry 16, 4848 (1977)).

(a): Aerobacter aerogenes enzyme, pH 9.0, $K_m$ (chorismate) = $1.3 \times 10^{-3}$ M (Chao and Berchtold, Biochemistry 21, 2778 (1982)).

The aldehyde is converted to the cyanohydrin under conditions sufficiently mild that lactonization is avoided,[30] and the bicyclic skeleton is constructed by selenocyclization and elimination.[31]

The double bond is elaborated into the allylic alcohol functionality via the epoxide, as described above for our shikimate synthesis. Strongly alkaline hydrolysis of the nitrile ester then provides the hydroxy diacid, predominantly as the exo isomer.

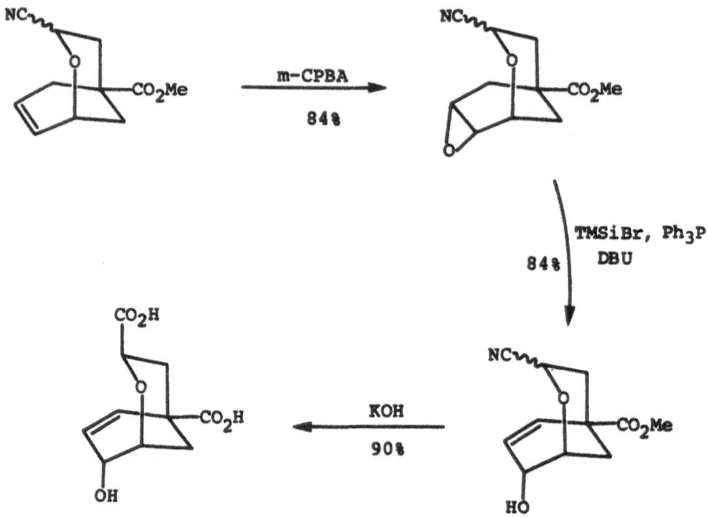
NC
CO2Me
m-CPBA
84%
NC
CO2Me
TMSiBr, Ph3P
DBU
84%
CO2H
CO2H
OH
KOH
90%
NC
CO2Me
HO

Introduction of the nitro group turned out to be quite a challenge. We learned in early experiments that the α-nitro ether moiety is quite labile hydrolytically.[32] We therefore devised the protecting scheme shown below, to allow us to deblock the bridgehead carboxyl and allylic hydroxyl under mild conditions and unleash the nitronate anion under alkaline conditions as a final step.

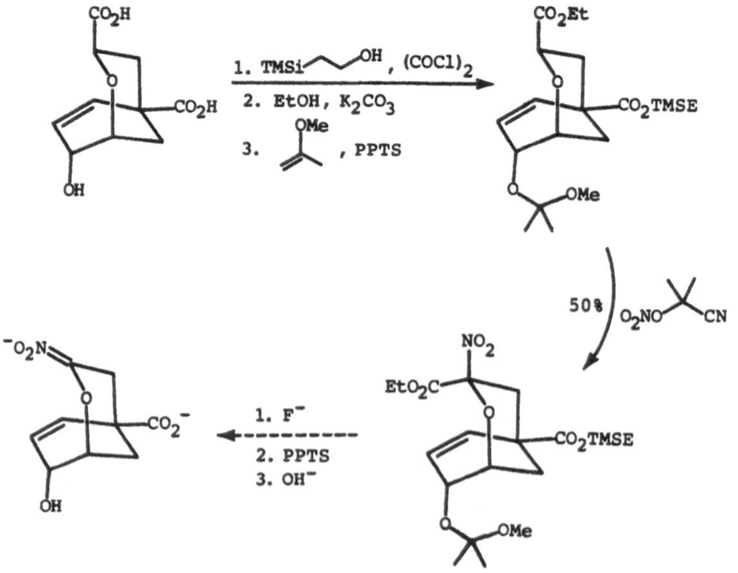
CO2H
CO2H
OH
1. TMSi OH, (COCl)2
2. EtOH, K2CO3
OMe
3. , PPTS
CO2Et
CO2TMSE
OMe
50%
O2NO CN
-O2N
CO2-
OH
1. F-
2. PPTS
3. OH-
NO2
EtO2C
CO2TMSE
OMe

Although we apparently were able to generate the desired nitronate by alcoholysis (retro-Claisen-type cleavage) of a nitro diester, attempted isolation led to the lactone 37 (R = ethyl) as the major product, arising from an exceedingly facile Nef reaction. Direct hydrolysis of a nitro monoester (R = H) was thwarted by cleavage first of the carbon-nitrogen linkage and formation of the keto-acid 38.

The structure of lactone 37 was verified by synthesis, and that of ketoacid 38 was determined spectroscopically.

Although nitronate 36 remains an elusive goal, we were able to apply the same strategy to synthesis of the hydrox-imino lactones 39. These materials are initially produced as a mixture of the Z and E isomers; however, they isomerize to the more stable E form fairly rapidly on standing.

Z-39                              E-39

These materials were examined as inhibitors of the Escherichia coli chorismate mutase/prephenate dehydroge-nase; however, none is a particularly potent inhibitor (Table 2). Although the oximino derivative E-39 has many of the features of our desired target 36, the N-hydroxyl group is not ionized at neutral pH (pK$_a$ ~ 12) and it is only a poor carboxylate mimic.

A number of other compounds were produced in the course of synthesis which turned out to be of interest. Alkaline hydrolysis of the bicyclic cyanohydrin produced, in addition to the exo isomer 40, a small amount of the endo diastereomer 41. These are "improved" versions of the Andrews et al. hydroxy diacids 28 and 32.[26] In addition, as a side product in the enolate nitration reaction, a dehydro derivative is produced, which can subsequently be deblocked to afford the sp$^2$-hybridized analog 42.

Table 2

| | $I_{50}/K_m$ | |
|---|---|---|
| | pH 7.5 ($K_m = 15$ μM) | pH 9.0 ($K_m = 240$ μM) |
| E-39 | > 300 | -- |
| 37 | 150 | -- |
| | -- | 6 |
| 38 | -- | 0.7 |

Whereas the <u>exo</u> isomer 40 is bound about as tightly as the saturated carbocycle 32 and the dehydro analog 42 is slightly better, the <u>endo</u> isomer 41 is an excellent inhibitor. In fact, with an $I_{50}$ of 1.5 X $10^{-7}$ <u>M</u> at pH 7.5, 41 is the most potent inhibitor yet reported for a chorismate mutase (Table 3).[33]

The dichotomy between the behavior of the <u>endo</u> isomers 28 and 41 clearly arises from the position of equilibrium between the chair and boat conformations of

$J_{ab}$ = 3 Hz, $J_{ac}$ = 7 Hz, $J_{bc}$ = 14 Hz

Table 3

|  | $I_{50}/K_m$ [a] |
|---|---|
| **28** | $> 100$ [b] |
| **32** | $3.5$ [b] |
| **40** | $3.9$ [c] |
| **42** | $1.1$ [c] |
| **41** | $0.010$ [c] |

a) Escherichia coli chorismate mutase/prephenate dehydrogenase, pH 7.5
b) Andrews et al., Biochemistry 16, 4848 (1977); $K_m =$
   $1.1 \times 10^{-4}$ M
c) Our work; $K_m = 1.5 \times 10^{-5}$ M

the bridging ring.  In the saturated carbocycle 28,
steric repulsion between the carboxyl group and the axial
hydrogen on the adjacent ring in the chair conformation
forces the upper ring into the boat conformation.   In
contrast, the axial hydrogen is absent from the allylic
ether 41, hence the steric interaction is reduced in
the chair conformation and it predominates.  The $^1$H NMR
spectrum provides clear evidence for the chair structure
of 41.  This placement of the bridging carbonyl group
over the lower ring is a crucial element in the effec-
tiveness of 41 as an inhibitor, in accord with our initial
premise.

In view of the varying behavior of chorismate
mutases under different conditions and from different
sources, we evaluated the best previously reported
chorismate mutase inhibitor, 1-adamantyl phosphonate 35,[27]
in direct comparison with 41.  Although the phosphonate

|  | pH 7.5 | pH 9.0 |
|---|---|---|
| Chorismate | $K_m$:  15 μM | 240 μM |
| 35 | $I_{50}$:  5.5 μM  (0.37)   $(I_{50}/K_m)$ | 1.2 μM  (0.005)   $(I_{50}/K_m)$ |
| 41 | $I_{50}$:  0.15 μM  (0.010) | 10 μM  (0.04) |

[E. coli chorismate mutase/prephenate dehydrogenase]

is a better inhibitor at pH 9, where binding of both chorismate and 41 fall off considerably, at the more normal pH of 7.5, 41 is significantly more potent. With an affinity for chorismate mutase some 100-fold better than that of the substrate, 41 still falls short of expectations for a transition state analog. However, it appears to be the best so far.

## ACKNOWLEDGMENTS

I would like to express my appreciation to Loretta A. McQuaid, Paul M. Chouinard, Charles R. Johnson, and Robert T. Lum, who carried out the research described above. We would like to thank Professor Jeremy R. Knowles for providing us with purified chorismate mutase-prephenate dehydrogenase used in the studies described in Tables 2 and 3 as well as an authentic sample of shikimate-3-phosphate. The financial support of Merck Sharp & Dohme, as well as the National Institutes of Health (grant no. GM-28965), is also gratefully acknowledged.

## REFERENCES

1.  HASLAM, E.   1974.   The Shikimate Pathway.   Wiley-
       Interscience, New York.
2.  GANEM, B.   1978.   From glucose to aromatics:   Recent
       development in natural products of the shikimic acid
       pathway.   Tetrahedron 34: 3353-3383.
3.  For a review of early syntheses of shikimate, see
       Reference 2.   For more recent syntheses, see Refer-
       ences 4 and 5 and references cited therein.
4.  BARTLETT, P.A., L.A. McQUAID.   1984.   Total synthesis
       of (±)-methyl shikimate and (±)-3-phopsphoshikimic
       acid.   J. Amer. Chem. Soc. 106: 7854-7860.
5.  MIRZA, S., A. VASELLA, 1984.   Synthesis of methyl
       shikimate and of diethyl phosphoshikimate from D-
       ribose.   Helv. Chim. Acta 67: 1562-1567.
6.  DANISHEFSKY, S., M. HIRAMA, N. FRITSCH, J. CLARDY.
       1979.   Synthesis of disodium prephenate and disodium
       epiprephenate.   Stereochemistry of prephenic acid
       and an observation on the base-catalyzed rearrange-
       ment of prephenic acid to p-hydroxyphenyllactic
       acid.   J. Amer. Chem. Soc. 101: 7013-7018.

7.  GRAMLICH, W., H. PLIENINGER. 1979. Total synthesis of
    disodium prephenate. II. Synthesis and stereochem-
    ical assignment of disodium prephenate. Chem. Ber.
    112: 1571-1584. For a recent synthesis of prephenate,
    see:
    RAMAGE, R., A.M. MACLEOD. 1984. J. Chem. Soc. Chem.
    Commun. 1008-1010.
8.  RASTETTER, W.H., T.J. RICHARD, M.D. LEWIS. 1978. 3,5-
    Dinitroperoxybenzoic acid, a crystalline, storable
    substitute for peroxytrifluoroacetic acid. J. Org.
    Chem. 43: 3163-3166.
9.  COBLENS, K.E., V.B. MURALIDHARAN, B. GANEM. 1982.
    Shikimate-derived metabolites. 12. Stereocontrolled
    total synthesis of shikimic acid and 6β-deuterio-
    shikimate. J. Org. Chem. 47: 5041-5042.
10. McKENNA, C.E., J. SCHMIDHAUSER. 1979. Functional
    selectivity in phosphate ester dealkylation with
    bromotrimethylsilane. J. Chem. Soc. Chem. Commun.
    739.
11. UHLMANN, E., W. PFLEIDERER. 1980. New improvement in
    oligonucleotide synthesis by use of the p-nitro-
    phenylethyl phosphate blocking group and its
    deprotection by DBU or DBN. Tetrahedron Lett.
    1181-1184.
12. OVERMAN, L.E., F.M. KNOLL. 1979. Palladium(II)-
    catalyzed rearrangement of allylic acetates.
    Tetrahedron Lett. 321-324.
13. GANEM, B., N. IKOTA, V.B. MURALIDHARAN, W.S. WADE,
    S.D. YOUNG, T.Y. YUKIMOTO. 1982. Total synthesis
    of (±)-chorismic acid. J. Amer. Chem. Soc. 104:
    6787-6788.
14. DIELS, O., P. FRITZSCHE. 1911. Zur Kenntnis der
    Azodicarbonsäureester. Chem. Ber. 44: 3018-3027.
15. TENG, C.Y., Y. YUKIMOTO, B. GANEM. 1985. Shikimate-
    derived metabolites. 14. Chiral synthesis of (-)-5-
    enolpyruvylshikimate-3-phosphate. Tetrahedon Lett.
    26: 21-24.
16. DAVIS, B.D., E.S. MINGIOLI. 1953. Aromatic biosyn-
    thesis. VII. Accumulation of two derivatives of
    shikimic acid by bacterial mutants. J. Bacteriol.
    66: 129-136.
17. KHORANA, H.G., A.R. TODD. 1953. Studies on phospho-
    rylation. XI. The reaction between carbodiimides
    and acid esters of phosphoric acid. A new method
    for the preparation of pyrophosphates. J. Chem.
    Soc. 2257-2260.

18. McGOWAN, D.A., G.A. BERCHTOLD. 1982. Total synthesis of racemic chorismic acid and (-)-5-enolpyruvyl-shikimic acid. J. Amer. Chem. Soc. 104: 7036-7041. HOARE, J.H., P.P. POLICASTRO, G.A. BERCHTOLD. 1983. Improved synthesis of racemic chorismic acid. Claisen rearrangement of 4-epi-chorismic acid and dimethyl 4-epi-chorismate. J. Amer. Chem. Soc. 105: 6264-6267.

19. SOGO, S.G., T.S. WIDLANSKI, J.H. HOARE, C.E. GRIMSHAW, G.A. BERCHTOLD, J.R. KNOWLES. 1984. Stereochemistry of the rearrangement of chorismate to prephenate: Chorismate mutase involves a chair transition state. J. Amer. Chem. Soc. 106: 2701-2703.

20. BARTLETT, P.A., P.M. CHOUINARD. 1983. Stereocontrolled synthesis of (E)- and (Z)-3-deuteriophosphoenolpyruvate. J. Org. Chem. 48: 3854-3855.

21. ROSE, I.A. 1975. Preparation of phosphoenolpyruvate and pyruvate specifically labeled with deuterium and tritium. Methods Enzymol. 41B: 110.

22. HILL, R.K., G.R. NEWKOME. 1969. Stereochemistry of chorismic acid biosynthesis. J. Amer. Chem. Soc. 91: 5893-5894.

23. WOLFENDEN, R. 1976. Transition state analog inhibitors and enzyme catalysis. Annu. Rev. Biophys. Bioeng. 5: 271-306.

24. STARK, G.R., P.A. BARTLETT. 1984. Design and use of potent, specific enzyme inhibitors. Pharmacol. Ther. 23: 45-78.

25. ANDREWS, P.R., G.D. SMITH, I.G. YOUNG. 1973. Transition-state stabilization and enzymic catalysis. Kinetic and molecular orbital studies of the rearrangement of chorismate to prephenate. Biochemistry 12: 3492-3498.

26. ANDREWS, P.R., E.N. CAIN, E. RIZZARDO, G.D. SMITH. 1977. Rearrangement of chorismate to prephenate. Use of chorismate mutase inhibitors to define the transition state structure. Biochemistry 16: 4848-4852.

27. CHAO, H.S., G.A. BERCHTOLD. 1981. Inhibition of chorismate mutase activity of chorismate mutase-prephenate dehydrogenase from Aerobacter aerogenes. Biochemistry 21: 2778-2781.

28. ALSTON, T.A., D.J.T. PORTER, H.J. BRIGHT. 1983. Enzyme inhibition by nitro and nitroso compounds. Accts. Chem. Research 16: 418-424.

29.  FLEET, G.W.J., P.J. HARDING.  1979.  Convenient
     synthesis of bis(triphenylphosphine)-copper(I)
     tetrahydroborate and reduction of acid chlorides
     to aldehydes.  Tetrahedron Lett. 975–978.
30.  EVANS, D.A., L.K. TRUESDALE.  1973.  Cyanosilylation
     of aldehydes and ketones.  A convenient route to
     cyanohydrin derivatives.  J. Chem. Soc. Chem.
     Commun. 55–56.
31.  NICOLAOU, K.C.  1981.  Organoselenium-induced cycliza-
     tions in organic synthesis.  Tetrahedron 37:
     4097–4109.
32.  AEBISCHER, B., J.H. BIERI, R. PREWO, A. VASELLA.
     1982.  Synthese von Ketosen durch Kettenverlängerung
     von 1-Desoxy-1-nitroaldosen.  Nucleophile Additionen
     und Solvolyse von Nitroaethern.  Helv. Chim. Acta
     65: 2251–2272.
33.  BARTLETT, P.A., C.R. JOHNSON.  1985.  An inhibitor of
     chorismate mutase resembling the transition state
     conformation.  J. Amer. Chem. Soc. 107: 7792–7793.

Chapter Six

INDOLEACETIC ACID, ITS SYNTHESIS AND REGULATION:  A BASIS
FOR TUMORIGENICITY IN PLANT DISEASE

TSUNE KOSUGE AND MARGARET SANGER

Department of Plant Pathology
University of California
Davis, California 95616

INTRODUCTION

     The multibranched shikimic acid pathway provides the
intermediates for the synthesis of the three amino acids
phenylalanine, tyrosine and tryptophan in microorganisms
and plants.  In plants, these three amino acids are precursors for a variety of secondary metabolites such as
alkaloids, coumarins, flavonoids, lignin precursors, indole
derivatives and numerous phenolic compounds (Fig. 1).  The
role of the aromatic amino acids in protein synthesis is
well known as is the role of indoleacetic acid in plant
development; however, the function of the various secondary
products is much less clear.  Various physiological roles
have been proposed including pest resistance, chromagens in
flowers and fruits, and precursors for the structural
component, lignin.

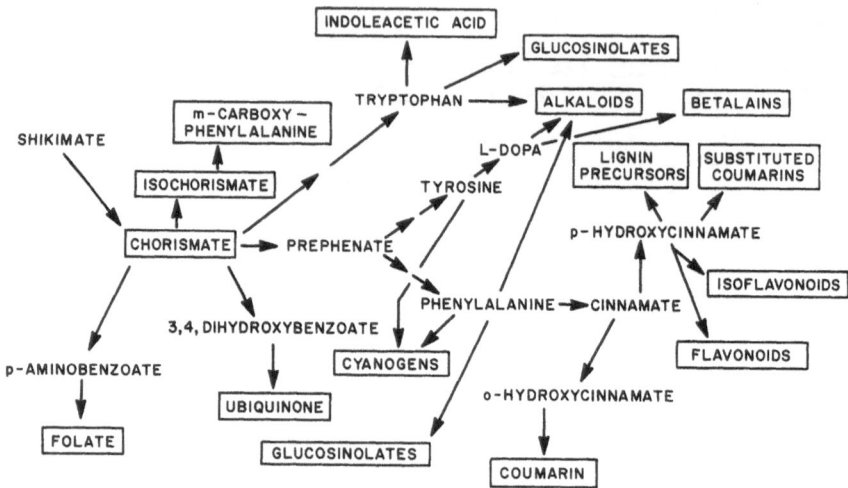

Fig. 1. A generalized scheme showing the kinds of secondary products that arise from the aromatic amino acids in higher plants. Several similarities are found in fungi and bacteria: some fungi produce alkaloids from tryptophan and lignin-like materials from phenylalanine. Plant pathogenic fungi produce cinnamate and para and meta hydroxy phenylacetate from phenylalanine. Certain bacteria produce antibiotics and fluorescent pigments from metabolites in the shikimate pathway.[1] Microorganisms are not known to produce coumarin, substituted coumarins, flavonoids and isoflavonoids.

In microorganisms, the shikimate pathway also provides precursors for a variety of secondary products including alkaloids, fluorescent pigments, antibiotics and phytohormones.[1] With the aid of molecular genetic approaches, it has been possible to define the roles of some of the secondary metabolites in the ecology of the organism involved. In the following discussion, we will discuss the role of indoleacetic acid (IAA) as a virulence factor in the interaction of a bacterium, Pseudomonas syringae pv savastanoi (P. savastanoi), and its hosts, oleander and olive. Virulence, in this case, is assessed by the production of tumor-like outgrowths by the plants in response to secretion of phytohormones by the pathogen during its growth in host tissue.

It is important to emphasize that production of a
virulence factor such as IAA is but one of the many attrib-
utes possessed by microorganisms that allow them to invade
and cause disease in plants.  Further, virulence concerns
those factors directly responsible for the formation and
severity of symptoms caused on plants; however, more than
one compound or factor may be responsible for the primary
symptoms of disease.  In the example that will be discussed
in this chapter, production of cytokinin, together with
IAA, is essential for full expression of virulence.

It is interesting to note that cytokinins and IAA
serve as microbial virulence factors in hosts which require
the same phytohormones for normal growth and development.
Therefore the basis of virulence (or tumorigenicity) rests
with the production and secretion by the pathogen of high
concentrations of cytokinins and IAA.  Production of phytohormones by the pathogen in invaded tissue is highly
localized and represents a source not regulated by the
plant systems.  As a consequence, both the physiological
concentrations of the individual phytohormones as well as
the normal phytohormone balance in the plant tissue are
altered to unnatural levels and cause the plant cells
surrounding the point of infection to proliferate in an
unorganized manner.

In keeping with the general theme of this volume, we
will limit the discussion to IAA, a virulence factor
arising as a secondary product of tryptophan metabolism
in the pathway.

THE RELATIONSHIP BETWEEN PHYTOHORMONE PRODUCTION AND TUMOR
FORMATION

Skoog and Miller[2] first described the effects of auxin
and cytokinin balances in promoting morphogenic responses
of normal tobacco callus tissue.  By addition of varying
concentrations of auxin and cytokinins to culture media,
they determined that high cytokinin-auxin ratios promoted
formation of shoots from the callus tissue, while intermediate ratios maintained the tissue in a rapidly prolif-
erating, undifferentiated state.  Low ratios promoted the
formation of roots.  Mounting evidence in recent years has
unequivocally indicated that the principles for phytohormone control of morphogenesis in plant callus cultures

elucidated by Skoog and Miller also apply to tumor formation
in plants.

    That phytohormone production is a basis for tumori-
genicity was first demonstrated by Braun[3] who found that
bacteria-free tobacco crown gall tumor tissue could grow
in culture without addition of auxin and cytokinins.  Normal
(untransformed) tobacco callus tissue would grow only on
media supplemented with the phytohormones.  Work in several
laboratories (as cited by Gelvin[4]) demonstrated that
tumorigenicity is associated with the presence of a large
plasmid called Ti in the crown gall pathogen, Agrobacterium
tumefaciens.  During the infection process, a region of
this plasmid, called T-DNA, is transferred from the
bacterium and integrated into the host genome.  Once
integrated, genes on the T-DNA are expressed and confer
the tumorigenic condition to host cells.  Subsequently,
Garfinkel and Nester[5] and others[4] constructed Tn 5 mutants
of A. tumefaciens which gave rise to leaf and root forming
tumors when inoculated into tobacco plants.  Mutations
that were associated with the production of morphologically
distinct tumors occurred in specific loci in the T-DNA.
The loci were designated tms 1 and 2, and tmr 4; inactivation of these loci conferred the unique tumor phenotypes.
When such tumors were analyzed for phytohormone content,
the cytokinin/auxin ratios were found to be highest in
tissues of leafy tumors, intermediate in wild type undifferentiated tumors and lowest in the rooty tumors.[6]  Thus
the principles elucidated by Skoog and Miller[2] on the role
of cytokinin/auxin ratios in morphogenic responses of normal
plant cells also apply to tumor formation on plants.


INDOLEACETIC ACID PRODUCTION AND VIRULENCE IN P. SAVASTANOI

    The association between IAA production and virulence
is also apparent in the interaction between the tumorigenic
bacterium, P. savastanoi, and its hosts, oleander and
olive.  Production of IAA confers virulence in P. savastanoi
because mutants deficient in IAA production fail to incite
the production of tumors on oleander plants.[7]  When genes
for IAA production are transferred from parental strains to
these mutants, both IAA synthesis and full virulence are
restored.[8,9]  Since loss of IAA production is not lethal to
the bacterium, IAA synthesis constitutes secondary metabolism of tryptophan in the bacterium.  In contrast to A.

tumefaciens, genetic information from P. savastanoi is not
transferred to its hosts; instead tumor formation by the
plant is a response to secretion of phytohormones by the
pathogen.

P. savastanoi produces IAA by the sequence of reactions
L-tryptophan ⟶ indoleacetamide ⟶ indoleacetic acid.
The two enzymes involved and their genetic determinants are:
tryptophan 2-monooxygenase (iaaM) and indoleacetamide
hydrolase (iaaH).[8,9,10] These genes (IAA genes) occur on
a plasmid, pIAA, in oleander strains and on the chromosome
of olive strains. The plasmid occurs in several sizes
ranging from 53 kb (pIAA) to 72 kb (pIAA2) in different
oleander strains.[11] Strains can be cured of pIAA readily by
selecting for resistance to α-methyltryptophan. Such
mutants are deficient in IAA synthesis and show attenuated
virulence. Loss of IAA production also occurs from dele-
tions of the IAA genes and inactivation of iaaM by IS
elements.[12]

Mechanisms which cause loss of IAA production in P.
savastanoi offer explanations for the spontaneous non-lethal
loss of virulence commonly observed in both plant pathogenic
bacteria and fungi. Expression of virulence in such
pathogens could be associated with the production of a
secondary metabolite.

As noted in Figure 2, tryptophan monooxygenase fulfills
the important function of diverting tryptophan into the
production of IAA; the enzyme occupies a key position not in
only regulating the production of IAA but also the diversion
of tryptophan into secondary metabolism. It is an FAD linked
enzyme with a monomer weight of 62 kd, and is strongly inhibited by its product, indoleacetamide ($K_i$ = 7 uM).[13] Because
the inhibition is competitive with respect to the substrate,
L-tryptophan, the cellular pool size of tryptophan is
important in regulating the flow of tryptophan into the
production of IAA. Thus, the production of the virulence
factor, IAA, depends upon the availability of the primary
metabolite, L-tryptophan, the production of which is regulated by feedback inhibition of anthranilate synthase by
tryptophan (Fig. 2). Therefore, mechanisms controlling
production and utilization of the amino acid affect expression of virulence. In the normal life cycle of the bacterium,
the priority for tryptophan utilization is protein synthesis
and secondarily IAA production. Control over the conversion

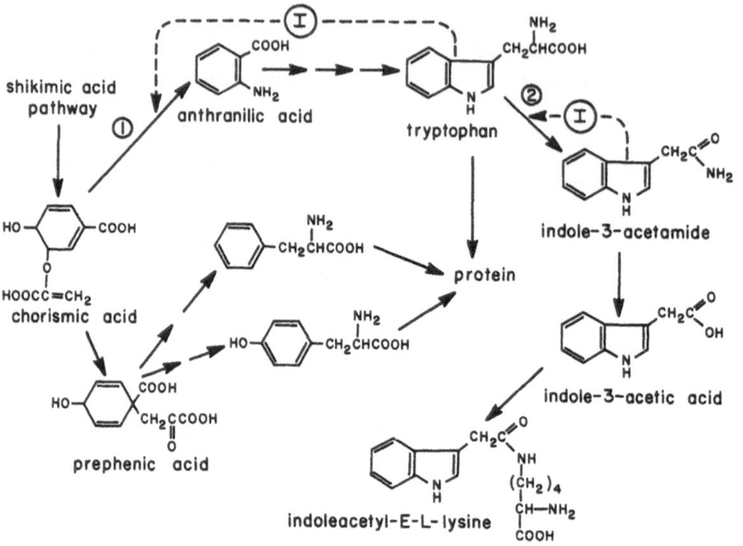

Fig. 2. The pathway for indoleacetic acid synthesis in P. savastanoi. Control of IAA synthesis occurs in part by inhibition of tryptophan monooxygenase by indoleacetamide. Because the inhibition is competitively reversed by the substrate, tryptophan, tryptophan pool size is important in modulating the inhibition by indoleacetamide.

Tryptophan pool size is regulated by feedback inhibition of anthranilate synthase by tryptophan. Other mechanisms, as yet unidentified, may regulate tryptophan and indoleacetic acid synthesis.

of tryptophan to IAA occurs from strong inhibition of tryptophan monooxygenase by indoleacetamide that is competitively reversed by the substrate, L-tryptophan.[13,14]

## CONTROL OF TRYPTOPHAN SYNTHESIS AFFECTS IAA PRODUCTION AND VIRULENCE

That the size of the tryptophan pool limits IAA production in wild type cells is evident since addition of an exogenous source of L-tryptophan boosts IAA production at least 10-fold.[7] One mutant, 2015-5, when compared with

its parent, 2015, accumulates 3 to 4-fold higher amounts of
IAA in culture without tryptophan supplementation but has
comparable levels of tryptophan monooxygenase.[7] When grown
with an exogenous source of L-tryptophan, IAA production
was increased but was comparable to the elevated amounts
accumulated by parental strains grown in the presence of
tryptophan. Thus, the super producer phenotype of mutant
2015-5 is expressed only under conditions of limited
(controlled) tryptophan supply. The intracellular concentration of tryptophan in this mutant was 35-fold higher
than that of its parent strain.

This mutant causes severe symptoms and in some cases
causes production of rooty tumors on oleander. The tumor
phenotype appears to be due to a lowered cytokinin/IAA
balance created by overproduction of IAA by the bacterium.
Although proficient in the production of IAA, the mutant
grows somewhat more slowly in oleander tissues compared
with its parent wild type.[7] Other mutants with very high
capacity for IAA production fail to grow in host tissue,
even though they multiply in culture at wild type rates
and accumulate 10-fold higher concentrations of IAA (L.
Glass and T. Kosuge, unpublished results). Apparently,
relaxed control of IAA production, in this case, reduces
fitness in this mutant. These mutants should provide
insights on the relationship between fitness and altered
regulation of IAA synthesis.

As noted above, tryptophan synthesis in this bacterium
is regulated, in part, by feedback inhibition of anthranilate synthase by L-tryptophan (Smidt and Kosuge,
unpublished results). Because regulation of the enzyme
is designed to control tryptophan pool size for efficient
utilization of cellular nutrients for biosynthetic activities and energy generation, mutations to relaxed control
of anthranilate synthase will result in increased tryptophan
and IAA production and, potentially, increased virulence.
However, relaxed control of tryptophan synthesis will reduce
the efficiency of carbon, nitrogen and energy utilization in
the bacterium and will reduce the capacity of the pathogen
to compete and to survive under nutrient limitation. Additional mechanisms controlling tryptophan and IAA synthesis
remain to be elucidated in this bacterium.

MOVABLE ELEMENTS INACTIVATE GENES FOR INDOLEACETIC ACID
SYNTHESIS

Movable elements such as IS-elements in bacteria are
commonly associated with changes in phenotype by insertional
inactivation of genes.[15]   In P. savastanoi this phenomenon is
commonly associated with loss of IAA production and virulence
in mutants that are selected for resistance to α-methyltryptophan.  Loss of virulence in three iaa⁻ mutants has been
attributed to movable elements IS-51 and IS-52 which insert
into and inactivate the monooxygenase gene, iaaM.[12]   Both
IS-51 and IS-52 have been sequenced (T. Yamada and T. Kosuge,
manuscript submitted).  Although the IS-elements differ
significantly in nucleotide sequences, they nevertheless
share the common property of transposition into iaaM causing
loss of virulence and IAA production.  The site of integration in iaaM is marked by the consensus sequence, purine-A-G,
in the target gene.  However, the integration is not at the
same location.  Both IS-51 and IS-52 have characteristics in
common with those found in other IS-elements; IS-51 has
inverted repeats of 26 bp at each end whereas IS-52 has
inverted repeats of 12 bp.

Mutations by insertional inactivation are responsible
for the change of phenotype in other plant pathogenic
bacteria.  Loss of tumorigenicity has been observed in A.
tumefaciens by spontaneous insertion of an IS-element in the
T-DNA.[5]  It should be recalled that Tn5 insertional inactivation of the T-DNA genes was used to isolate mutants which
either were avirulent or caused formation of morphologically
altered tumors on tobacco different hosts of Agrobacterium.


CONVERSION OF IAA TO INDOLEACETYL LYSINE

Certain strains of P. savastanoi convert IAA to
indoleacetyl-ε-L-lysine (Fig. 2).  Because IAA synthesis
is responsible for tumorigenicity and virulence in P.
savastanoi, it follows that conjugate synthesis can reduce
the pool size of free IAA and reduce virulence in the
pathogen.  Such a role is seen for the system which catalyzes the conversion of IAA to IAA-lysine in oleander
strains of P. savastanoi.[16,17]  The enzyme catalyzing the
synthesis of IAA-lysine from IAA requires ATP and a divalent
metal cation such as $Mg^{+2}$ or $Mn^{+2}$.  The genetic determinant
(iaaL) for the enzyme is borne on pIAA but is not part of

the iaa operon (Fig. 3).  It is interesting that olive
isolates, unlike their counterparts from oleander, are
unable to produce IAA-lysine.

IAA-lysine synthetase may help control the pool size
of free IAA since oleander strains typically accumulate less
IAA in culture when compared with olive strains (L. Glass
and T. Kosuge, unpublished results).  IAA-lysine is not the
end product of IAA metabolism, however; oleander strains
convert the compound to its N-acetyl derivative[18] and to
other compounds which remain to be identified.[17]

## IAA PRODUCTION IN OTHER TUMORIGENIC SYSTEMS

Because production of phytohormones is a basis for
tumorigenicity in both crown gall tumor tissue and in
oleander knot, questions were raised about the similarities
between the two systems (Fig. 3).  Characterization of gene
2 has shown that it codes for an amidohydrolase converting

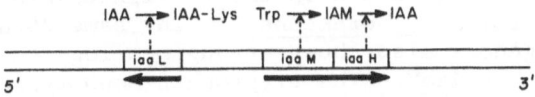

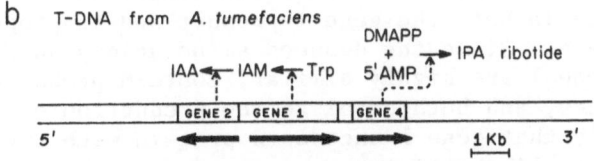

Fig. 3.  A model of A. tumefaciens T-DNA and the IAA region
of pIAA1 from P. savastanoi.  The direction of transcription
of T-DNA and P. savastanoi genes are shown by arrows; the
proposed catalytic activities of the gene products are
indicated as reactions converting tryptophan (trp) to
indoleacetamide (IAM) (gene 1) and to indoleacetic acid
(IAA) (gene 2).[19,20,21]  Gene 4 encodes a transferase catalyzing the production of isopentenyladenine ribotide (IPA
ribotide) from dimethylallylpyrophosphate (DMAPP) and 5'
AMP.[22,23]

indole-3-acetamide (IAM) into IAA.[19],[20]  This observation led
to the suggestion that IAA in tumors might be synthesized
from tryptophan by the pathway involving 2 steps previously
demonstrated in P. savastanoi, and that gene 1 might code for
an enzyme converting tryptophan into IAM.  It has been shown
that tobacco tissue transformed with a vector containing
gene 1, but without other T-DNA genes, accumulated high
levels of IAM (1500 pMol/g.fr.wt.).  Normal tissue contained
comparatively little IAM (1.0 pMol/g.fr.wt.).[21]  However,
conversion of tryptophan into IAM has not been successfully
demonstrated using cell-free enzyme preparations from tumor
tissue.

    Significant homology occurs between nucleotide sequences
of iaaM and iaaH of P. savastanoi and those of genes 1 and 2
of A. tumefaciens T-DNA.  Gene 1 and iaaM have about 50%
perfect matches in deduced amino acid sequence; comparison
of the open reading frames for these genes indicate that
gene 1 codes for a polypeptide which is 198 amino acids
longer and about 22,000 daltons heavier than that encoded
by iaaM.[24]  Comparisons of the deduced amino acid sequences
of iaaH and gene 2 indicate a lesser homology of about 27%
for perfect amino acid matches.  Moreover, a sequence of 24
amino acids has been found to be homologous between gene 1
and the FAD binding domain in the FAD-linked hydroxybenzoate
hydroxylase from P. fluorescens.[25]  The same 24 amino acid
sequence of gene 1 also shows strong homology to a 24 amino
acid sequence in iaaM.  The tryptophan monooxygenase encoded
by iaaM is known to have an FAD cofactor, and it seems
likely that this amino acid sequence comprises an FAD-
binding site in both the gene 1 product and tryptophan
monooxygenase.  Since the deduced amino acid sequences of
iaaM and gene 1 are highly similar, contain probable FAD-
binding sites, and both catalyze the production of IAM, it
seems likely that gene 1 encodes a protein with tryptophan
monooxygenase-like activity.

    The tryptophan monooxygenase encoded by iaaM confers
resistence to the toxic tryptophan analog, 5-methyltryptophan
(5MT) and certain other ring substituted tryptophan analogs
but not to the side-chain substituted α-methyltryptophan in
P. savastanoi.[13]  Resistance is related to the capacity of
tryptophan monooxygenase to use 5MT as a substrate and
convert it into the less toxic 5-methylindolyl-3-acetamide.
The enzyme cannot use α-methyltryptophan as a substrate and
subsequently cannot confer resistance to this toxic anti-

metabolite.  The response of crown gall tumor cells to
tryptophan analogs is similar to that seen in P. savastanoi
cells.  Transformed tobacco cells containing gene 1 are
resistant to 5MT and other ring-substituted tryptophan
analogs and sensitive to α-methyltryptophan.  Normal, non-
transformed cells and transformed cells lacking gene 1 are
sensitive to tryptophan analogs.  5MT resistance is
correlated to the presence of gene 1 in tumor cells.  The
metabolic basis for tryptophan analog resistance in tumor
cells has not yet been fully characterized.  However, the
similar pattern of tryptophan analog resistances correlating
to iaaM in P. savastanoi and gene 1 in A. tumefaciens
suggests additional relationships between these two genes,
their products, and the production of IAA in these tumorigenic diseases.

The foregoing evidence suggests that the genes for IAA
synthesis had a common origin in the two tumorigenic systems.
Perhaps even more intriguing is the question of the role of
the iaa operon in bacteria in general.  A number of plant
pathogenic bacteria not associated with gall formation on
their hosts nevertheless have DNA sequences showing strong
homology with the iaaM and iaaH.  If these genes are functional in those bacteria, what selection pressure maintains
the iaa genes in these bacteria?  Perhaps they perform the
function of detoxifying toxic tryptophan analogues that are
produced as antibiotics by competing epiphytic bacteria.

SUMMARY

In this chapter, the role of the shikimic acid pathway
in virulence of a plant pathogen is discussed using indoleacetic acid (IAA) production by P. savastanoi as a model.
A product of the shikimic acid pathway, IAA is a virulence
factor for the bacterium, and is involved in tumor formation
on its hosts, oleander and olive.  The bacterium produces
IAA from tryptophan by the sequence of reactions:
L-tryptophan ⟶ indoleacetamide ⟶ indoleacetic acid.
The enzymes concerned and their genetic determinants are:
tryptophan 2-monooxygenase (iaaM) and indoleacetamide
hydrolase (iaaH).  The two genes, iaaM and iaaH which are
virulence genes for the bacterium, occur on a plasmid, pIAA,
in oleander strains and on the chromosome of olive strains
of the pathogen.  The virulence gene product, tryptophan
monooxygenase, fulfills the important role of diverting

tryptophan into the production of IAA; thus, the enzyme not only helps confer virulence but also regulates the production of IAA and the diversion of the primary metabolite, L-tryptophan, into secondary metabolism. The enzyme is a flavoprotein and is product inhibited by indoleacetamide ($K_i$ = 7 uM); the inhibition is competitive with respect to the substrate, L-tryptophan. Control of IAA production and virulence therefore occurs at the step catalyzed by tryptophan monooxygenase.

The availability of tryptophan is also a determining factor in the formation of indoleacetic acid; the pool size of tryptophan is regulated in part by tryptophan feedback inhibition of anthranilate synthase.

Two IS-elements, IS-51 and IS-52, resident in the bacterium have been isolated and sequenced. By integration into iaaM, they are responsible for loss of IAA production and virulence in several mutants that were isolated by selection for α-methyltryptophan resistance. Insertional inactivation of virulence genes by transposable elements similar to IS-51 and IS-52 may be the basis of the spontaneous loss of virulence commonly observed in other plant pathogenic bacteria.

Certain strains of P. savastanoi convert IAA to its lysine conjugate; the gene for synthesis of the lysine conjugate occurs on pIAA but it is not part of the IAA operon. Conversion of IAA to its lysine conjugate may affect the free pool size of IAA and the amount secreted into host tissue.

Both iaaM and iaaH have been sequenced; on the basis of both nucleotide and deduced amino acid sequences, the coding sequences of iaaM show significant homology with the sequences of the open reading frame of tms 1, a gene encoding IAA production in crown gall T-DNA; less, but nevertheless significant homology occurs between the coding sequences of iaaH and the open reading frame of tms-2 of crown gall T-DNA. The results suggest that the genes for IAA production in P. savastanoi and crown gall T-DNA have a common origin.

REFERENCES

1. LEISINGER, T., R. MARGRAFF. 1979. Secondary metabolites of the fluorescent Pseudomonads. Microbiol.
   Rev. 43: 422-442.
2. SKOOG, F., C.O. MILLER. 1957. Chemical regulation of
   growth and organ formation in plant tissues cultured
   in vitro. Soc. Exp. Biol. Symp. 11: 118-131.
3. BRAUN, A.C. 1958. A physiological basis for autono-
   mous growth of the crown gall tumor cell. Proc.
   Natl. Acad. Sci. USA 44: 344-349.
4. GELVIN, S.B. 1984. Plant tumorigenesis. In Plant-
   Microbe Interactions - Molecular and Genetic
   Perspectives. (T. Kosuge, E.W. Nester, eds.),
   MacMillan Inc., New York, pp. 243-377.
5. GARFINKLE, D.J., E.W. NESTER. 1980. Agrobacterium
   tumefaciens mutants affected in crown gall tumorigenesis and octopine catabolism. J. Bacteriol.
   144: 732-743.
6. AKIYOSHI, D.E., R.O. MORRIS, R. HINZ, B.S. MISCHKE,
   T. KOSUGE, D.J. GARFINKEL, M.P. GORDON, E.W. NESTER.
   1983. Cytokinin/auxin balance in crown gall tumors
   is regulated by specific loci in the T-DNA. Proc.
   Natl. Acad. Sci. USA 80: 407-411.
7. SMIDT, M., T. KOSUGE. 1978. The role of indole-3-
   acetic acid accumulation by alpha methyl tryptophan-
   resistant mutants of Pseudomonas savastanoi in gall
   formation on oleanders. Physiol. Plant Pathol. 13:
   203-214.
8. COMAI, L., T. KOSUGE. 1980. Involvement of plasmid
   deoxyribonucleic acid in indoleacetic acid synthesis
   in Pseudomonas savastanoi. J. Bacteriol. 143: 950-
   957.
9. COMAI, L., T. KOSUGE. 1982. Cloning and characterization of iaaM, a virulence determinant of
   Pseudomonas savastanoi. J. Bacteriol. 149: 40-46.
10. COMAI, L., T. KOSUGE. 1983. The genetics of indole-
    acetic acid production and virulence in Pseudomonas
    savastanoi. In Molecular Genetics of the Bacteria-
    Plant Interactions. (A. Puhler, ed.), Springer-
    Verlag, Berlin, pp. 363-366.
11. COMAI, L., G. SURICO, T. KOSUGE. 1982. Relation of
    plasmid DNA to indoleacetic acid production in
    different strains of Pseudomonas syringae pv
    savastanoi. J. Gen. Microbiol. 128: 2157-2163.

12.  COMAI, L., T. KOSUGE. 1983. Transposable element that
     causes mutations in a plant pathogenic Pseudomonas
     sp. J. Bacteriol. 154: 1162-1167.
13.  HUTCHESON, S.W., T. KOSUGE. 1985. Regulation of
     3-indoleacetic acid production in Pseudomonas
     syringae pv savastanoi. J. Biol. Chem. 260: 6281-
     6287.
14.  KOSUGE, T., M.G. HESKETT, E.E. WILSON. 1966. Microbial
     synthesis and degradation of indole-3-acetic acid.
     I. The conversion of L-tryptophan to indole-3-
     acetamide by an enzyme system from Pseudomonas
     savastanoi. J. Biol. Chem. 241: 3738-3744.
15.  CHANDLER, M., D.J. GALAS. 1985. Studies on the
     transposition of IA1. In Plasmids in Bacteria.
     (D.R. Helinski, S.N. Cohen, D.B. Clewell, D.A.
     Jackson, A. Hollaender, eds.), Plenum Press, New
     York, pp. 53-77.
16.  KOSUGE, T., L. COMAI, N.L. GLASS. 1983. Virulence
     determinants in plant-pathogen interactions. In
     Plant Molecular Biology. (R. Goldberg, ed.), Alan
     R. Liss, Inc., New York, pp. 167-177.
17.  HUTZINGER, O., T. KOSUGE. 1968. Microbial synthesis
     and degradation of indole-3-acetic acid. III. The
     isolation and characterization of indole-3-acetyl-
     ε-L-lysine. Biochemistry 7: 601-605.
18.  EVIDENTE, A., G. SURICO, N.S. IACOBELLIS, G. RANDAZZO.
     1985. Isolation and structural characterization
     of α-N-acetyl-indole-3-acetyl-ε-L-lysine: a new
     metabolite of indole-3-acetic acid from Pseudomonas
     syringae pv savastanoi. Phytochemistry 24: 1499-1502.
19.  SCHRODER, G., S. WAFFENSCHMIDT, E.W. WEILER, J.
     SCHRODER. 1984. The T-region of Ti plasmids codes
     for an enzyme synthesizing indole-3-acetic acid.
     Eur. J. Biochem. 138: 387-391.
20.  THOMASHOW, L.S., S. REEVES, M.F. THOMASHOW. 1984.
     Crown gall oncogenesis: evidence that a T-DNA
     gene from the Agrobacterium Ti plasmid pTiA6
     encodes an enzyme that catalyzes synthesis of
     indoleacetic acid. Proc. Natl. Acad. Sci. USA 81:
     5071-5075.
21.  VAN ONCKELEN, H., P. RUDELSHEIM, D. INZE, A. FOLLIN,
     E. MESSENS, S. HOVEMANS, J. SCHELL, M. VAN MONTAGU,
     J. DeGREEF. 1985. Tobacco plants transformed with
     the Agrobacterium tumefaciens T-DNA gene 1 contain
     high amounts of indoleacetamide. FEBS Lett. 181:
     373-376.

22.  AKIYOSHI, D.E., H. KLEE, R.M. AMASINO, E.W. NESTER,
     M.P. GORDON. 1984. T-DNA of _Agrobacterium
     tumefaciens_ encodes an enzyme of cytokinin
     biosynthesis. Proc. Natl. Acad. Sci., USA 81:
     5994-5998.
23.  BARRY, G.F., S.G. ROGERS, R.T. FRALEY, L. BRAND. 1984.
     Identification of a cloned cytokinin biosynthetic
     gene. Proc. Natl. Acad. Sci., USA 81: 4776-4780.
24.  YAMADA, T., C.J. PALM, B. BROOKS, T. KOSUGE. 1985.
     Nucleotide sequences of _Pseudomonas savastanoi_
     indoleacetic acid genes show homology with
     _Agrobacterium tumefaciens_ T-DNA. Proc. Natl. Acad.
     Sci., USA 82: 6522-6526.
25.  KLEE, H., A. MONTOYA, F. HORODYSKI, C. LICHTENSTEIN,
     D. GARFINKEL, S. FULLER, C. FLORES, J. PESCHON,
     E.W. NESTER, M.P. GORDON. 1984. Nucleotide
     sequence of the tms genes of the pTi A6NC octopine
     Ti plasmid: two gene products involved in plant
     tumorigenesis. Proc. Natl. Acad. Sci., USA 81:
     1728-1732.

Chapter Seven

HYDROXYBENZOIC ACIDS AND THE ENIGMA OF GALLIC ACID

EDWIN HASLAM

Department of Chemistry
The University of Sheffield
Sheffield   S3 7HF
United Kingdom

INTRODUCTION

"Esters and/or glycosides of a variety of substituted benzoic acids are found in all higher plants."  Examining the veracity of this statement[1] brings one in mind of Oscar Wilde's aphorism concerning truth - that it is never pure and rarely simple.  Certainly $C_6$–$C_1$ phenolic acids occur widely in plants and also microorganisms.  Some plant species are particularly rich in these compounds - for example[1] the leaves of winter-green (Gaultheria procumbens), extracts of which upon alkaline hydrolysis yield p-hydroxybenzoic (2), salicylic (4), protocatechuic (6), 2,3-dihydroxybenzoic (5), gentisic (9), vanillic (7) and syringic (8) acids (Fig. 1).  Although metabolites containing these functionalities are widely distributed in angiosperms (in the botanical sense), according to the extent literature, their occurrence is nevertheless sporadic and often appears to constitute something of a taxonomic speciality.  In this respect the most familiar example is probably the willow family (Salicaceae) in

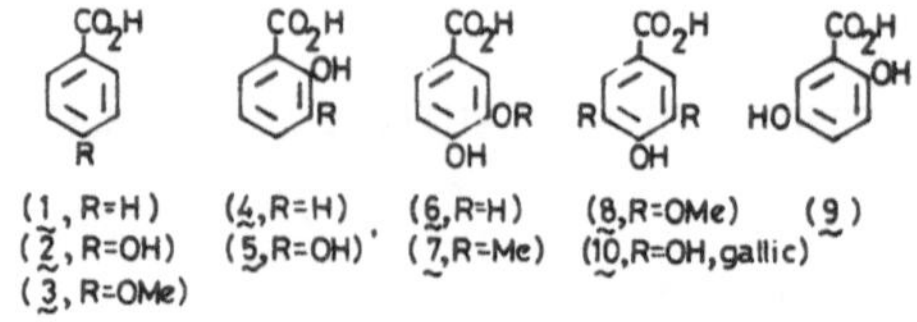

Fig. 1.  Shikimate derived hydroxybenzoic acids in plants.

which derivatives of salicylic acid, such as the ω-
salicylsalicin metabolites (11-13) isolated from Salix
purpurea bark,[2] are found.  Hydroxybenzoic acids, their
analogues, and significantly benzoic acid (1) itself are
also consistently located esterified to both terpenoid and
alkaloid structures throughout the plant kingdom.  The
metabolites (Fig. 2, 11-25) typically illustrate the
breadth and diversity of this mode of occurrence.[3-8]  In
view of the later biosynthetic discussions particular note
should be made of the β-aminophenylalanine ester group in
taxol (23).

     In microorganisms p-hydroxybenzoic (2), salicylic (4)
and 2,3-dihydroxybenzoic (5) acids are all synthesised via
chorismic acid[9,10] in metabolism which is purposeful in
character (Fig. 3).  p-Hydroxybenzoic acid (2) is thus an
intermediate in the biosynthesis of the ubiquinones[9,10] and
2,3-dihydroxybenzoyl serine (26), as its cyclic trimer
enterochelin, is intimately involved in iron transport in
some bacteria.[9,10]  Protocatechuic acid (6) in contrast is
probably formed as a shunt-metabolite in certain mutants of
Aerobacter aerogenes which are blocked in the biosynthesis
of the aromatic amino acids.[11]

     Whatever qualification or interpretation one may place
upon the opening quotation, in higher plants gallic acid
(10) and its metabolites provide clearer support for the
statement both in the manner of their occurrence and their
distribution than do any of the other $C_6$-$C_1$ acids already
alluded to.  The occurrence of gallic acid has been noted
in some 20 or so plant families[12,13,14] and one freshwater
alga.[15]  In particular tissues, e.g. plant galls, metabolites of gallic acid often accumulate in substantial
quantity.  Indeed, as they occur in extracts of oak-galls,
gallic acid and its derivatives constitute a chemical
reagent of considerable antiquity used in the analysis of

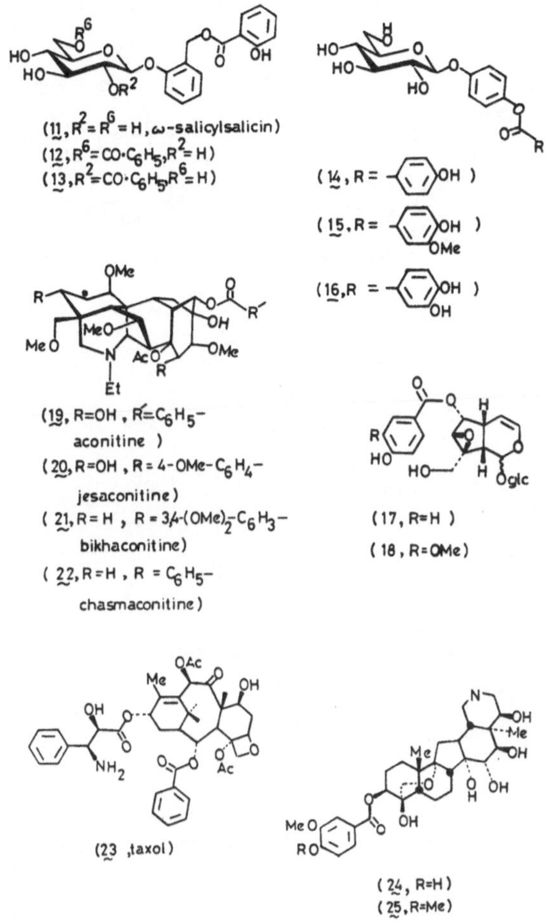

(11, $R^2 = R^6 = H$, $\omega$-salicylsalicin)
(12, $R^6 = CO \cdot C_6H_5$, $R^2 = H$)
(13, $R^2 = CO \cdot C_6H_5$, $R^6 = H$)

( 14, R = benzene ring with OH )

( 15, R = benzene ring with OH, OMe )

( 16, R = benzene ring with OH, OH )

(19, R=OH , R'=$C_6H_5$-
   aconitine )
(20, R=OH , R = 4-OMe-$C_6H_4$-
   jesaconitine)
( 21, R=H , R = 3,4-(OMe)$_2$-$C_6H_3$-
   bikhaconitine)
( 22, R=H , R = $C_6H_5$-
   chasmaconitine)

(17, R=H )
(18, R=OMe)

(23 ,taxol)

( 24, R=H)
(25, R=Me)

Fig. 2.    Modes of occurrence of hydroxybenzoic acids in plants.

mineral waters and as a component of invisible ink as early as the sixteenth century.  However, over and above all these characteristics, it is the variety and range of forms in which it is found in plants which distinguishes gallic acid from the other hydroxybenzoic acids.  It thus occurs in ester forms which vary from the very simple mono-esters such as β-D-glucogallin (27), theogallin (28) and the flavan-3-ol-gallates (29, 30) - all found in young tea shoots (Camellia sinensis) - to the complex polyesters

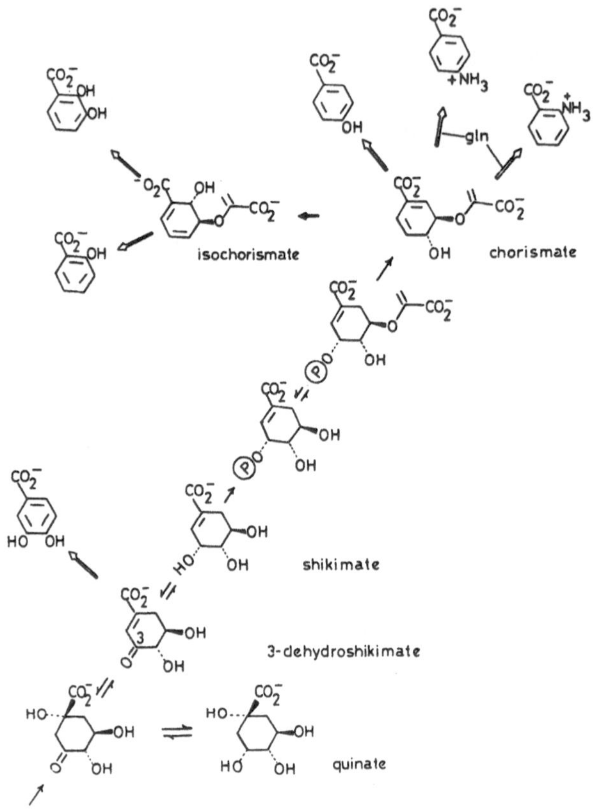

Fig. 3.  Shikimate derived hydroxybenzoic acids in micro-organisms.

with D-glucose whose molecular weights extend to 3000 and which together with the proanthocyanidins constitute the so-called vegetable tannins[16] of the earlier chemical and botanical literature.  These patterns of occurrence should also be contrasted with those of the various hydroxycin-namic acids (p-coumaric, caffeic, ferulic, sinapic).  These acids are ubiquitous in higher plants but they are normally only found as mono- and occasionally bis-esters with polyols (e.g. chlorogenic acid, 31).

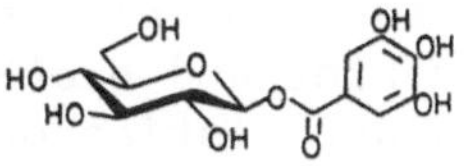

(27)

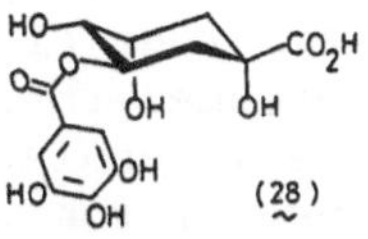

(28)

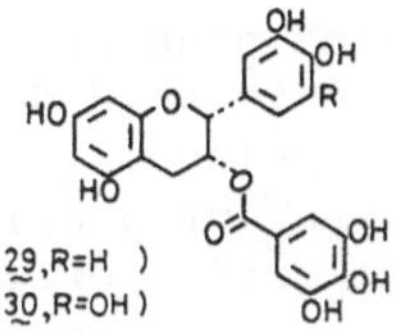

(29, R=H )
(30, R=OH )

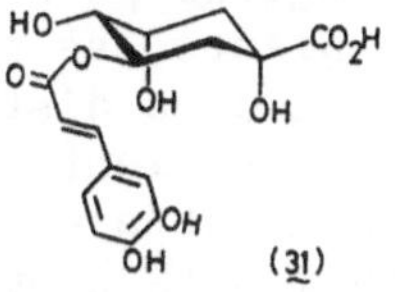

(31)

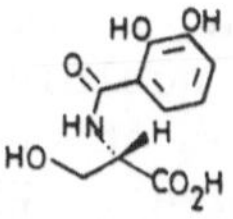

(26)

Although gallic acid is almost invariably located in plant tissues in ester form - with for example sugars, polyols, glycosides and other phenols[12] - its metabolism in glycosidic form provides an interesting variation on this pattern.   One of the most fascinating biological phenomena in plants is that of leaf movement such as is witnessed in the sensitive plant Mimosa pudica.   The chemical agonists of phytodynamics - turgorins - which elicit leaf movement after perception of an external stimulus do so by causing a drastic change in the hydrostatic internal pressure of the vacuoles of parenchymal cells in the motile organs.   The South African plant Acacia karoo folds its leaves together pairwise when it reacts nyctinastically and it then has the appearance of a sleeping mimosa plant.   Two periodic leaf-movement factors isolated from A. karoo are the sulphate esters (32 and 33) of 4-O-β-D-glucopyranosyl gallic acid. The glucoside (32) has also been detected in Gleditsia triacanthos, Robinia pseudacacia, Oxalis sp., Albizia julibrissin and Abutilon grandiflorum and, along with the related glucuronide (34) from M. pudica.[17]

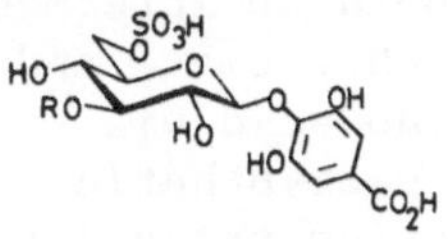

(32 , R=H )
(33 , R=SO₃H)

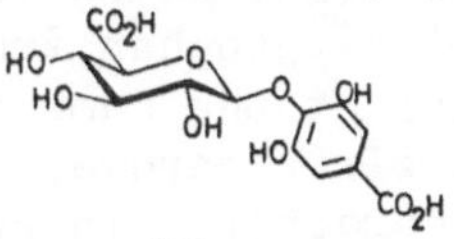

(34)

turgorins

The concepts of purpose and function are essential to
advances in the study of living matter and biology since it
is generally assumed that the characteristics of an organism,
whether they be molecular or morphological, have been
selected because of their functional advantage.  Such
considerations are often a stimulus to the understanding
of complex biological phenomena.  The metabolism of gallic
acid in higher plants, as compared to that of the hydroxybenzoic acids, remains distinctive and presents something of
an enigma.  In many plant species gallic acid derivatives
proliferate.  Some attain a relative molecular size (> 1000)
which endows them with properties apparently unique amongst
secondary metabolites of plant or microbial origin.  These
characteristics of ready and reversible complexation with
proteins, polysaccharides, nucleic acids and nitrogenous
bases (vide infra) make them objects well worthy of scientific study in their own right.  What is the particular
function of this form of phenolic metabolism in higher
plants is a question which remains unanswered.  It is one
to which attention must inevitably move in future years if
a rational explanation of the purpose of secondary metabolism in general is to be uncovered.


BIOSYNTHESIS

In higher plants and microorganisms single biosynthetic
pathways generally exist to the major metabolites of the
$C_6-C_3$ class such as phenylalanine and tyrosine and (in
plants) the various hydroxycinnamic acids.  The principal
exception to this generalisation is in the blue-green
bacteria and in some Pseudomonas sp. in which the arogenate
pathway is operative.[18]  It is therefore of some significance that a dichotomy of pathways may be utilised in
nature for the synthesis of the corresponding $C_6-C_1$ hydroxy
acids.  Biosynthesis may thus occur[9,10] from an acyclic
intermediate or by transformation of precursors which
already contain an aromatic ring.  The situation is most
clearly exemplified by that of the $C_6-C_1$ amino acid
anthranilic acid (35) which is both an intermediate in the
biosynthesis of tryptophan from chorismate and in the oxidative degradation of the same amino acid via kynurenine[9,10]
(Fig. 4).  The first pathway is biosynthetic, the second
catabolic in character.  In practice the existence of this
duality of pathways has led, in particular instances, to
apparently conflicting evidence and conclusions.  For the

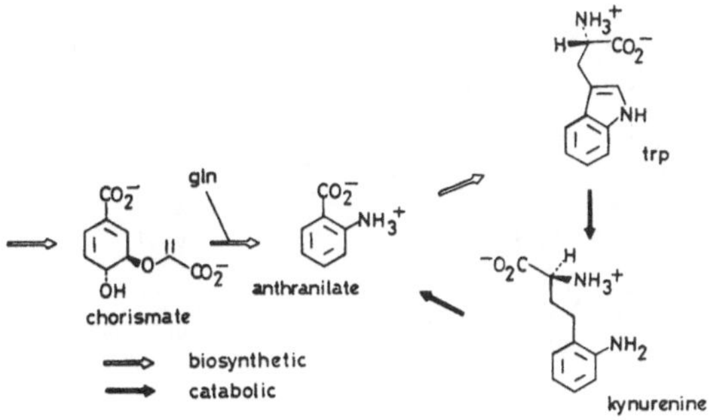

Fig. 4. Anabolic and catabolic pathways to the $C_6-C_1$ acid, anthranilic acid.

formation of the hydroxybenzoic acids in plants the data has been critically assessed by Zenk.[19]

Synthesis of several hydroxybenzoic acids in micro-organisms has been demonstrated to occur via aromatisation of acyclic intermediates in the shikimate pathway – notably 3-dehydroshikimic acid, chorismic acid and isochorismic acid[10,20-23] (Fig. 3). In each case the phenolic hydroxyl groups derive from oxygen atoms in the precursor and by analogy with anthranilic acid (35, Fig. 4) such pathways may be considered to be biosynthetic. However in higher plants alternative routes to these metabolites also exist and by the same token these may be envisaged as essentially catabolic in nature. Thus the shortening of the side-chain of a $C_6-C_3$ compound (phenylalanine or a cinnamic acid) or hydroxylation of a less highly substituted benzoic acid can lead to the desired metabolite. Here the phenolic groups originate principally by hydroxylation (introduction of oxygen) of the aromatic ring.. The various alternatives are illustrated in the case of p-hydroxybenzoic acid (2), Figure 5: isotopic tracer work suggests that in this,[24] and in analogous cases, the favored pathway in plants to the $C_6-C_1$ acid is through the breakdown of a $C_6-C_3$ substrate. A mechanism for the degradation of cinnamic acids to the $C_6-C_1$ derivatives was proposed by Zenk[25] similar to the β-oxidation of fatty acids involving cinnamoyl-CoA esters as intermediates. Some doubt has however been cast on this hypothesis, notably by Kindl and his associates.[26] If only phenylalanine and tyrosine can serve as substrates for the

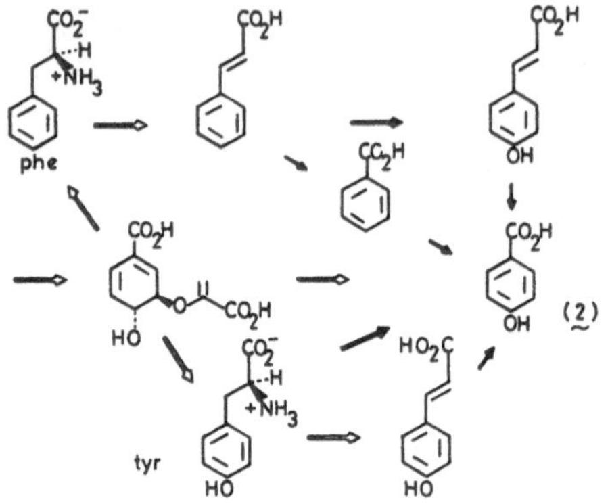

Fig. 5.  Pathways to p-hydroxybenzoic acid in plants.

chain shortening reactions, as suggested by Kindl, it is
interesting to speculate on the mechanism of such conversions in the light of the natural occurrence of such
metabolites as taxol (23) and taxine (36) which are both
derivatives of the unusual β-aminophenylalanine structure.

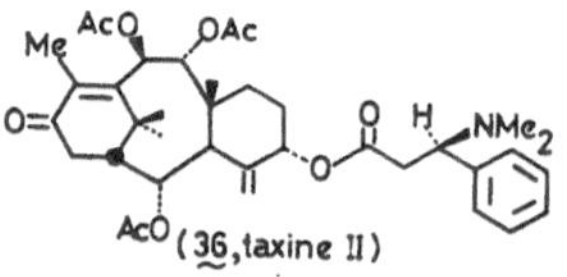

(36,taxine II)

The latter may be envisaged as an intermediate in the degradation of phenylalanine.  In clostridia catabolic breakdown
of lysine (37) occurs[27] by initial conversion to β-lysine
(38), (mediated by a 2,3-aminomutase) and thence via 3-keto-
5-aminohexanoate to acetate and butyrate.  An analogous
catabolic sequence commencing with phenylalanine would lead
to benzoic acid (as its CoA ester) and acetoacetate with
the β-aminophenylalanine (39) as an intermediate (Fig. 6).

Against this backcloth it is perhaps not surprising to
learn, that, despite its distinctive position in the overall
patterns of plant phenol metabolism, ambiguity still
surrounds the biosynthesis of gallic acid.  Several pathways
have been proposed.  Zenk[28] formulated a conventional pathway (Fig. 7, a) from L-phenylalanine to 3,4,5-trihydroxy-
cinnamic acid followed by β-oxidation to give gallic acid,

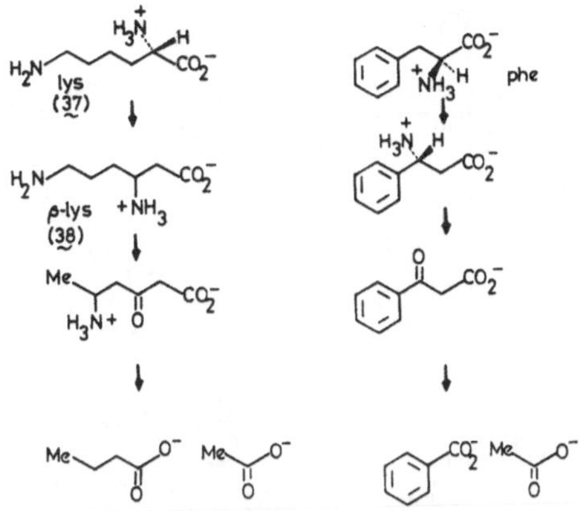

Fig. 6. Amino-acid catabolism by amino transfer ($\alpha \to \beta$): a putative pathway to benzoic acid from L-phenylalanine.

10, in experiments in Rhus typhina (Sumach, Anacardiaceae), and Neish and Towers[29] favored a variation of this theme (Fig. 7, b). Both of these pathways postulate that gallic acid is derived, overall, by oxidative shortening of the $C_3$ side chain and hydroxylation of the aryl ring, of one of the end products (L-phenylalanine) of the shikimate pathway. In contrast the third pathway,[30-35] for which experimental evidence has been obtained with the mold Phycomyces blakesleeanus and with the plants R. typhina, C. sinensis, Acer and Geranium sp. suggests that gallic acid is a shunt or overflow metabolite of the shikimate pathway derived by direct dehydrogenation of the intermediate 3-dehydroshikimate (Fig. 7, c). This pathway predicts that not only the carbon skeleton but also the three phenolic groups originate directly from the alicyclic precursor. In support of this type of mechanism is the observation (N. Amrhein, private communication) that the herbicide glyphosate, which acts as a competitive inhibitor of the enzyme 5-enolpyruvyl shikimate phosphate synthetase, stimulates gallic acid production when administered in limiting amounts to certain plant tissues.

In terms of the experimental observations the case of gallic acid is noteworthy for, among the hydroxybenzoic acids, it remains the only example in higher plants where the weight of evidence suggests its formation directly from

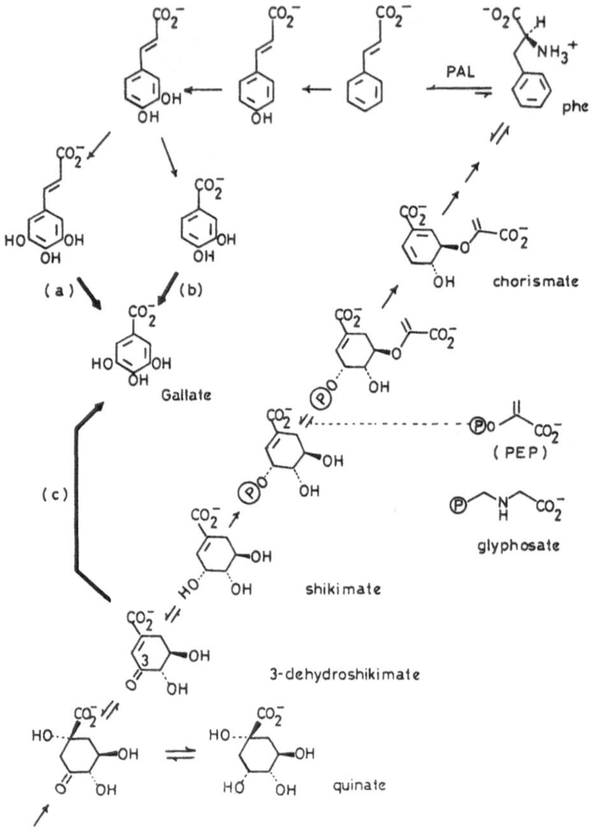

Fig. 7.  Pathways of biosynthesis of gallic acid in plants.


an alicyclic precursor.  Juxtaposed with the substantive
differences which have been noted in the range and pattern
of metabolites, this lends further circumstantial support
to the unique, enigmatic position which gallic acid is
thought to hold amongst the hydroxybenzoic acids in plants.

    Comparatively little is yet known concerning the
biosynthesis of the various derivatives of gallic acid,
such as the esters and depsides which occur in plants.
Gross[36,37] recently commenced a study of the enzymology of
gallic acid esters.  Cell-free extracts of fresh oak leaves
(Quercus robur) were found to catalyse the esterification

of gallic acid and UDP-glucose. The product of the reaction
was identified as β-D-glucogallin (27). This first intermediate is then sequentially esterified (possibly via the
participation of a species such as galloyl coenzyme A) to
give finally β-penta-O-galloyl-D-glucose (40). This intermediate, based upon the chemical intuition common to all
biogenetic hypothesis, is then believed to play a pivotal
role in the biosynthesis of both the so-called gallotannins
and ellagitannins in many, but not all, plant species[12,14,38,39] which metabolize gallic acid.

A variation on the sequential acylation of β-D-
glucogallin (27) to give (40) as suggested by Gross might
be one in which acylation occurs preferentially at the
anomeric center and is then followed by acyl migration to
the appropriate position of the D-glucopyrose ring (e.g.
1-galloyl → 6-galloyl → 1,6-digalloyl → etc). Such a
mechanism would explain the frequent occurrence in plants
of galloyl esters in which the anomeric center is free.


GALLIC ACID METABOLISM

The initial enthusiams which were generated in the
early years of the century by the isolation of polyphenols
in a crystalline form from plants gradually waned as workers
realized the enormous inherent complexity of many extracts.
Studies of plant polyphenols became one of the untidy
corners of organic chemistry. It was the advent of impres-
sive new methods of separation and structure determination
which gave a new impetus to research and to the comparatively recent rapid advances in our knowledge of plant
polyphenols. The vastly expanded armamentarium of physical
methods which the organic chemist now has at his command to
solve structural problems has, at the same time, brought
about a reappraisal of its objectives. Structure determination of itself no longer remains the major goal.
Increasingly attention focusses on the way in which natural
products themselves participate in the dynamics of living
systems. Nowhere has this trend been more evident than

with plant polyphenols - in fields as diverse as mammalian
pharmacology and ecology.  Within the compass of this
review no attempt will be made to give a comprehensive
account of these burgeoning developments but rather particular facets will be brought to the reader's attention
which highlight recent advances and some of their subtle-
ties.

Structure Determination

     Gallic acid is almost invariably found in plant
tissues, associated with sugars, polyols, glycosides and
other phenols, in ester form.[12,14]  Methods of structure
determination have been amply documented[12,40-43] in recent
years.  Attention is here directed to some aspects of the
structural analysis of D-glucose derivatives.

     Reliable mass measurements are a fundamental prerequi-
site for structural elucidation of a natural product.  The
last decade has seen a number of attempts to overcome the
specific problems associated with mass-spectrometry of
polyphenols, namely the requirement that the sample must be
presented to the instrument in the gas phase before it is
ionized.  Fast atom bombardment[44,45] (FAB) of solids is a
new technique for obtaining high quality mass-spectra of
molecules which are polar, thermally unstable and of relatively high molecular weight such as peptides, saccharides,
nucleotides and polyphenols.  Sample volatilization is not
necessary and thermal effects are thus avoided.  The method
employs as a key feature the phenomenon of 'ion sputtering'
from solutions (mainly glycerol) using a beam of fast
neutral atoms, typically Ar of 2-8 keV, as the primary
particles.  Mass-spectra are characterized by high pseudo-
molecular ion sensitivity giving $(M+H)^+$ in positive ion
spectra and $(M-H)^-$ in the negative ion case.  The method
promises to be a boon in the analysis of natural polyphenol
structures.  A typical example of its use (in collaboration
with R. Self, AFRC - Norwich) is in the analysis of the
astringent polyphenolic principle of several plants such
as Rosa sp., Filipendula ulmaria, Tellima grandiflora[42]
(Fig. 8).  It has a molecular weight of 1874 and contains
29 phenolic groups.  In addition to giving an accurate
molecular mass, structurally significant fragment ions were
also observed.

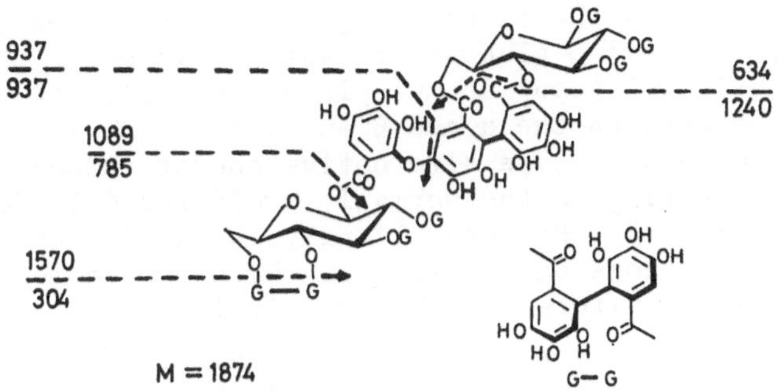

Fig. 8. Fast atom bombardment (FAB) mass spectrometry of complex polyphenols: phenol from Rosa sp. and Filipendula ulmaria.

[1]H and [13]C NMR spectroscopy with spin decoupling provides a complete structural analysis for many natural galloyl-D-glucopyranose derivatives. When the anomeric center is acylated the galloyl group almost invariably adopts the β-configuration (however vide infra) and the sugar adopts the C-1 ([4]C[1]) conformation.[12,42,43] Immediate recognition of these two features is given by the characteristic low field doublet for H-1 (δ 6.39 – 5.7 ppm TMS, J = 9.5 Hz) in the [1]H NMR spectrum. The position of this signal is often a measure of the index of galloylation in the D-glucose series.[12] Galloyl ester groups are distinguished by a series of 2 proton singlets ($d_6$ acetone, δ 6.95 – 7.25 ppm TMS). The position of the [13]C resonances due to the galloyl ester carbonyl groups is solvent sensitive but the relative intensities of these signals and their multiplicity provides an additional measure of the extent of the esterification of the glucose nucleus by gallic acid. Resonances due to the carbon atoms of the aryl rings of the galloyl ester groups are also readily recognized and identified. In polygalloyl esters (e.g. β-1,2,3,6-tetra-O-galloyl-D-glucose, Fig. 9) the [13]C resonances due to C-2 and C-3 in the individual galloyl groups are not resolved at 25.15 MHz and generally occur as singlets at δ 109.8 – 110.6 and 145.9 – 146.3 (ppm TMS) respectively. Conversely the [13]C resonances due to C-1 and C-4, just as the carbonyl carbon atoms, of the various galloyl groups are invariably resolved at this same operating frequency. They likewise provide an

accurate index of the number of galloyl ester groups in the
molecule.  Galloyl-D-glucose derivatives in which the
anomeric hydroxyl group is unacylated not only set up an
equilibrium between the α and β anomeric forms in solution
but they may also assume alternative conformations of the
glucopyranose ring to the normally preferred C-1 ($^4C_1$)
conformation.[46,47]  Structural analysis is, therefore, by
contrast with those derivatives in which the anomeric
position is substituted, much more difficult.

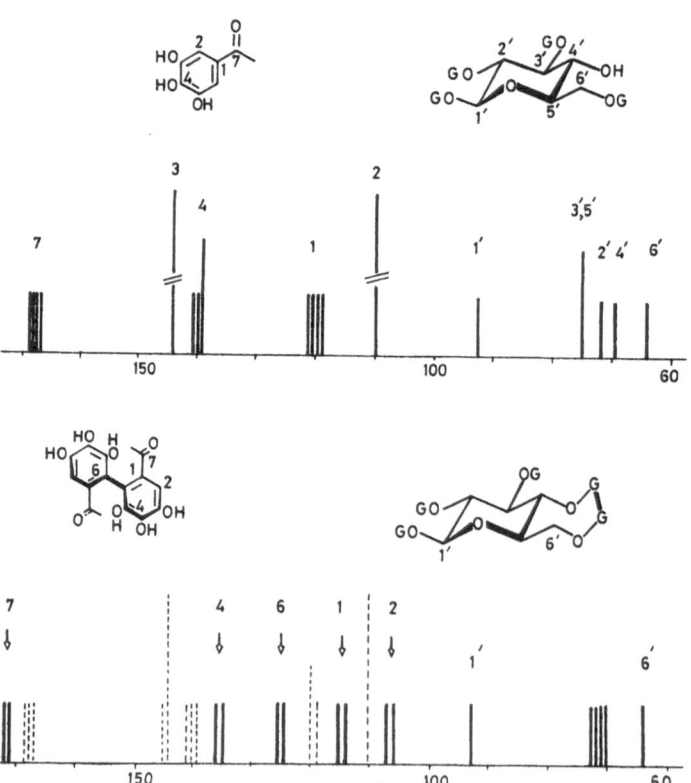

Fig. 9.  $^{13}C$ NMR spectroscopy of galloyl and hexahydroxydiphenoyl esters.

In his taxonomic surveys of the 1950's Bate-Smith[13] drew particular attention to the occurrence of ellagic acid (41) in the acid hydrolysates of dicotyledonous plant extracts. This, he inferred, was derived from metabolites which contained the hexahydroxydiphenoyl group (42) bound in ester form (Fig. 10). Such ester groupings are readily detected by an old but very distinctive test, the Proctor-Paessler reaction,[12] in which an instantaneous carmine-rose color formed with nitrous acid changes through green and purple to indigo-blue. Although there is no firm biosynthetic proof it is generally assumed, following the hypothesis of Schmidt and Mayer,[38] that esters of this type are generated by oxidative coupling of suitably placed galloyl ester groups in an appropriate precursor (Fig. 10). The formation of a diphenoyl ester group in this manner has a number of consequences. The closure of a large numbered ring containing two cis (E) double bonds reduces conformational flexibility and gives the molecule much greater rigidity. Where the bridging of the D-glucose residue occurs 1,6;2,4 or 3,6 this constrains the D-glucopyranose residue to adopt a thermodynamically unfavorable conformation.[43,47] One specific chirality is imposed on the hexahydroxydiphenoyl ester group as it is formed and this chirality is determined by the need of the hexahydroxydiphenoyl ester to bridge particular positions in the alcohol portion (invariably D-glucose) of the substrate molecule. The absolute configuration of corilagin (43) has been determined by reference to the twisted biphenyl system of the cyclo-octane lignan schizandrin (44)[48,49] (Fig. 10). The chirality of the other hexahydroxydiphenoyl esters may be determined by measurements of circular dichroism (CD) and comparison with 43. The CD spectra of hexahydroxydiphenoyl esters are characterized by a distinctive couplet centered at $\sim$ 200 - 210 nm with a positive or negative maximum at 228 - 238 nm. The $\Delta\varepsilon$ values are approximately incremental for the number of hexahydroxydiphenoyl ester groups in the molecule.[12,48,49]

The presence of a hexahydroxydiphenoyl group in a polyphenolic metabolite is readily defined by spectroscopic means.[12] In the $^1$H NMR spectra the two aryl protons of the hexahydroxydiphenoyl ester group appear as singlets generally upfield (d$_6$ acetone, $\delta$ 6.3 - 6.8 ppm TMS) from the two proton singlet of a galloyl ester group. Correspondingly several distinctive $^{13}$C NMR characteristics aid the recognition of the hexahydroxydiphenoyl ester group. The

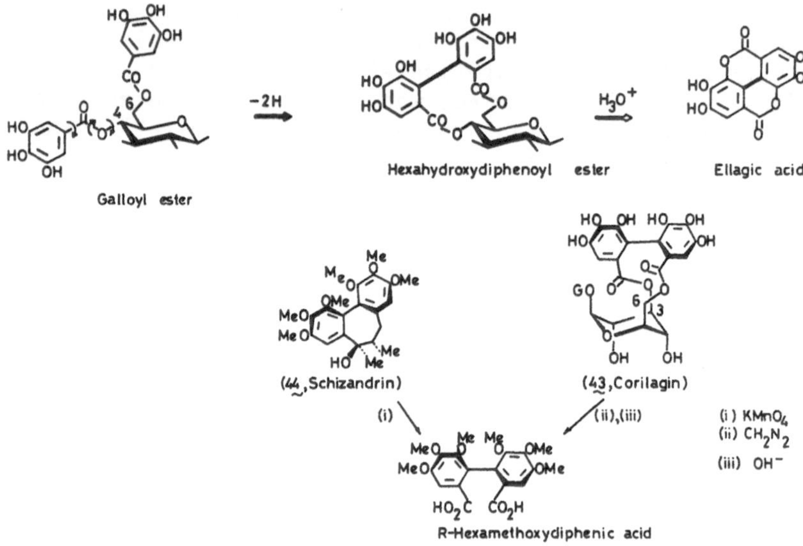

Fig. 10.   Biosynthesis of hexahydroxydiphenoyl esters.

signals due to the carbonyl carbon atoms appear downfield from TMS relative to those of galloyl esters and likewise the signals due to C-1, C-2 and C-6 are diagnostic for the hexahydroxydiphenoyl ester group.  Indeed the number of such groups in a metabolite may usually be discerned from the multiplicity of singlets in the $^{13}$C NMR spectrum arising from C-6 or the carbonyl carbon atoms or the aryl proton singlets in the $^1$H NMR.  Some facets of these distinctive NMR features of galloyl and hexahydroxydiphenoyl esters are illustrated in Figures 9, 11, 12 and 13.

Biogenetically one important and perhaps unexpected variation of the hexahydroxydiphenoyl ester group found in a limited number of metabolites is the dehydro hexahydroxydiphenoyl ester (45) first identified by Schmidt[50,51] and later noted in other metabolites of the families Geraniaceae, Euphorbiaceae and Aceraceae.  Isomerization occurs in solution and leads to an equilibrium mixture of internal hemi-acetal forms (45a, 45b).[12,40,43]  The attainment of the equilibrium is characterised by a distinctive but complex series of changes in both the $^1$H and $^{13}$C NMR spectra.[12,40,43]

(45)          (45 a)          (45 b)

     A critical feature of the assignment of structure to
polyesters of the type noted above has been the location
of the various different types of ester functionality on
the D-glucopyranose molecule.  Previous methodology has
relied heavily on specific chemical and/or enzymic degradation and empirical arguments based on chemical shift data.
A recent development (in association with K. Spencer, B.E.
Mann and R. Martin, University of Sheffield) now makes such
assignments directly possible from NMR measurements.  The
concept is illustrated for the case of β-penta-O-galloyl-
D-glucose (40).  The $^1$H NMR spectrum displays a series of
two proton singlets for each of the five galloyl ester
groups (Figs. 11, 12); the proton-coupled $^{13}$C NMR spectrum
an analogous sequence of five signals for the individual
carbonyl carbon atoms (Fig. 13) - four quartets ($J_{C-H} \sim$
5 Hz) and a multiplet.  In the $^1$H NMR spectrum the signals
from the protons of the D-glucopyranose ring may be associated unequivocally with the individual protons H-1, H-2,
H-3 etc., by conventional proton decoupling techniques.
Use is then made of the characteristic three-bond long
range C-H coupling to establish, via a two dimensional NMR
pulse sequence, the connectivities between individual
protons on the D-glucopyranose, the carbonyl carbon atoms
and the protons of each galloyl ester group.  In this way
the association of each galloyl ester group two proton
singlet (Fig. 12, $\sim \delta$ 6.95 - 7.25) and each carbonyl carbon
atom signal (Fig. 13, $\sim \delta$ 166 - 168) may be specifically
associated with a particular galloyl ester group on the
D-glucopyranose core.  (Parenthetically it is personally
very satisfying to record that this particular result was
predicted on purely empirical grounds 20 years earlier!)
Extension of this method to a range of hexahydroxydiphenoyl
and related esters has permitted complete assignments of
structure (vide infra).

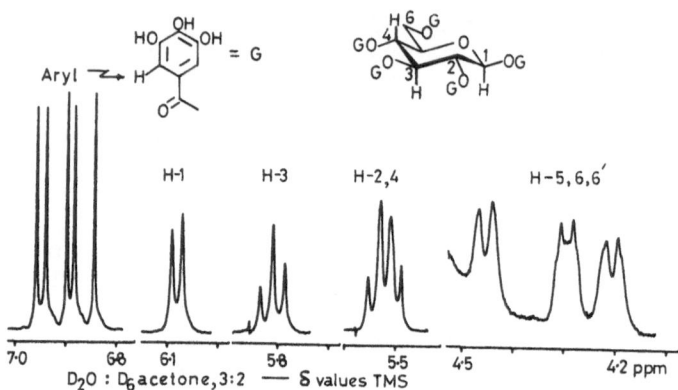

Fig. 11.   ¹H NMR spectroscopy of β-pentagalloyl-$\underline{\underline{D}}$-glucose.

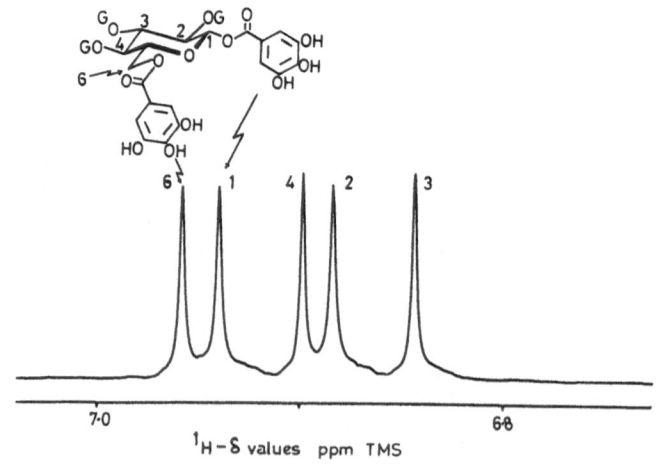

Fig. 12.   ¹H NMR spectroscopy of β-pentagalloyl-$\underline{\underline{D}}$-glucose.

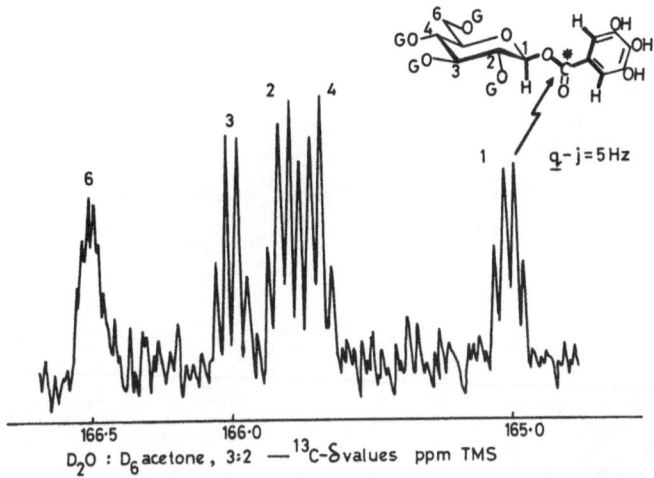

Fig. 13.   $^{13}$C NMR spectroscopy of β-pentagalloyl-$\underline{D}$-glucose.


## Biogenetic Relationships

Using phenolic biosynthesis in the leaves of plants as a point of reference the metabolism of gallic acid for many plants begins and ends with their ability to synthesize esters of gallic acid with $\underline{D}$-glucose, various other sugars, polyols, phenols and phenolic glycosides.[12,14]  For a great many other plants, however, this is not the limit of the range of metabolites associated with gallic acid and particular note is made here of those plants able to biosynthesize β-penta-$\underline{O}$-galloyl-$\underline{D}$-glucose (40, Fig. 14). This intermediate appears to mark a biosynthetic watershed for many plants and from it other biosynthetic pathways subsequently diverge to give the gallotannins and various ellagitannins.[12,38]  Biogenetic relationships have been suggested to link these diverse metabolites to the pivotal intermediate, 40.  Valuable as they are as a means of showing underlying affinities and relationships, it is important to stress that they are nevertheless based largely on patterns of occurrence and the chemical intuition common to all biogenetic hypotheses.  In this context, however, and supporting the focal position which β-penta-$\underline{O}$-galloyl-$\underline{D}$-glucose (40) is deemed to occupy in many plants,[12] it is interesting to note that callus tissue of Quercus robur metabolizes 40 in small amount along with

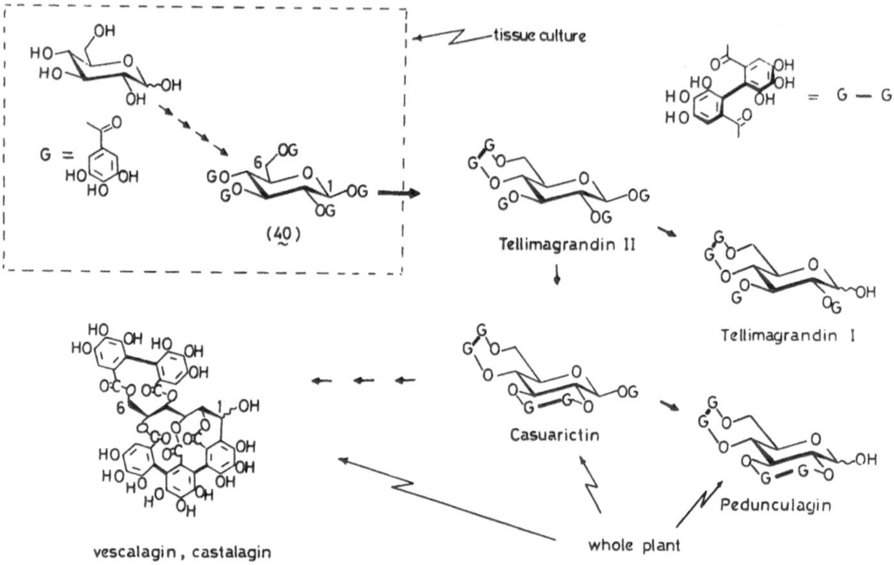

Fig. 14.  Polyphenol metabolism in tissue cultures of Quercus sp.

gallic acid and pyrogallol (Fig. 14).  The tissue culture
produces none of the many other derivatives of gallic acid
and hexahydroxydiphenic acid which leaves and other tissue
of the fully differentiated plant Q. robur synthesizes
(Fig. 14).[12,14]  Once again, as in other areas of secondary
metabolism, chemical and morphological specialization go
hand in hand.  Under the conditions of culture the callus
tissue only expresses a part of the genetic information
necessary for secondary metabolite production.

The patterns of metabolism of β-pentagalloyl-D-glucose
(40) which have been discerned in the leaves of higher plants
are broadly divisible into three major groupings (Fig. 15,
2A, B, C).  In the first of these additional galloyl groups
are added as m-depsides to give the typical gallotannins
(2A).  In the second (S)-hexahydroxydiphenoyl esters are
elaborated by oxidative coupling of vicinal galloyl ester
groups (4, 6 and 2, 3) attached to D-glucopyranose in its

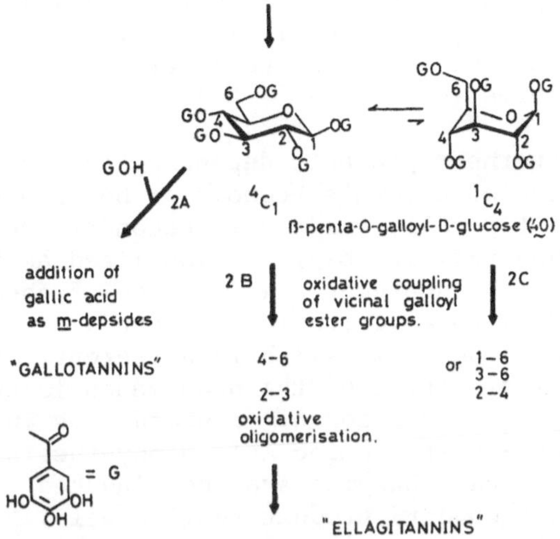

Fig. 15. Metabolism of β-pentagalloyl-$\underline{D}$-glucose in plants: the overall patterns.

most stable $^4C_1$ (1-C) conformation (2B). This group may be further sub-divided on the basis of the possession of enzymes able to bring about simply 4, 6 or 4, 6 and 2, 3 coupling and on the nature of the biosynthetic end-products (vide infra). According to present evidence a rather smaller group of plants adopts a further metabolic variation in which oxidative coupling of adjacent galloyl ester groups occurs 1, 6; 2, 4 or 3, 6 in a $\underline{D}$-glucopyranose precursor which itself adopts the less favorable $^1C_4$ (C-1) conformation or a skew-boat version (2C). For its routine metabolic processes nature appears to prefer the conformationally most stable sugars, and among the $\underline{D}$-hexoses the three most stable ones glucose, mannose, and galactose are very widely distributed. Those with higher free energies occur rarely, if ever. It is a point of some curiosity, therefore, that in this particular form of oxidative metabolism of galloyl esters of $\underline{D}$-glucose transformations apparently occur with the galloyl-$\underline{D}$-glucopyranose precursor in an energetically unfavorable chair ($^1C_4$; C-1) or related skew-boat conformation. The difference in free energy between the $^1C_4$ (C-1) and $^4C_1$ (1-C) forms of $\underline{D}$-glucopyranose is calculated as + 5.95 kcal $mol^{-1}$ (24.8 kJ $mol^{-1}$) and this difference may in part explain why oxidative coupling of galloyl ester groups via the $^{4}C_1$ (C-1) form (class 2B) of the D-glucopyranose precursor is much more widely encountered in plants than that via the alternative $^{1}C_4$ (1-C) or skew-boat forms (class 2C).

These patterns represent, depending on one's outlook, classic examples of nature's economy or her biosynthetic prodigality so typically evident in secondary metabolism. One key compound (gallic acid) is visualized as being formed. This then undergoes a wide range of chemical modifications leading to a veritable cornucopia of secondary metabolites, each one only slightly different from the next. One plausible explanation of the phenomenon is that it results from an accumulation of intermediates in the shikimate pathway of aromatic amino acid biosynthesis at some stage during growth. Enzymes are then induced for the synthesis of a secondary product (gallic acid) from intermediates in this pathway as a means of adjusting to changing circumstances. Such a proposition would suggest that it is the processes of secondary metabolism rather than the products (metabolites) which are important.

Depside metabolites. The ability to produce metabolites in which additional molecules of gallic acid are esterified to a pre-existing galloyl ester in the form of meta-depsides (Fig. 15, 2A) is limited to a comparatively few plant families - principally those of the Rhoideae tribe of the Ancardiaceae. In the earlier literature[52] many of these products were grouped together under the generic term 'gallotannin'. Although the principal form in which gallic acid is found in depside form is that based on β-pentagalloyl-D-glucose (40),[12,14] other esters of this type have been encountered in which the depsidically linked galloyl groups are attached to β-1,3,4,6-tetra-O-galloyl-D-glucose (Turkish gallotannin in galls of Quercus infectoria), 3,4,5-tri-O-galloylquinic acid (Caesalpinia spinosa) and 2,6-di-O-galloyl - 1,5-anhydro-D-glucitol (Acer saccharinum, A. tartaricum, A. ginnale). A novel method of analysis based on the application of $^{13}C$ NMR was recently discovered independently by Nishioka[53] and Haslam.[41] This work suggests that the polyphenolic ester based on 40 is typically, as originally suggested by Emil Fischer,[54] a heterogeneous mixture of esters varying from β-pentagalloyl-D-glucose (40) itself to various isomers of deca- and undecagalloyl-D-glucose. The proportion of each type determines the final composition of the gallotannin.

Using $^{13}$C NMR the formation of the additional m-depside
linkages was shown to occur preferentially to those galloyl
ester groups attached to C-2 or C-3, C-4 and C-6 of the D-
glucopyranose[41,53] (Fig. 15).

Hexahydroxydiphenoyl esters (2B, 2C). A very widely
distributed metabolic fingerprint is that in which β-penta-
O-galloyl-D-glucose (40) - it is assumed - is further
transformed by oxidative coupling of pairs of adjacent
galloyl ester groups (2, 3 and 4, 6) on the β-D-glucopyra-
nose core (Fig. 15, 2B). Some details of this pattern of
metabolism were hinted at in earlier work by Hillis and
Siekel,[46] by Jurd,[55] by Wilkins and Bohm,[56] and Schmidt.[57]

Plants whose phenolic metabolism places them within
this category furnish, along with β-penta-O-galloyl-D-
glucose (40) one or more of the hexahydroxydiphenoyl esters
(45-49, tellimagrandin II, tellimagrandin I, casuarictin,
potentillin and pedunculagin, Fig. 16) as key phenolic
metabolites (Haslam et al.[12,42]). The ratio of these metabolites varies markedly from plant to plant, even within the
same plant family. Parallel with this variation there is a
concomitant change in the nature of higher molecular weight
polyphenolic metabolites which are formed by intramolecular
C-O oxidative coupling of the metabolites 45-48. It is
noticeable that the major biosynthetic thrust in these
plants is towards the formation of these 'dimers' and
'trimers'. Thus for example in the plant family Rosaceae
the leaves of Rosa canina and F. ulmaria metabolize substantial quantities of the phenolic esters (45-46); however the
principal phenolic metabolites of these tissues are the
'dimers' rugosin D, (50) and rugosin E, (51) (T2B and T2A

Valoneic acid

$(45) + (45) \xrightarrow{-2H} (50; R = β\text{-}OG; T_{2B}, \text{Rugosin D})$

$(45) + (46) \xrightarrow{-2H} (51; R = α, β\text{-}OH; T_{2A}, \text{Rugosin E})$

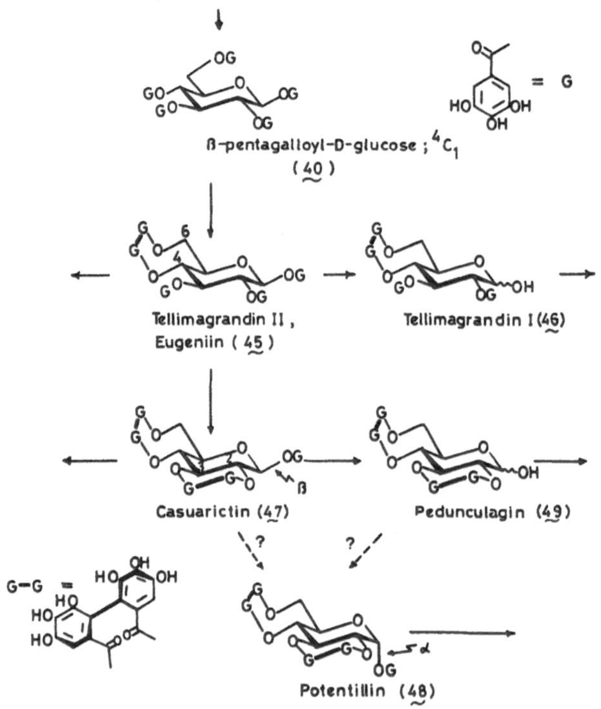

Fig. 16.   Metabolism of β-pentagalloyl-D-glucose in plants: class 2B metabolites.

in references 12, 14) formed presumably by oxidative C-O coupling of the precursors tellimagrandins I and II, (45) and (46). It is noteworthy that the two glucose residues are linked by the formation of a valoneic acid residue. Correspondingly in the same plant family Rubus idaeus, Rubus fructicosus, Geum rivale and Potentilla sp. metabolize substantial quantities of the 'dimer', sanguin H-6, (52) ($T_1$ in the references 12, 14) alongside its presumed

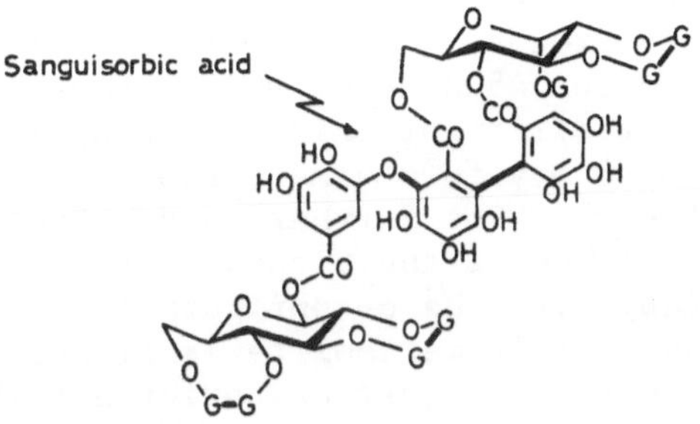

Sanguisorbic acid

$$(47) + (48) \xrightarrow{-2H} (52; T_1, \text{Sanguin-H6})$$

precursors casuarictin (47) and potentillin (48). In this polyphenol the two glucose residues are linked by a sanguisorbic acid residue – isomeric with valoneic acid and formed similarly, it is presumed, by C-O oxidative coupling (Haslam[12],[14]). Finally amongst polyphenols of the 'dimeric' form may be noted gemin A (53) from Geum japonicum (Okuda et al.[58]). In this structure units of tellimagrandin II (45) and potentillin (46) are linked by a third type of structural fragment a dehydrodigallic acid unit (Mayer[59]).

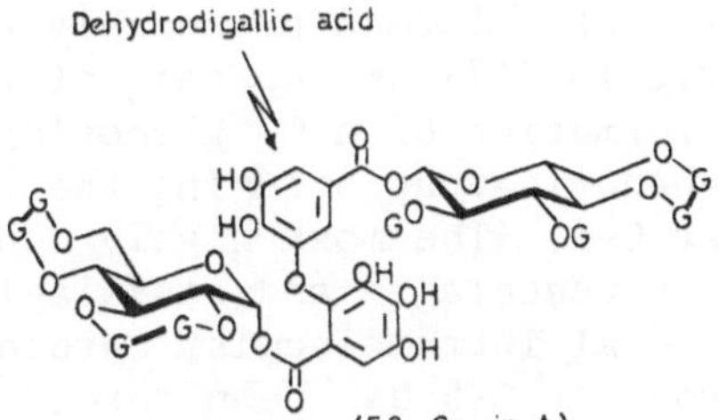

Dehydrodigallic acid

$$(45) + (48) \xrightarrow{-2H} (53; \text{Gemin A})$$

[G and G—G as in (45-49)]

Further examples of this type of polyphenolic structure in
which molecular complexity and size are increased by oxidative coupling of two or more glucose derivatives (45-49) by
formation of new C-O bonds have been described.[60-62]

It is noteworthy that whilst all plants in this category (Group 2B, Fig. 15) possess enzymes (it is presumed)
to facilitate C-C coupling of the 4 and 6 galloyl ester
groups on 40 to give 45 (Fig. 16), many do not possess the
means to bring about the subsequent coupling of the 2 and
3 galloyl ester groups.  Members of this class thus form
dimers from species 45 and 46, whilst those which possess
the additional facility of oxidative coupling of the ester
groups at 2 and 3 may form dimers from all the potential
precursors (45-49).  One further, and as yet unexplained,
note is the presence of the bishexahydroxydiphenoyl ester
potentillin (48) with the α-configuration at C-1 of the
glucose residue.  This metabolite is almost invariably
found alongside the β-anomer casuarictin (47) and is
presumably of some crucial biogenetic significance.  It is
the only example, to date, of an α-glucoside detected
amongst the 'monomeric' galloyl and hexahydroxydiphenoyl
ester metabolites.

A very interesting series of variations on the
patterns of metabolism discussed above occurs in members
of the plant families Fagaceae, Casuarinaceae, Stachyura-
ceae, Juglandaceae and Myrtaceae (Mayer et al.[63,64] and
Okuda et al.[65]), amongst others.  The ability to bring
about both 4,6- and 2,3- oxidative coupling of galloyl
ester groups in 40 is retained by plants in these families.
However the principal phenolic metabolites are not those of
additional C-O oxidative coupling as outlined but are
'monomeric' products, formed presumably from pedunculagin
(49) or casuarictin (47), by opening of the pyranose ring
of the sugar, formation of a C- glycosidic link to the
hexahydroxydiphenoyl group bridging the 2,3- positions and
galloylation at C-5.  The most highly condensed polyphenols
of this type are vescalagin and castalagin (Fig. 14) in
which an additional intramolecular carbon-carbon bond to
the galloyl group at C-5 has been formed.  The creation in
these molecules of three C-C biphenyl linkages and a C-
glycosidic bond results in the formation of relatively
inflexible propeller shaped species which contrast quite
markedly with their presumed biosynthetic precursor, β-
penta-O-galloyl-D-glucose (40), which has a perfectly

flexible disc-like shape.  This indeed appears to be a
common feature of all the polyphenolic metabolites -
grouped under the headings 2B and 2C (Fig. 15) - namely
the progressive formation of more highly condensed and
conformationally less flexible structures.

The biosynthetic origins of these highly condensed
metabolites such as vescalagin and castalagin (Fig. 14),
with the unique open-chain D-glucose structure, is something of a mystery.  Similarly the biogenetic significance
of the α-glucoside, potentillin (48) and the invariable
occurrence among all the metabolic fingerprints of the
metabolites unacylated at the anomeric center (e.g. 46
and 49) remains unexplained.  A speculative scheme which
links all these biogenetic uncertainties is shown in
Figure 17.  It presumes that acylation (galloylation) -
deacylation occurs preferentially at the anomeric center
of D-glucopyranose via an enzyme process similar to that
adopted in the fatty acid synthetases (vide supra).

According to present evidence[12,14] a rather smaller
group of plants - principally members of the plant families
Cercidiphyllaceae, Ericaceae, Onagraceae, Combretaceae,
Nyssaceae, Aceraceae, Punicaceae, Simaroubaceae and
Gerniaceae - adopts a further metabolic variation in which
oxidative coupling of adjacent galloyl ester groups occurs
1,6; 3,6 or 2,4 from a D-glucopyranose precursor which
itself adopts the less favorable $^1C_4$ (C-1) or an intermediate skew-boat conformation (Figs. 15, 18).  Typical of
the metabolites of this class is geraniin (54).[40,43]  The
dehydrohexahydroxydiphenoyl ester group (45), present in

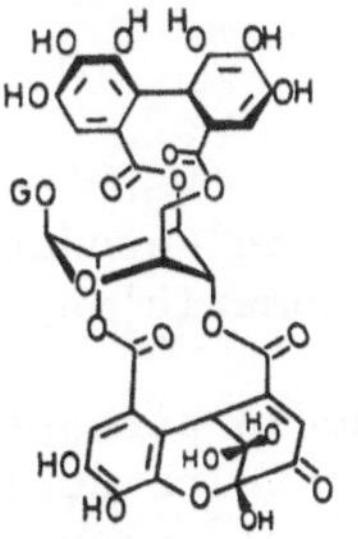

(54, Geraniin)

the structure of geraniin, or one of its transformation
products, is often another distinctive feature of metabolites of this class.  In contrast to the group 2B (vide
supra), no evidence has yet been obtained among group 2C

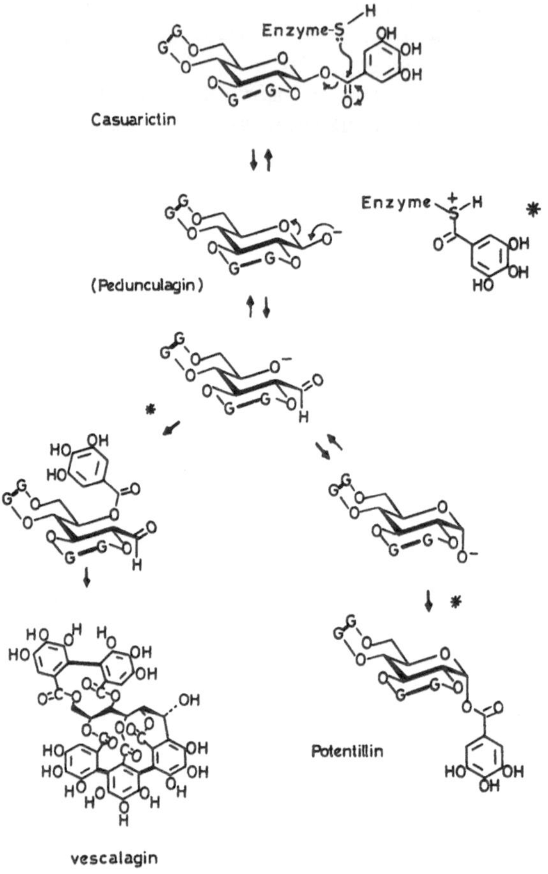

Fig. 17.  Biosynthesis of vescalagin (castalagin):  a putative scheme.

metabolites to show that 'dimeric' products are formed by oxidative coupling of the 'monomeric' species.

Finally brief comment should perhaps be made on the overall pattern of metabolites which derive in a large number of plants from the pivotal intermediate β-penta-O-galloyl-D-glucose (40).  The pathways leading to a great many of the biosynthetic end-products are clearly catabolic in character (Fig. 19) - dehydrogenation is the common means of chemical transformation of 40.

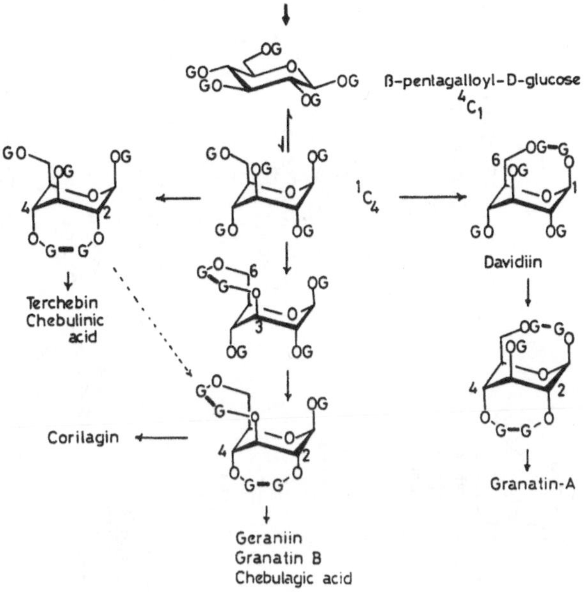

Fig. 18.  Metabolism of β-pentagalloyl-$\underline{D}$-glucose:  class 2C
metabolites.


## Polyphenol Complexation

Polyphenols (syn vegetable tannins[16,52] – proanthocyanidins and esters of gallic acid, vide supra) constitute
one of the most distinctive groups of higher plant secondary
metabolites.  Their uniqueness lies not only in their
molecular size and polyphenolic character but also in their
ability to complex strongly with proteins, carbohydrates,
nucleic acids and alkaloids.  Studies of these properties
are not only of intrinsic scientific interest .... a scientist who confesses to an interest in polyphenols is
invariably asked one question – why do plants form them?
.... but they are of considerable practical significance.
The characteristics of many plant products – their taste,
palatability, nutritional value, pharmacological and toxic
effects and their microbial decomposition – are substantially influenced by the polyphenols they contain.

Recent work[66,67] has delineated some of the critical
facets of polyphenol structure necessary for enhanced

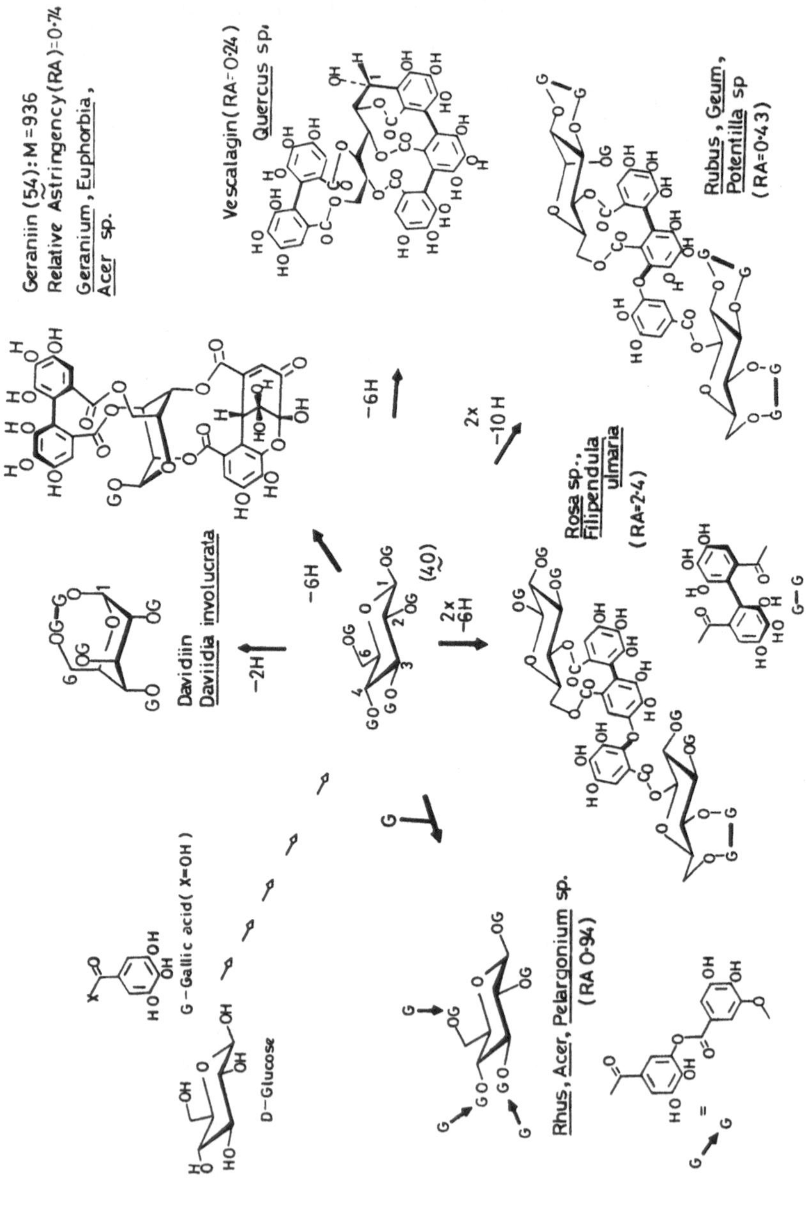

Fig. 19. Metabolism of β-pentagalloyl-D-glucose: biosynthetic 'end products'.

complexation with proteins:

(i)     Complexation is dependent on protein type and pH. Association is essentially a surface phenomenon, maximized at or near the isoelectric point of the protein and occurs principally via the intermediacy of hydrogen bonds and hydrophobic interactions.

(ii)    Molecular size of the polyphenol.  In, for example, the galloyl-$\underline{D}$-glucose series the efficacy of binding increases exponentially in the series tri → tetra → penta ($M_R$-636, 788 and 940).

(iii) Conformation flexibility of the polyphenol.  When conformational restraints are placed on the polyphenolic substrate, its capacity to complex, whatever its molecular size, is dramatically reduced (e.g. Rubus dimer, polymeric proanthocyanidins).  Complimentarity leading to enhanced complexation between the polydentate ligand (polyphenol) and the receptor (protein) is maximized by conformational mobility in both components (Fig. 19).

Collectively these results fully complement those of Hagerman and Butler[68,69] who showed that proline-rich and conformationally mobile proteins have high affinities for polyphenols and that, on occasion, such proteins may be preferentially precipitated in presence of other proteins. Complementarity between the polydentate ligand (polyphenol) and the receptor (protein) is maximized by conformational flexibility in both components.  The results also show that there is a very wide variability in the protein-complexing capabilities of the polyphenolic 'end-products' of gallic acid metabolism (Fig. 19).  There is no clearly consistent pattern nor is there evident any discernable correlation between the apparent 'metabolic cost' to the plant of its synthesis of a particular polyphenol and that polyphenol's astringency, i.e. its ability to complex with protein.


CONCLUSION

It is a precept of modern chemical ecology that energy is unlikely to be 'wasted' in the production of secondary metabolites unless there is some compensating adaptive advantage to the organism in question.[70]  The biosynthesis of polyphenols involves seemingly a substantial input of

energy and it might be expected, on these grounds, that
there was a considerable compensating gain to plants which
contain polyphenols. If, as Rhoades has suggested,[71] there
has been positive selection for the production of such
metabolites, this does not appear to be based solely, if
at all, on the capacity of these polyphenols to bind to
protein. While it may well be argued that polyphenol
complexation with protein is not, as has been assumed, the
critical property of polyphenols operative in plant defense,
these observations suggest the need for a reappraisal, at
the very least, of the current hypotheses in plant-
herbivore interactions. The present evidence[72] suggests
that although polyphenolic synthesis, such as has been
outlined, may well confer an advantage, a secondary benefit,
on the plant and may be the basis of selective pressures,
it does not clearly support the proposition that the
purpose of polyphenol metabolism is to specifically generate agents for the plant's defence. Protection would be
a consequence of polyphenol formation, not a cause.

Circumstantial evidence suggests that the complexation
of polyphenols with proteins is modified by carbohydrates,
other phenols and nitrogen-containing metabolites. Correspondingly the influence of polyphenols on the properties of
plant materials - for example the changes in astringency of
fruits as they ripen or in storage - is substantially
modified.

Polyethylene oxides and amyloses ($M_R$ > 4000) readily
complex with polyphenols but quantitative studies have been
severely limited by the availability of water-soluble polysaccharides with clearly defined molecular characteristics.
Semi-quantitative studies show that the association of
polyphenols with polysaccharides is - in contrast to that
with proteins - broadly independent of pH. Molecular size
and flexibility are likewise critical factors but, significantly, where the polysaccharide can sequester the
hydrophobic aryl residues of the polyphenol - holes in a
crystal lattice (cellulose) or hydrophobic cavities (amylose
and polysaccharide gels) - then complexation is substantially enhanced. Open, flexible, filamentous polysaccharides, such as the 1-α-6-dextrans conversely bind phenolic
substrates very poorly. It is interesting to note that
model 'polysaccharide holes' - in the form of the α- and
β- cyclodextrins - can sequester the aryl residues of
certain polyphenols in the core of the molecule. In doing

so they effectively compete with peptides and other
nitrogen-containing metabolites for polyphenolic substrates.
Quite clearly the study of such non-covalent molecular
interactions of polyphenols is likely to provide a fascinating field of study in the next few years, not only
because of its practical significance but also because of
its theoretical implications to biology as a whole.


ACKNOWLEDGMENT

The author thanks numerous colleagues who have collab-
orated in various aspects of this work but in particular
Dr. T.H. Lilley (University of Sheffield).


REFERENCES

1.  TOWERS, G.H.N.  1964.  Metabolism of phenolic acids in
        higher plants and microorganisms.  In Biochemistry
        of Phenolic Compounds.  (J.B. Harborne, ed.),
        Academic Press, London and New York, pp. 249-294.
2.  PEARL, I.A., S.F. DARLING.  1970.  Phenolic extractives
        of Salix purpurea bark.  Phytochemistry 9: 1277-
        1281.
3.  CHALLICE, J.S., A.H. WILLIAMS.  1968.  Phenolic
        compounds of the genus Pyrus I.  The occurrence of
        flavones and phenolic acid derivatives of 3,4-
        dihydroxybenzyl alcohol 4-glucoside in Pyrus
        calleryana.  Phytochemistry 7: 119-130.
4.  CHALLICE, J.S.  1973.  Phenolic compounds of the genus
        Pyrus VI.  Distribution of phenols in Pyrus stem.
        J. Sci. Food Agric. 24: 285-293.
5.  BOBBITT, J.M., D.W. SPIGGLE, S. MAHBOOB, W. von
        PHILLIPSBORN, H. SCHMID.  1962.  Catalpa glycosides
        II.  The structure of catalposide.  Tetrahedron
        Lett., 521-529.
6.  WEINGES, K., P. KLOSS, W.D. HENKELS.  1972.  Picroside
        II, ein neues 6-vanilloyl catalpol aus Picrorhiza
        kurooa.  Liebig's Ann. Chem. 759: 173-182.
7.  PELLETIER, S.W., L.H. KEITH.  1970.  Diterpene alka-
        loids from Aconitum, Delphinium and Garrya species:
        the $C_{19}$-diterpene alkaloids.  In The Alkaloids.
        (R.H.F. Manske, ed.), Vol. 12, Academic Press,
        London and New York, pp. 1-134.

8.  WANI, M.S., H.L. TAYLOR, M.E. WALL, P. COGGON, A.T.
    McPHAIL. 1971. Antitumour agents VI. The isola-
    tion and structure of taxol, a novel antileukemic
    and antitumour agent from Taxus brevifolia. J. Am.
    Chem. Soc. 93: 2325-2327.

9.  HASLAM, E. 1974. The Shikimate Pathway, Butterworths,
    London, 316 pp.

10. WEISS, U., J.M. EDWARDS. 1980. The Biosynthesis of
    Aromatic Compounds. Wiley-Interscience, New York -
    Chichester, 728 pp.

11. GIBSON, F., C.H. DOY, A.J. PITTARD. 1962. Possible
    relationship between o-dihydroxyphenols and trypto-
    phan synthesis by Aerobacter aerogenes. Biochim.
    Biophys. Acta 57: 290-295.

12. HASLAM, E. 1982. The metabolism of gallic acid and
    hexahydroxydiphenic acids in higher plants.
    Fortschr. Chem. Org. Naturst. 41: 1-46.

13. BATE-SMITH, E.C. 1962. The phenolic constituents
    of plants and their taxonomic significance. I.
    Dicotyledons. Linn. Soc. London Bot. J. 58: 95-
    173.

14. HADDOCK E.A., R.K. GUPTA, S.M.K. AL-SHAFI, K. LAYDEN,
    E. HASLAM, D. MAGNOLATO. 1982. The metabolism of
    gallic acid and hexahydroxydiphenic acid in plants.
    Biogenetic and molecular taxonomic considerations.
    Phytochemistry 22: 1049-1062.

15. NISHIZAWA, M., T. YAMAGISHI, G. NONAKA, I. NISHIOKA,
    M. RAGAN. 1985. Gallotannins of the freshwater
    green alga Spirogyra sp. Phytochemistry 24:
    2411-2413.

16. HASLAM, E. 1981. Vegetable tannins. In The Biochem-
    istry of Plants. (E.E. Conn, ed), Vol. 7, Academic
    Press, London and New York, pp. 527-554.

17. SCHILDKNECHT, H. 1983. Turgorins, hormones of the
    endogenous daily rhythms of higher organized
    plants - detection, isolation, structure, synthesis
    and activity. Angew. Chem. Int. Ed. Engl. 22:
    695-710.

18. JENSEN, R.A., D.L. PIERSON. 1975. Evolutionary impli-
    cations of different types of microbial enzymology
    for L-tyrosine biosynthesis. Nature 254: 667-671.

19. ZENK, M.H. 1971. Metabolism of prearomatic and
    aromatic compounds in plants. In Pharmacognosy and
    Phytochemistry. (H. Wagner, L. Horhammer, eds.),
    Springer-Verlag, Berlin-Heidelberg-New York, pp.
    314-346.

20. FLOSS, H.G., M.H. ZENK, D.K. ONDERKA, M. CARROLL, K.H.
    SCHARF. 1971. Stereochemistry of the enzymatic
    and non-enzymatic conversion of 3-dehydroshikimate
    into protocatechuate. J. Chem. Soc., Chem. Commun.,
    765-766.
21. O'BRIEN, I.G., F. GIBSON. 1970. The structure of
    enterochelin and related 2,3-dihydroxybenzoyl-
    serine conjugates from Escherichia coli. Biochim.
    Biophys. Acta 215: 393-402.
22. RATLEDGE, C. 1966. Biosynthesis of salicylic acid
    by Mycobacterium smegmatis. In Biosynthesis of
    Aromatic Compounds. (G. Billek, ed.), Pergamon,
    Oxford, pp. 61-66.
23. MARSHALL, B.J., C. RATLEDGE. 1972. Salicylic acid
    biosynthesis in Mycobacterium smegmatis. Biochim.
    Biophys. Acta 264: 106-116.
24. ZENK, M.H., G. MULLER. 1964. Biosynthese von p-
    Hydroxybenzoesaure und andere Benzoesauren in
    hoheren Pflanzen. Z. Naturforsch. 198: 398-405.
25. ZENK, M.H. 1979. Recent work on cinnamoyl-CoA
    derivatives. In Biochemistry of Plant Phenolics.
    (T. Swain, J.B. Harborne, C.F. van Sumere, eds.),
    Plenum Press, London and New York, pp. 139-176.
26. KINDL, H., W. LOFFELHARDT. 1975. The conversion of
    L-phenylalanine into benzoic acid on the thylakoid
    membrane of higher plants. Hoppe-Seyler's Z.
    Physiol. Chem. 356: 21-28.
27. BARKER, H.A. 1981. Amino-acid degradation by anae-
    robic bacteria. Ann. Rev. Biochem. 50: 23-40.
28. ZENK, M.H. 1964. Zur Frage der Biosynthese von
    Gallussaure. Z. Naturforsch. 19B: 83-84.
29. NEISH, A.C., G.H.N. TOWERS, D. CHEN. S.Z. EL-BASYOUNI,
    R.K. IBRAHIM. 1964. The biosynthesis of hydroxy-
    benzoic acids. Phytochemistry 3: 485-492.
30. KNOWLES, P.F., R.D. HAWORTH, E. HASLAM. 1961.
    Gallotannins 4. The biosynthesis of gallic acid.
    J. Chem. Soc., 1854-1859.
31. CONN, E.E., T. SWAIN. 1961. Biosynthesis of gallic
    acid in higher plants. Chem. and Ind., 592-593.
32. CORNTHWAITE, D.C., E. HASLAM. 1965. Gallotannins 9.
    The biosynthesis of gallic acid in Rhus typina.
    J. Chem. Soc. (C), 3008-3011.
33. DEWICK, P.M., E. HASLAM. 1969. Phenol biosynthesis
    in higher plants - gallic acid. Biochem. J. 113:
    537-542.

34.  SAIJO, R.  1983.  Pathway of gallic acid biosynthesis
     and its esterification with catechins in young tea
     shoots.  Agric. Biol. Chem. 47: 455-460.
35.  BRUCKER, W., M. HASHEM.  1962.  Zur Phenolkarbonsaure-
     bildung von Phycomyces aus Shikimisaure - $^{14}$C.
     Flora 152: 57-67.
36.  GROSS, S.R.  1982.  Synthesis of galloyl coenzyme A
     thiol ester.  Z. Naturforsch. 37C: 778-783.
37.  GROSS, S.R.  1983.  Synthesis of mono-, di- and tri-
     galloyl-β-D-glucogallin dependent galloyl
     transferases from oak leaves.  Z. Naturforsch.
     38C: 519-523.
38.  SCHMIDT, O.Th., W. MAYER.  1956.  Naturliche
     Gerbstoffe.  Angew. Chem. 68: 103-105.
39.  SCHMIDT, O.Th., J. SCHULZ, H. FEISSER.  1967.  Die
     Gerbstoffe der Myrabolanen.  Liebig's Ann. Chem.
     706: 187-197.
40.  OKUDA, T., T. YOSHIDA, T. HATANO.  1982.  Constituents
     of Geranium thunbergii.  J. Chem. Soc., Perkin
     Trans. I, 9-14.
41.  HASLAM, E., E.A. HADDOCK, R.K. GUPTA, S.M.K. AL-SHAFI,
     D. MAGNOLATO.  1982.  The metabolism of gallic acid
     and hexahydroxydiphenic acid in plants.  Part 1.
     Naturally occurring galloyl esters.  J. Chem. Soc.
     Perkin Trans. I, 2515-2524.
42.  HASLAM, E., R.K. GUPTA, S.M.K. AL-SHAFI, K. LAYDEN.
     1982.  The metabolism of gallic acid and hexahy-
     droxydiphenic acid in plants.  Part 2.  Esters of
     (S)-hexahydroxydiphenic acid with D-glucopyranose
     ($^{4}C_1$).  J. Chem. Soc., Perkin Trans. I, 2525-2534.
43.  HASLAM, E., E.A. HADDOCK, R.K. GUPTA.  1982.  The
     metabolism of gallic acid and hexahydroxydiphenic
     acid in plants.  Part 3.  Esters of (R) and (S)-
     hexahydroxydiphenic acid and dehydrohydroxydiphenic
     acid with D-glucopyranose ($^{1}C_4$ and related conforma-
     tions).  J. Chem. Soc., Perkin Trans. I, 2535-2545.
44.  SURMAN, D.J., J.C. VICKERMAN.  1981.  Fast atom
     bombardment quadruple mass spectrometry.  J. Chem.
     Soc., Chem. Commun., 324-325.
45.  BARBER, M., R.S. BORDOLI, R.D. SEDGWICK, A.N. TYLER.
     1981.  Fast atom bombardment of solids (FAB): A
     new ion source for mass spectrometry.  J. Chem. Soc.,
     Chem. Commun., 325-327.
46.  HILLIS, W.E., M. SIEKEL.  1970.  Hydrolysable tannins
     of Eucalyptus delegatensis wood.  Phytochemistry
     9: 1115-1128.

47.  JOCHIMS, J.C., G. TAIGEL, O.Th. SCHMIDT. 1968.
     Protonresonanz- Spectrum und Konformations
     bestimmung einiger naturlicher Gerbstoffe.
     Liebig's Ann. Chem. 717: 169-185.
48.  OKUDA, T., T. YOSHIDA, T. HATANO, T. KOGA, N. TOH,
     N. KURIYAMA. 1982. Circular dichroism of hydro-
     lysable tannins-I. Ellagitannins and gallotannins.
     Tetrahedron Lett. 23: 3937-3940.
49.  OKUDA, T., T. YOSHIDA, T. HATANO, T. KOGA, N. TOH,
     N. KURIYAMA. 1982. Circular dichroism of
     hydrolysable tannins-II. Ellagitannins and
     gallotannins. Tetrahedron Lett. 23: 3941-3944.
50.  SCHMIDT, O.Th., R. SCHANZ, R. ECKERT, R. WURMB.
     1967. Brevilagin I. Liebig's Ann. Chem. 706:
     131-153.
51.  SCHMIDT, O.Th., R. SCHANZ, R. WURMB, W. GROEBKE.
     1967. Brevilagin 2. Liebig's Ann. Chem. 706:
     154-168.
52.  HASLAM, E. 1966. Chemistry of Vegetable Tannins,
     Academic Press, London and New York, 179 pp.
53.  NISHIOKA, I., G. NONAKA, M. NISHIZAWA, T. YAMAGISHI.
     1982. Tannins and related compounds. Part 5.
     Structure of Chinese gallotannin. J. Chem. Soc.
     Perkin Trans. 1, 2963-2968.
54.  FISCHER, E. 1919. Untersuchungen uber Depside und
     Gerbstoffe, Springer-Verlag, Berlin, 541 pp.
55.  JURD, L. 1958. Plant polyphenols III. The isolation
     of a new ellagitannin from the pellicle of the
     walnut. J. Am. Chem. Soc. 80: 2249-2252.
56.  WILKINS, C.K., B.A. BOHM. 1976. Ellagitannins from
     Tellima grandiflora. Phytochemistry 15: 211-214.
57.  SCHMIDT, O.Th., L. WURTELE, A. HARREUS. 1965.
     Pedunculagin, eine 2,3:4,6-Di-(-) hexahydroxydi-
     phenoyl - glucose aus Knoppern. Liebig's Ann.
     Chem. 690: 150-162.
58.  OKUDA, T., T. YOSHIDA, M. USMAN-MEMON, T. SHINGU.
     1982. Structure of gemin A, a new dimeric ellagi-
     tannin having $\alpha$- and $\beta$-glucose cores. J. Chem.
     Soc., Chem. Commun, 351-353.
59.  MAYER, W. 1954. Dehydrodigallusaure. Liebig's Ann.
     Chem. 578: 34-44.
60.  OKUDA, T., T. YOSHIDA, Y. MARUYAMA, M. USMAN-MEMON,
     T. SHINGU. 1982. Gemin B and C, dimeric ellagi-
     tannins from Geum japonicum. Chem. Parm. Bull.
     30: 4245-4248.

61. OKUDA, T., T. HATANO, K. YAZAKI, N. OGAWA. 1982.
    Rugosin A, B and C and praecoxin A, tannins having
    a valoneoyl group. Chem. Pharm. Bull. 30: 4230-
    4234.

62. OKUDA, T., T. HATANO, N. OGAWA. 1982. Rugosin D. E,
    R and G, dimeric and trimeric hydrolysable tannins.
    Chem. Pharm. Bull. 30: 4234-4237.

63. MAYER, W., F. KULLMAN, G. SCHILLING. 1971. Die
    Struktur des Vescalins. Liebig's Ann. Chem. 747:
    51-59.

64. MAYER, W., H. SEITZ, J.C. JOCHIMS, K. SCHAUERTS, G.
    SCHILLING. 1971. Struktur des Vescalagins.
    Liebig's Ann. Chem. 751: 60-68.

65. OKUDA, T., T. HATANO, T. YOSHIDA, K. YAZAKI, M.
    ACHIDA. 1982. Ellagitannins of the Casuarinaceae,
    Stachyuraceae and Myrtaceae. Phytochemistry 21:
    2871-2874.

66. McMANUS, J.P., K.G. DAVIS, J.E. BEART, S.H. GAFFNEY,
    T.H. LILLEY, E. HASLAM. 1985. Polyphenol inter-
    actions. Part 1. Introduction; some observations
    on the reversible complexation of polyphenols with
    proteins and polysaccharides. J. Chem. Soc.,
    Perkin Trans. II, 1429-1438.

67. BEART, J.E., T.H. LILLEY, E. HASLAM. 1985. Poly-
    phenol interactions. Part 2. Covalent binding
    of procyanidins to proteins during acid-catalyzed
    decomposition; observations on some polymeric
    proanthocyanidins. J. Chem. Soc., Perkin Trans.
    II, 1439-1443.

68. HAGERMAN, A.E., L.G. BUTLER. 1981. The specificity
    of proanthocyanidin protein interactions. J. Biol.
    Chem. 256: 4494-4497.

69. HAGERMAN, A.E., L.G. BUTLER. 1980. Condensed tannin
    purification and characterization of tannin-
    associated proteins. J. Agric. Food Chem. 28:
    947-952.

70. FEENY, P.P. 1976. Plant apparency and chemical
    defence. Recent Adv. Phytochem. 10: 1-40.

71. RHOADES, D.F. 1979. Evolution of plant chemical
    defence against herbivores. In Herbivores -
    Their Interactions with Secondary Plant Metabolites.
    (G.A. Rosenthal, D.H. Janzen, eds.), Academic
    Press, London and New York, pp. 3-54.

72. BEART, J.E., T.H. LILLEY, E. HASLAM. 1985. Plant
    polyphenols - secondary metabolism and chemical
    defence - some observations. Phytochemistry 24:
    33-38.

Chapter Eight

LIGNANS: SOME PROPERTIES AND SYNTHESES

ANDREW PELTER

Department of Chemistry
University College of Swansea
Swansea SA2 9PP
United Kingdom

INTRODUCTION

Lignans are widely distributed plant products which are dimers of phenylpropanoids derived from shikimic acid.[1] Particularly common are the dimers derived from two units at the coniferyl alcohol oxidation level. Rather confusingly, the term 'neolignan'[2] has been coined to describe some dimers formally derived from the dimerisation of propenyl- and allylphenols. In fact, nothing is known of the biosynthesis of these compounds and many so-called 'neolignans' may well have their oxidation levels adjusted after, rather than before, dimerisation. However, many compounds, not ββ-linked, occur among 'neolignans' leading to a wide structural diversity which perhaps justifies the sub-grouping. The carbon skeletons of ββ-linked lignans, together with some naturally occurring rearrangement products and their most common oxidation levels (substituents are almost always H, OH, OMe, OCOR) are shown in Figure 1 and these compounds are the subject of this chapter. The carbon frameworks of the 'neolignans' are shown in Figure 2.

It should be noted that 'mixed' compounds, in which a phenylpropanoid unit is found attached to other commonly

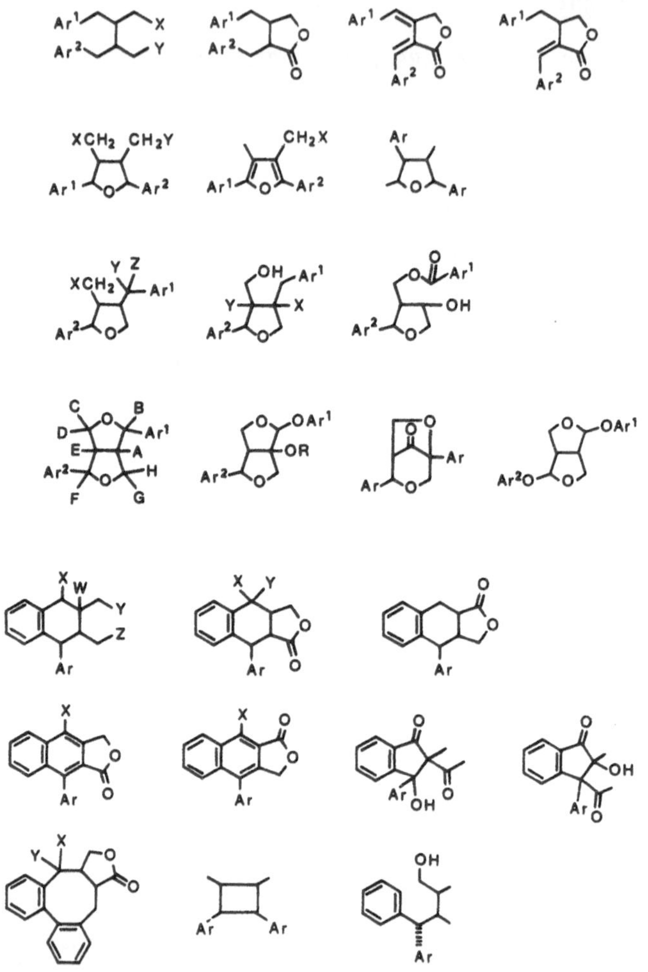

Fig. 1.   The carbon framework of some ββ-linked lignans.

occurring plant products such as flavonoids, coumarins and
xanthones are being found in increasing numbers (Fig. 3).
The physiological properties of these compounds are
extremely interesting.   The flavonolignan silybin and its
derivatives are used for treatment of hepatic disorders[3] as
well as phalloidin poisoning, while some coumarinolignans
such as cleomiscosin A are cytoxic.[4,5]   The hypotensive
alkaloids of the ephedradine and orantine series, chaeno-

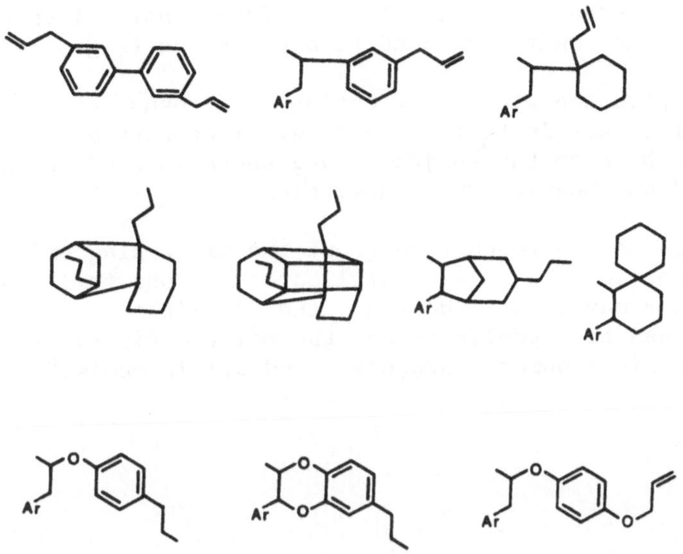

Fig. 2.   The carbon framework of some neolignans.

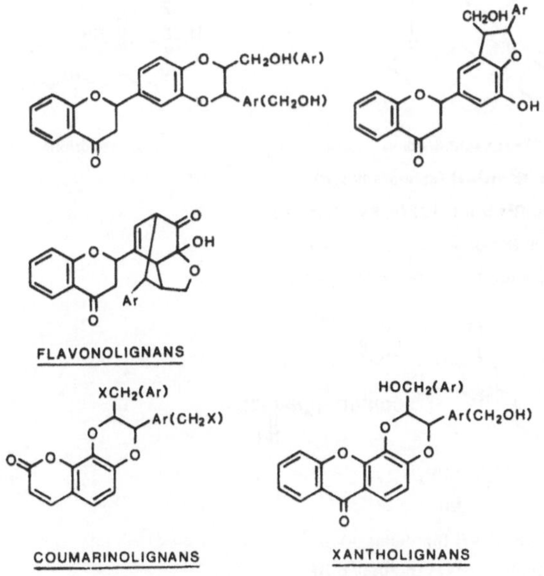

FLAVONOLIGNANS

COUMARINOLIGNANS          XANTHOLIGNANS

Fig. 3.   Compounds possessing 'lignan' structures.

rhine and verbascenine and the antifungal hordatines all
have a lignan unit connected to a polyamine (Fig. 4).[6]

Despite the wide distribution of lignans and the widespread interest in their physiological properties, there is
only one book on the subject,[1] and there is need for an
extended and heavily revised version.

Lignans are found throughout the plant kingdom[7] and it
is not possible to make generalizations about families in
which they may occur.  However, the 'neolignans' appear to
be confined to Magnoliales and the related Piperales, which
also contain monomeric propenyl- and allylphenols.[8]

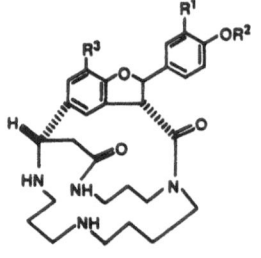

$R^1 = R^2 = R^3 = H$ (ephedradine A, orantine)

$R^1 = OMe$, $R^2 = R^3 = H$ (ephedradine B)

$R^1 = OMe$, $R^2 = Me$, $R^3 = H$ (ephedradine C)

$R^1 = R^2 = H$, $R^3 = OMe$ (ephedradine D)

$R^1 = H$, $R^2 = Me$, $R^3 = H$ (O-methylorantine)

Chaenorhine

$R^1 = H$, $R^2 = H$ (hordatine A)

$R^1 = OMe$, $R^2 = H$ (hordatine B)

Fig. 4.  Alkaloids which contain a 'lignan' structure.

BIOSYNTHESIS OF LIGNANS

The biosynthesis of lignans has received little direct study. It is probable that lignans derive from the general carbon flow towards lignin,[9],[10] although lignans are produced stereospecifically under enzymic control. This contrasts with lignin, a random chemical structure with no stereospecificity imposed during biosynthesis.[11] Nevertheless, without any hard evidence, reasonable schemes have been proposed for the biosynthesis of lignans based on analogies with the oxidative processes thought to be utilised for lignin production.[12],[13]

The production of ferulic acid, 1, and sinapic acid, 2, from p-coumaric acid, 3, is well documented[14] as is the reduction of their coenzyme A esters first to the corresponding alcohols[15] and then to the alkenes[16],[17] (Scheme 1).

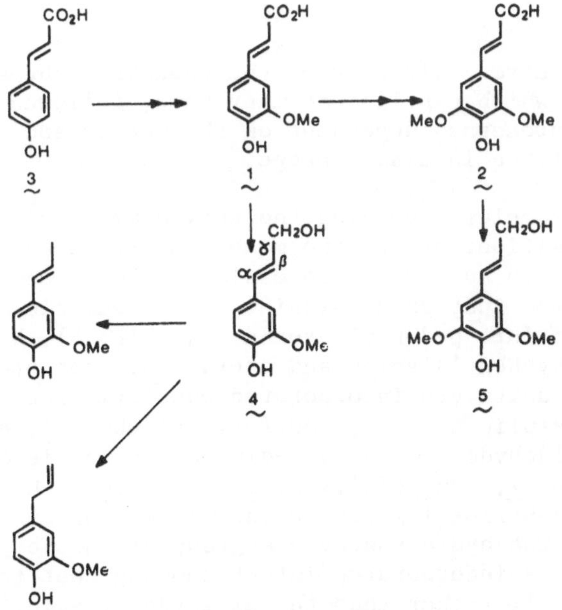

Scheme 1

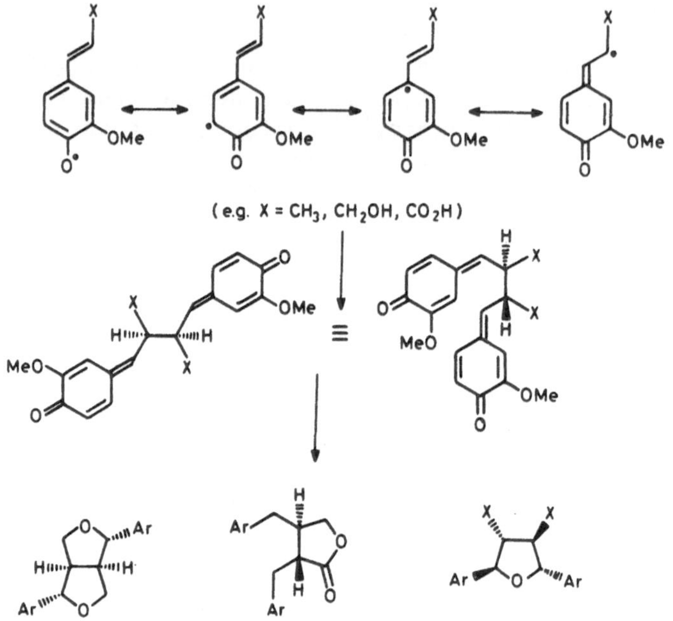

Scheme 2


    One electron oxidation of the phenols (Scheme 2) leads
to radicals which can link in many ways, followed by a wide
variety of processes dependent on the regio- and stereochemistry of the initial linkage.

    Some examples involving the very common initial linking
of the β positions of the two monomers are shown in Scheme
3.  Evidence to back such proposals is slowly accumulating.
Ayres[18] showed that phenylalanine and p-coumaric acid are
incorporated into podophyllotoxin, 6, (Fig. 5), but that
3,4-dihydroxyphenylalanine and acetate are not precursors.[19]
The monomer units are incorporated equally.  The incorporation of ferulic acid, 1, coniferyl alcohol, 4, and
coniferyl aldehyde but not 3,4-dimethoxycinnamic acid
into arctiin, 7, and phillyrin, 8, underlines the need
for a free 4-hydroxyl group in the monomer precursor for
lignans.[20]  The hydroxymethylene group of coniferyl
alcohol, 4, is incorporated intact showing that the monomer
is the alcohol, rather than the aldehyde or acid.[20]  Phenylalanine, ferulic acid, 1, and sinapyl alcohol, 5, are

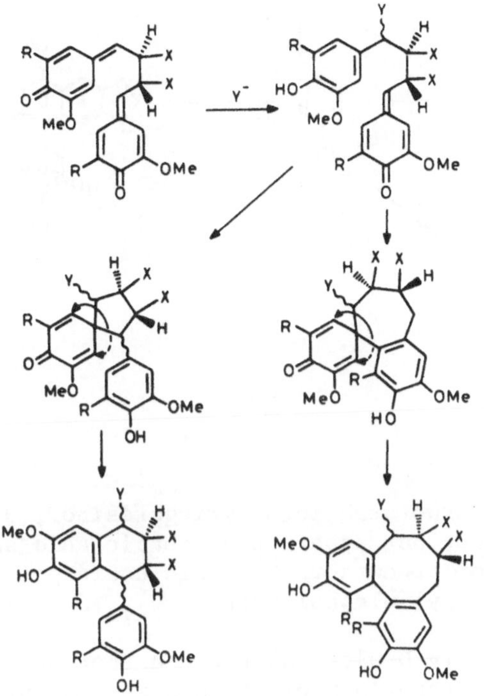

R = H or OMe
Y = H or OH

Scheme 3

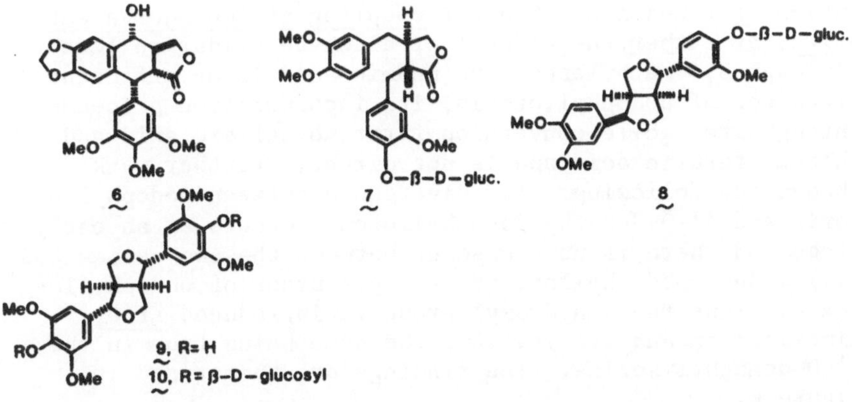

Fig. 5.  Lignans that have received biosynthetic attention.

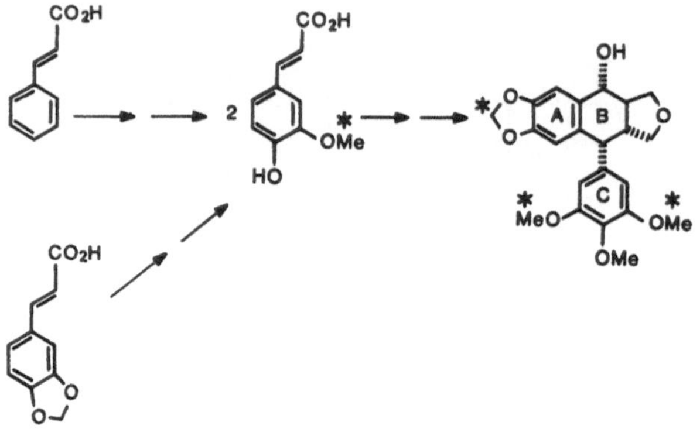

Scheme 4

efficiently incorporated into syringaresinol, 9, and
liriodendrin, 10, but apparently sinapic acid and coniferyl
alcohol are poor precursors.[21]  This could point to coupling of two sinapyl alcohol units.

Very recently Dewick has begun a thorough investigation of podophyllotoxin biosynthesis in Podophyllum
hexandrum.  Cinnamic and ferulic acids are each incorporated equally into both halves of the molecule, the carbon
of the O-methyl group of ferulic acid being retained by
both ring A and ring C (Scheme 4).[22]

Furthermore, sinapic acid and 3,4,5-trimethoxycinnamic
acid are not incorporated, suggesting adjustment of substitution of rings A and C after coupling of two equivalent
units, at either the alcohol or aldehyde oxidation level.
Although 3,4-methylenedioxycinnamic acid is an efficient
precursor of podophyllotoxin, its incorporation proceeds
through its facile conversion first to caffeic acid and
then to ferulic acid and is not direct.  Further work[23]
showed the following:  (i) divergence between podophyllotoxin and 4'-O-demethylpodophyllotoxin occurs at an early
stage and there is no crossover between the two series and
(ii) 4-deoxypodophyllotoxin is a precursor of podophyllotoxin.  Thus the 4-hydroxyl group is introduced after
dimerisation and cyclisation, the same being true in the
4'-O-demethyl series.  The findings are summarised in
Figure 6.

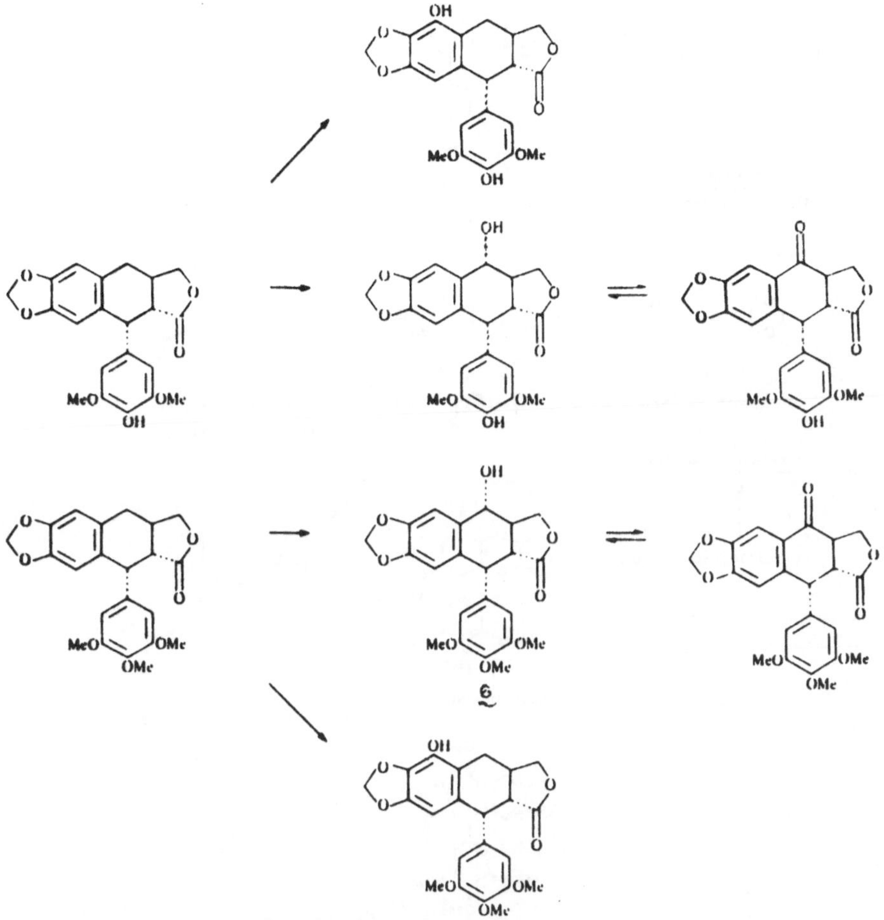

Fig. 6.  Proposed biosynthetic interrelationships for
Podophyllum lignans.

Recent unpublished work by W.M. Kamil and P.M. Dewick,
who I thank for permission to use these results, yields
fascinating information about the ring closure process
(Scheme 5).  Whereas yatein and its cis-isomer are
precursors of podophyllotoxin, most surprisingly the functionalized compounds anhydropodorhizol and podorhizol are
not.  This suggests a quinone methide intermediate with a
central function.  It is tempting to see matairesinol, not
yet tested, as a possible precursor and branching point for
all of the podophyllotoxin lignans.

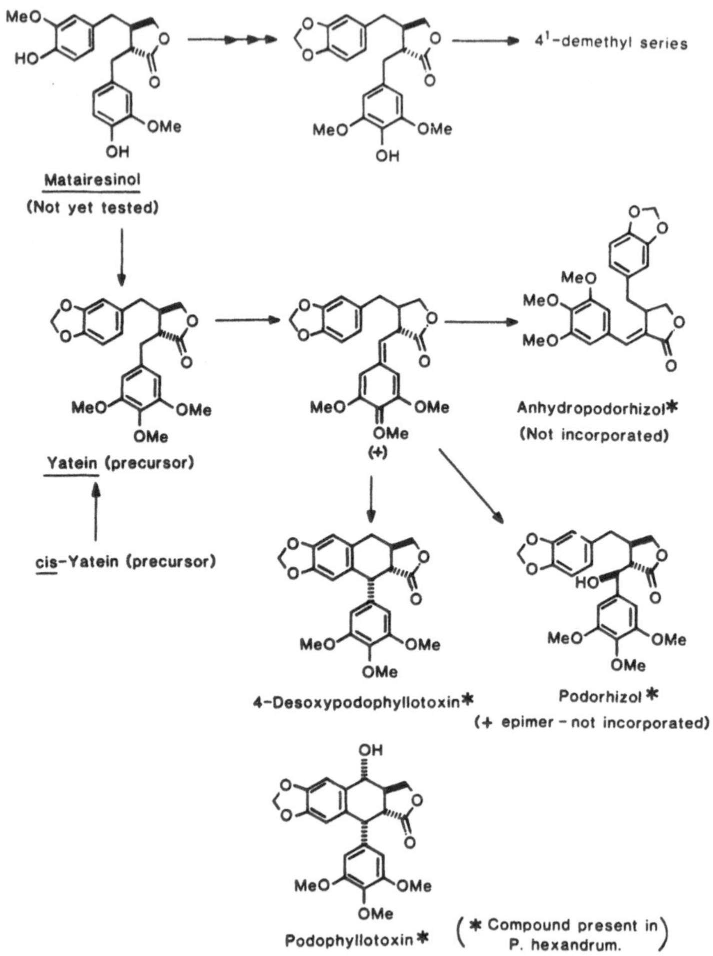

MeO
HO
Matairesinol
(Not yet tested)

→→ (benzodioxole structure) → 4¹-demethyl series

MeO   OMe
OH

Yatein (precursor)

cis-Yatein (precursor)

MeO   OMe
OMe

→ MeO   OMe
OMe
(+)

→ Anhydropodorhizol*
(Not incorporated)
MeO
MeO
MeO

4-Desoxypodophyllotoxin*
MeO   OMe
OMe

Podorhizol *
(+ epimer – not incorporated)
HO
MeO   OMe
OMe

OH
Podophyllotoxin*
MeO   OMe
OMe

( * Compound present in
P. hexandrum. )

Scheme 5

## CATABOLISM OF LIGNANS

Little is known of the catabolism of lignans. The
knowledge that is available is derived from studies on
dimers used as models for various bonds in lignin, and
as yet is highly fragmentary. The following discussion
is based partly on unpublished work by J.M. Palmer who
is thanked for permission to use Figure 7 and Schemes 7
and 8. Surprisingly, it appears that lignin, with its

multiplicity of linkages, is degraded by a single enzyme
which has been isolated from the white-rot fungus <u>Phanero-
chaete chrysosporium</u>[24,25] and shown to be present in
another such fungus, <u>Coriolus versicolor</u>.[26]  The enzyme
exhibits a broad specificity for phenolic and, more
importantly, for non-phenolic aromatic substrates.
Depending on the nature of the substituents in the aromatic
ring it can affect many of the previously identified reactions concerned with lignin biodegradation.  These are:
C-C bond cleavage in side chains, oxidation or hydroxylation
of benzylic methylene groups, hydroxylation of olefinic
groups in styrenes and oxidation of benzylic alcohols.  The
fact that it can also oxidise phenolic substrates recon-
ciles many earlier observations on the role of phenol-
oxidising enzymes in lignin biodegradation.[27-30]  However
horseradish peroxidase and the various fungal tyrosinases
and laccases do not have the same unique capacity as
ligninase to oxidise non-phenolic substrates.  This enzyme,
though previously termed an 'H$_2$O$_2$-dependent oxygenase'[24] or
a 'diarylpropane oxygenase'[25] would seem to be more
correctly classed as a peroxidase.[31]  The enzyme, which
works at the unusual optimum pH of 2.5, is like other
peroxidases but is unique in being able to generate an
oxyferyl centre of sufficiently high redox potential so
that it can oxidise both phenolic and non-phenolic
substrates.  With non-phenolic dimers it behaves in an
identical fashion to high potential chemical one-electron
transfer reagents such as CAN and Fe(III) trisphenantho-
line, and can function under argon.  It therefore cannot
be an oxygenase.  There is however an absolute requirement
for hydrogen peroxide.

It has been shown that one electron oxidation of
benzylic alcohols by chemical oxidants first yields a
cation radical that may degrade by C-H or C-C bond cleavage
(Scheme 6).[32]

The lignan <u>11</u> can behave analogously to yield <u>12</u>, <u>13</u>
and <u>14</u> (Fig. 7).

A seeming O-demethylation follows a similar pathway
of production of a cation radical, addition of hydroxyl and
loss of methoxyl to give the corresponding phenol (Scheme
7, Y = OMe).  Reduction of the cation radical and expulsion
of the methoxyl group could lead to its replacement by
hydrogen.

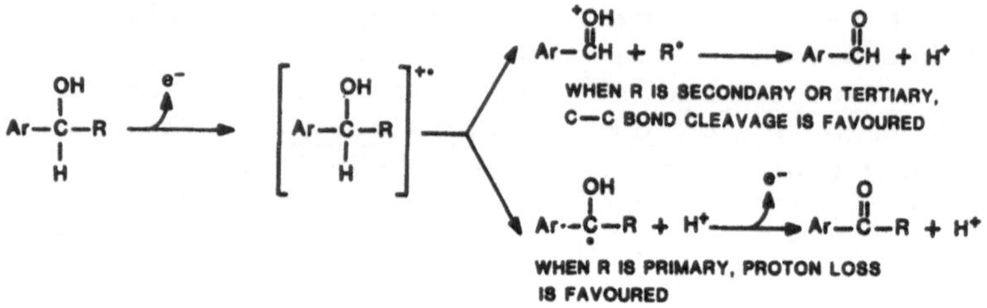

WHEN R IS SECONDARY OR TERTIARY, C—C BOND CLEAVAGE IS FAVOURED

WHEN R IS PRIMARY, PROTON LOSS IS FAVOURED

Scheme 6

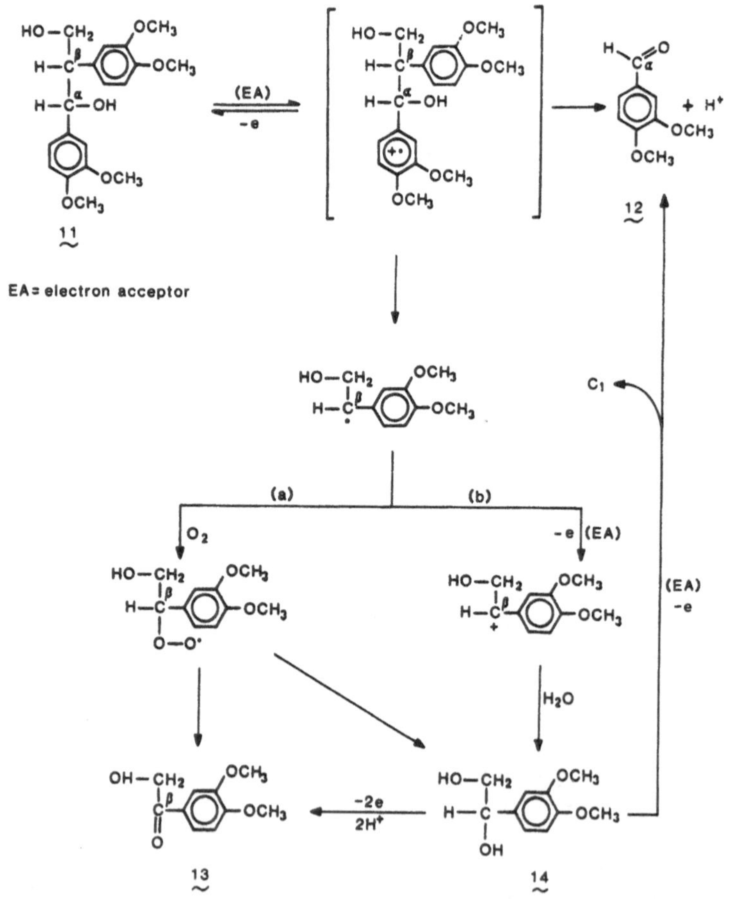

(EA)
−e

EA = electron acceptor

11

12

(a)   (b)

$O_2$

−e (EA)

$C_1$

(EA)
−e

$H_2O$

−2e
$2H^+$

13

14

Fig. 7. Proposed mechanism of oxidation of a 1,2-diaryl propane substructure by a one-electron transfer mechanism.

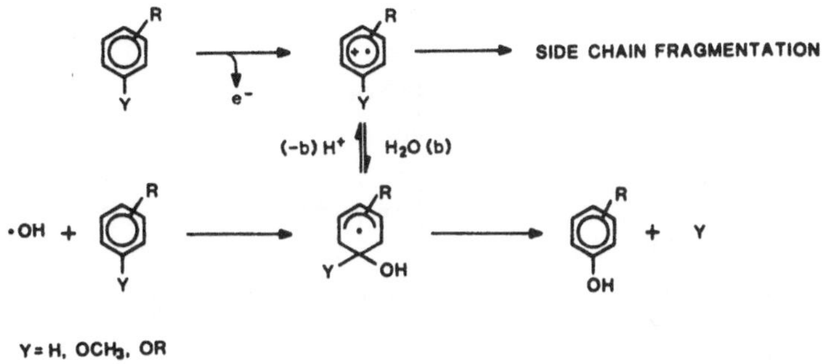

Y = H, OCH₃, OR

Scheme 7


Degradation of the side chain of 4-substituted phenols is readily accommodated as in Scheme 8. Other types of lignan substrate will lead to fascinating results when treated with ligninase.

Enterolactone, 15, is a mammalian lignan, found particularly in pregnant human females, which possibly affects cell division in the early stages of pregnancy.[33,34] It may be derived by removal of two 4-hydroxyl groups from dietary matairesinol 16 or the related secoisolariciresinol[35] or possibly two 4-methoxyl groups from 17 followed by demethylation (Scheme 9).


PHYSIOLOGICAL ACTIVITIES OF LIGNANS

It is not possible in this article to give detailed consideration to the various physiological activities of lignans. Fortunately, a comprehensive review[36] has recently appeared. Most of the known types of activity of lignans are listed in Table I. Although the list looks large it is, in fact, minimal. The activity one finds is a function of the activity one looks for, and no series of lignans has ever been comprehensively tested. Podophyllotoxin and its derivatives find clinical use in the treatment of human cancers[37,38] and virally induced genital warts.[39] The effects of 2,6-diaryl-3,7-dioxabicyclo[3.3.0]octane lignans on sickled erythrocytes[40] is of particular interest as is the antiarrythmic action of justicidin C (neojusticin B).[41]

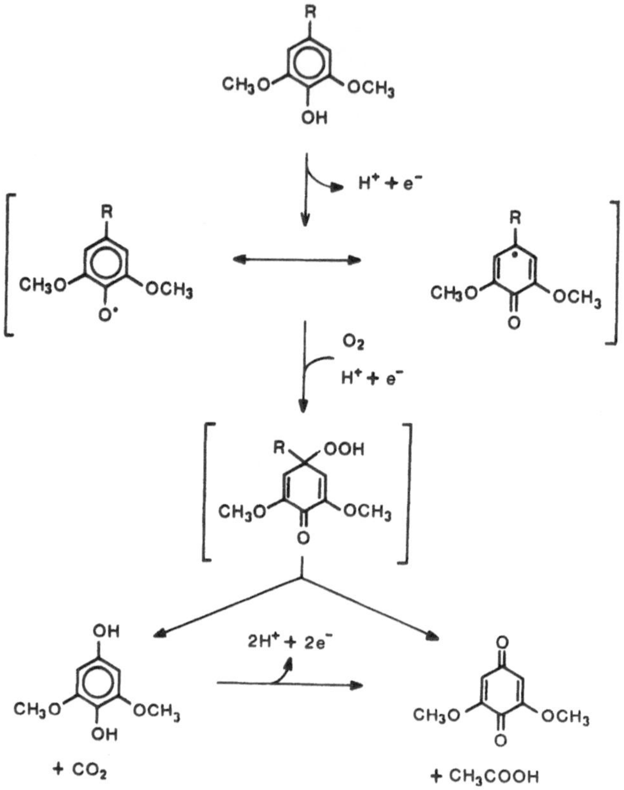

Scheme 8

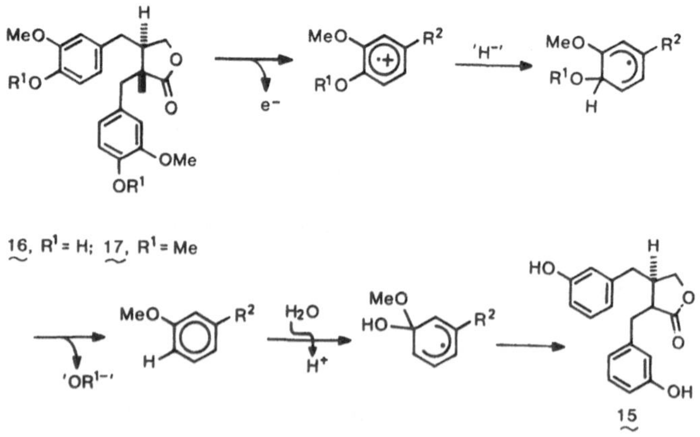

Scheme 9

Table 1.   Physiological effects of lignans.

| | |
|---|---|
| Antitumour activity | Cathartic effects |
| Antimitotic activity | Allergenicity |
| Antiviral activity | Hypotensive activity |
| Cytotoxicity | Insecticidal activity |
| Reversal of sickling/crenation of erythrocytes | Piscicidal activity |
| | Germination inhibition |
| Hepatoxic protection | Antimicrobial activity |
| C.N.S. activity | Fungistatic activity |
| Cardiovascular activity | Influence on nucleic acids |
| Antifertility agents | Influence on enzymes |
| Stress protection/antifatigue effects | |

Perhaps it is not reasonable to look for a common
factor linking the molecular basis for the action of lignans
over such a range of activities.  However, almost all of the
plant ββ-linked lignans contain two aromatic rings linked by
four carbon atoms, and the aromatic rings always contain at
least two adjacent oxygen functions.  The stripping of the
protecting groups from such systems would release o-quinols
which, of course, could interfere with natural redox
systems.  This is particularly facile with methylenedioxy
groups, as we have seen,[22] and this may explain the preva-
lence of such groups in physiologically active lignans.
However, methoxyl groups can also be replaced with hydroxyl
groups to lead to quinols.

At the moment, such generalisations are without
scientific foundation, but remain tempting when one examines
lignans such as the justicidins and related compounds, which
are simple arylnaphthalenes with no stereochemical complications (Fig. 8).  The tests on these closely related compounds
have been carried out in different laboratories mainly on
the basis of folk medicine.

Justicidins A and B, 16, have extraordinary piscicidal
activity,[42] as also have justicidins C, 17, D and E, 18.[43]
Clearly neither the configuration of the lactone ring nor
the presence of the 4-methoxy or the 6,7-dimethoxy groups
are necessary for this activity.  Justicidins A and B have
not been tested for insecticidal activity directly but show

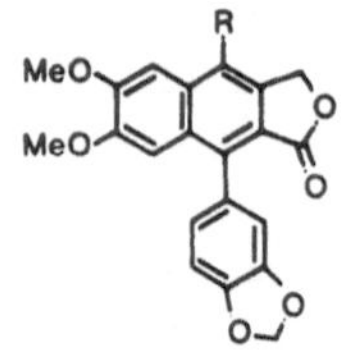

R = OMe, Justicidin A
R = H, Justicidin B

16

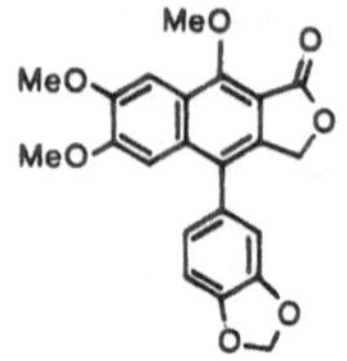

Justicidin C

17

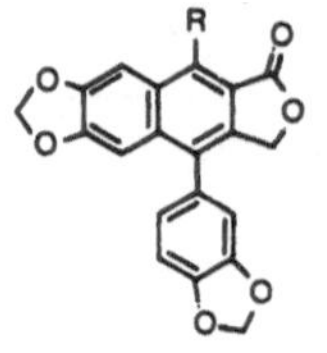

R = OMe, Justicidin D
R = H, Justicidin E

18

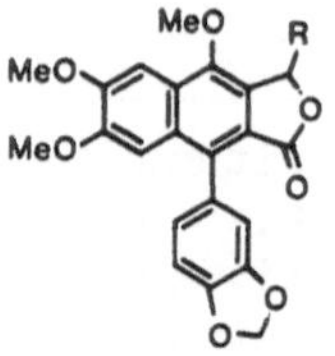

R = OMe, Justicidin P

19

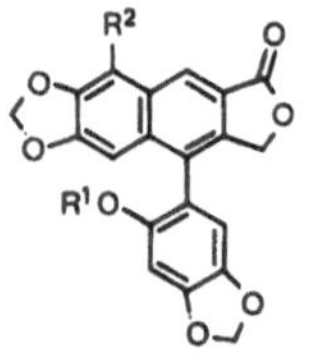

R¹ = H, R² = OMe, Prostalidin A
R¹ = Me, R² = OMe, Prostalidin B
R¹ = H, R² = H, Prostalidin C

20

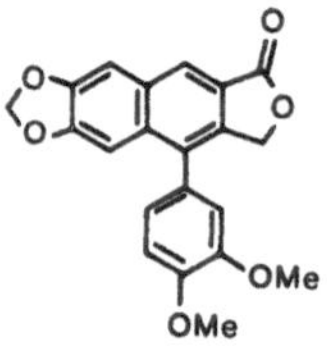

Retrochinensin

21

Fig. 8.   Related arylnaphthalene lignans.

synergistic activity.[44]   Justicidin B is active in the PS
system against NCI murine P388 lymphocytic leukemia.[45]
Justicidin C exhibits antiarrythmic activity[41] while
Justicidin P, 19 (R = OMe) intermediate between the
'normal' and 'retro'-lactone series, has not been tested
for piscicidal activity but is an excellent insecticide and
anti-feedant.  It is also reported as an antiviral agent
against Herpes virus Types I and II, human rhinovirus HRV-2
and vesicular stomatitis (VSV) virus.[42]  The compound is
active at much lower concentrations than 'antivir' and shows
a broad spectrum of activity.  The group R in 19 can be
varied widely (OH, OR¹, SR¹, F) and activity retained.  In
some cases cytotoxicity is evidenced but not at antiviral
levels.  The prostalidins, 20, and retrochinensin, 21, have
rather weak antidepressant activity.[46]  It would be interesting to test all of these compounds and related ones for
all the activities found.  In any case all they have in common
is a 4-arylnaphthalene ring system substituted such that
the production of o-quinols is possible.  Even the planar
naphthalene ring is unnecessary for antiviral activity as
podophyllotoxin and 4-deoxypodophyllotoxin are known to

act against Herpes simplex Type I and II, influenza and
vaccinia viruses.[47]

So much attention has been paid to podophyllotoxin
and its derivatives, that the details cannot be considered
here. Etoposide (VP-16), 21, and teniposide (VM-26), 22,
are used clinically in the treatment of human cancers and
their action has been compared with that of podophyllotoxin, 6, (Fig. 9). It suffices to say that, although
podophyllotoxin is closely related to VP-16 and VM-26,
uniquely they affect cells by entirely different modes of
action.[37,38] Podophyllotoxin behaves like colchicine, 24,
and β-peltatin, 23, in binding to tubulin, although colchicine and podophyllotoxin occupy adjacent sites rather than
the same site.[38] VP-16 and VM-26 are inactive in the inhibition of microtubule formation but inhibit DNA synthesis,
probably by causing DNA strand breakage.[48] The inhibition
occurs in vivo but not with purified DNA.[49] The drugs are
particularly active against small-cell lung cancer,
testicular cancer and malignant lymphoma. Structure-
activity relationships have been explored for podophyllotoxin, and an axial 1-aryl group, together with a trans-
fused lactone ring which holds the reduced ring rigid, have

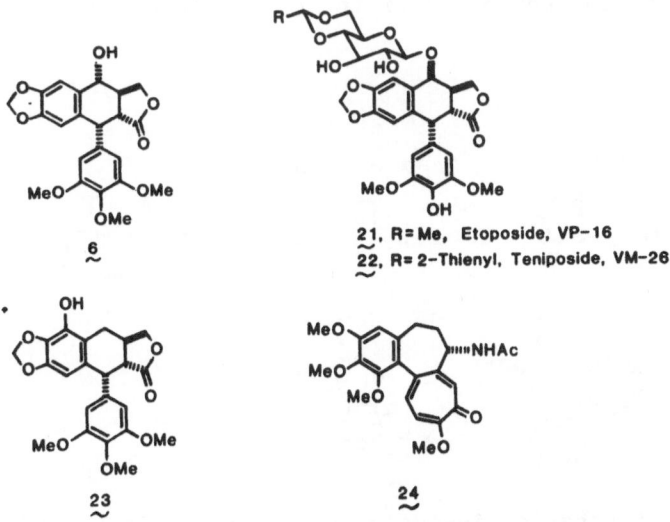

21, R = Me, Etoposide, VP-16
22, R = 2-Thienyl, Teniposide, VM-26

6

23

24

Fig. 9.  Compounds with antitumour activity.

been shown to be necessary.  This may of course have less
to do with the molecular basis of action than the specificity of adsorption and transport.


## CHEMICAL SYNTHESIS OF LIGNANS

The topic as a whole has been reviewed.[50]  In this
review the general synthetic approach, which is used in
our laboratory and is applicable to all $\beta\beta$-linked lignans,
will be described.

Our analysis (Fig. 10) consists of dissecting these
compounds into two Ar-$C_1$ units plus a central $C_4$ unit.[51]
The synthetic problem then consists of picking $C_4$ units
and two Ar-$C_1$ units, such that they can be joined in a
regiospecific manner, at the required oxidation level and,
if possible, with the required stereospecificity.  Scheme
10 stresses the need for sequential, controllable introduction of the two $C_6$-$C_1$ units if unsymmetrical lignans are to
be built up, and also the versatility of the charge distribution possible on the $C_4$-unit.  Our first targets were the
2,6-diaryl-3,7-dioxabicyclo[3.3.0]octane lignans,[52] which
are abundant and which contain compounds with a wide
variety of reported physiological activities.  We had
previously isolated and characterised (Fig. 11) a number
of such compounds with a varying number of hydroxyl groups
attached to the periphery of the central bicyclic system.[53]

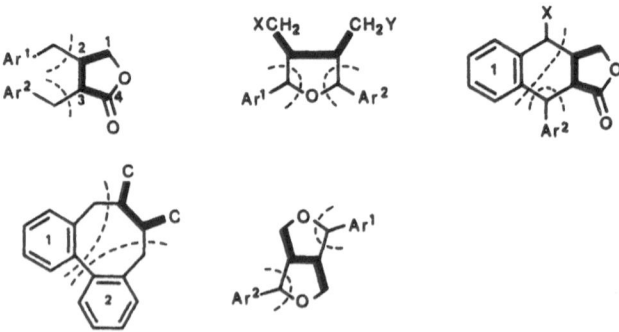

Fig. 10.  Synthetic dissection of lignans into $C_4$ and $C_6$-$C_1$
units.

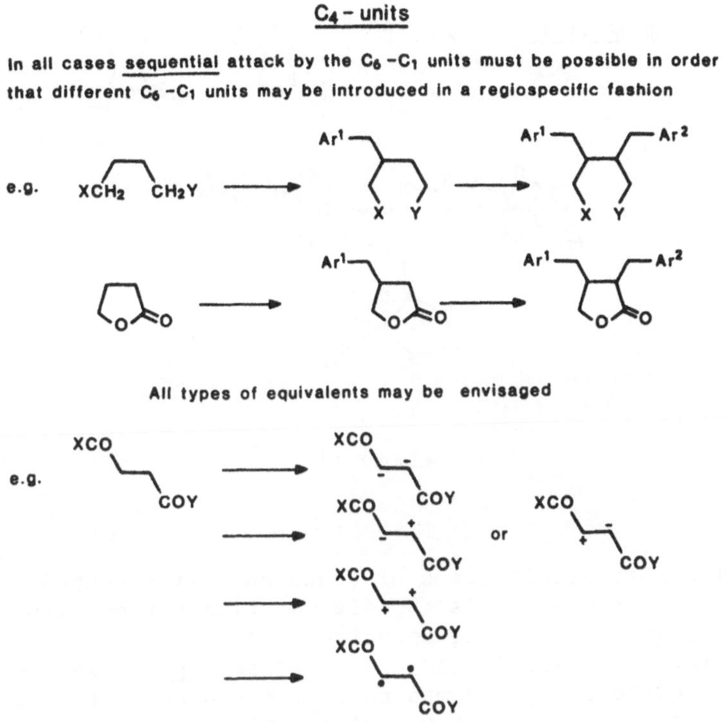

## $C_4$ – units

In all cases <u>sequential</u> attack by the $C_6$ –$C_1$ units must be possible in order that different $C_6$ –$C_1$ units may be introduced in a regiospecific fashion

e.g.

All types of equivalents may be envisaged

e.g.

Scheme 10

Paulownin        Gummadiol        Arboreol

Wodeshiol        4,8-Dihydroxy sesamin

( Ar = piperonyl)

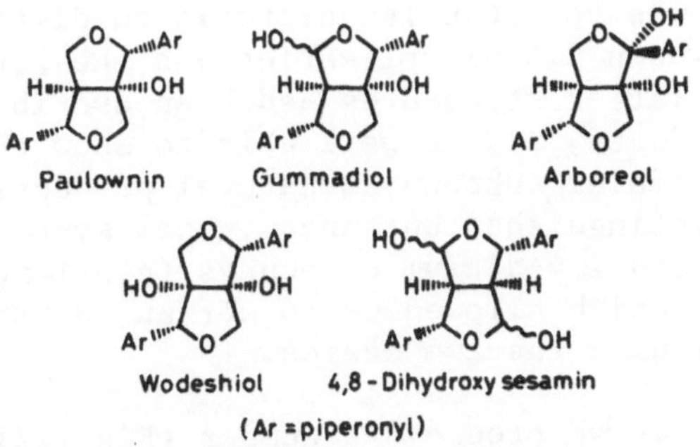

Fig. 11.   Hydroxy substituted 2,6-diaryl-3,7-dioxabicyclo [3.3.0]octane lignans.

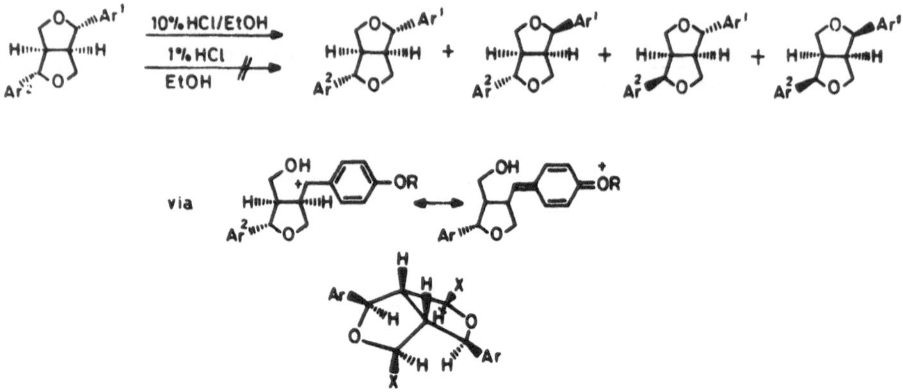

Scheme 11

The most stable class of these butterfly-shaped
molecules (Scheme 11) is the diequatorial series with the
two aryl groups pointing away from the concave part of the
molecule.  However, in the unsymmetrical epi-series there
are two different epi-compounds, of comparable stability
and hence, not only must two different aryl groups be introduced but they must each be placed with defined and
predictable regio- and stereochemistry.  Many unsymmetrical
epi-compounds are known.  The dia-series are unstable with
respect to the rest of the series and only rarely encountered.  Note that mild acid does not equilibrate the aryl
groups.

We were motivated further by a matter of some importance, which was to establish criteria to distinguish
between the known 2,6-diaryl series and the 2,4-diaryl
series.  The latter structures had been ascribed to various
important products, and we were able to show that in fact
this series remains unknown as natural products.  Scheme 12
shows an undistinguished but unequivocal synthesis which
yielded 2,6- and 2,4-diaryl compounds (confirmed by X-ray
analysis) and which allowed us to set up $^1$H NMR criteria
for distinguishing the two series.

In this way we produced a number (Fig. 12) of symmetrical diequatorial lignans but could not take the process
further.  A rather more satisfactory route to the unnatural

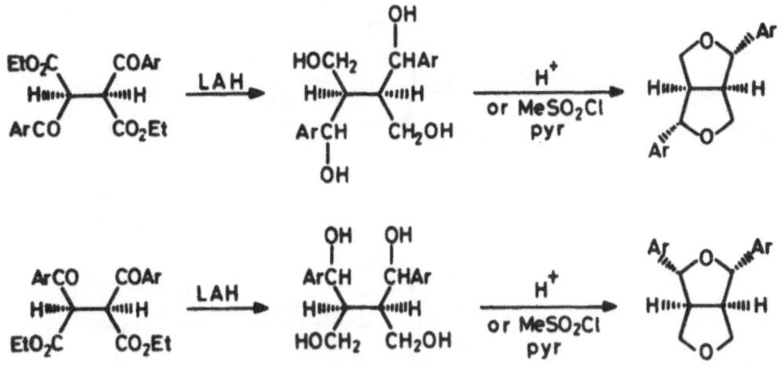

Scheme 12

a.  Ar = 4-hydroxy-3-methoxyphenyl
b.  Ar = 3,4-dimethoxyphenyl (veratryl)
c.  Ar = 4-benzyloxy-3-methoxyphenyl
d.  Ar = 3,4-methylenedioxyphenyl (piperonyl)
e.  Ar = phenyl

Fig. 12.  2,6- And 2,4-diaryl-3,7-dioxabicyclo[3.3.0]
octane lignans.

lignans which also yields the cyclobutane lignans, is shown
in Scheme 13.

At this point we approached the problem in a more
rational and general fashion, employing a $C_4$ unit as a
potential, sequentially reactive dianion (Scheme 14).

Our potential dianion, 1-0-methyl-2-methylthiosuccinic
acid, 25, readily produced from commercially available
2-thiolsuccinic acid, is shown in Scheme 15.  The intermediate spirothiolactone, 26, is a solid that can be made
in multigram quantities and stored.  The methylthio group
is present to help dianion formation at C-2 specifically
and then to block C-2 for later attack.  We also hoped that
it would help guide the attack at C-3 so that the incoming
group is cis to the carbomethoxyl group, not trans as is

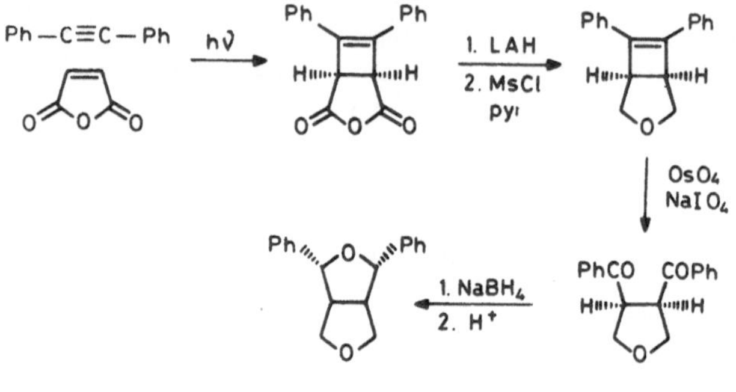

Scheme 13

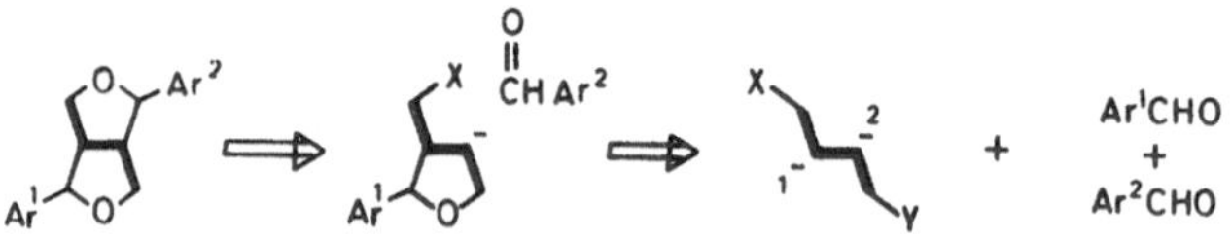

Need a C₄-unit tuned to give a dianion that reacts
sequentially with two separate aldehyde units

Scheme 14

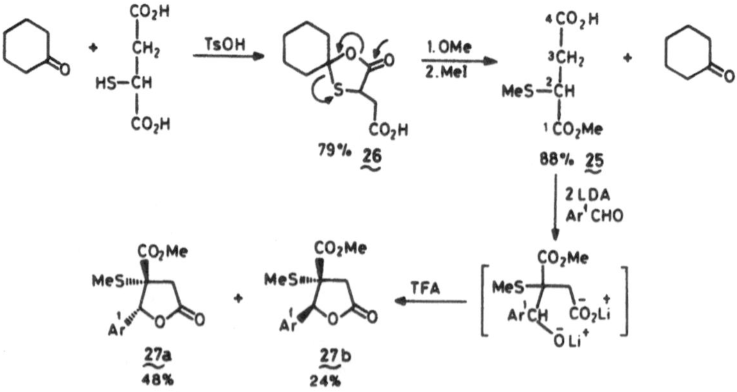

Scheme 15

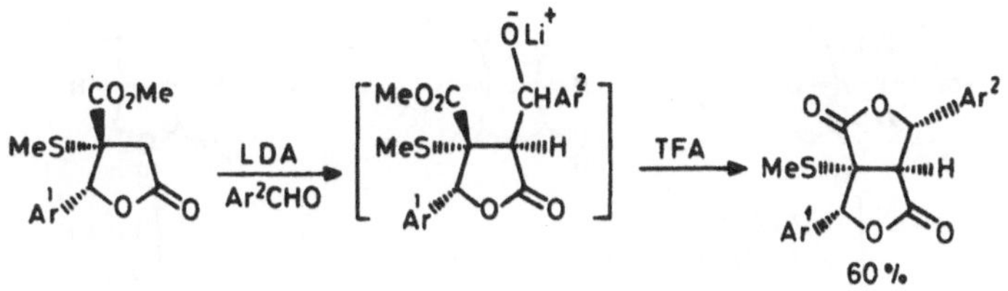

Scheme 16

normal in the unsubstituted systems.  Attack on the dianion
gave a mixture of two isomeric γ-lactones, readily separated by crystallisation.  The major isomer, 27a, shown by
NMR to have the methylthio- and aryl groups cis, was
treated with LDA and then a different aromatic aldehyde.
Acidification gave the required dilactone, 28 (Scheme 16).

     Thus the introduction of both $C_6$-$C_1$ groups is regio-
specific and also stereospecific at the $C_4$ unit, the
methylthio group acting as hoped.  Hence we obtained the
lignan nucleus containing sufficient functionality for
additional experimentation.  At this stage, the stereochemistry at C-6 is not really established, our normal NMR
criteria being in some doubt due to the presence of the
methylthio group.  All our efforts to remove the methylthio
group from 28 were in vain (Scheme 17), but DIBAL followed
by equilibration with 0.2% methanolic HCl, which we know
does not equilibrate the aryl groups, gave a single
compound, 29.  Normally the equilibrated methoxy groups
would be diequatorial, but we now know that we have an
8-axial methoxyl group, due to the adjacent 1-methylthio
functionality.  However at the time we could only be sure
of the relative stereochemistry of the groups on C-1 and
C-2.  We were now able to remove the methylthio group,
equilibrate with mild aqueous acid to give the naturally
occurring 4,8-dihydroxylignans, reduce and close the ring
to form the diequatorial lignans, 30.  The equilibration
proved somewhat difficult when methyledioxy groups were
present on the aryl rings.  Moreover, reduction with LAH
unfortunately led to a ring opened compound with possibi-
lities of isomerisation and ring closure to 2,4- rather
than 2,6-diaryl compounds.

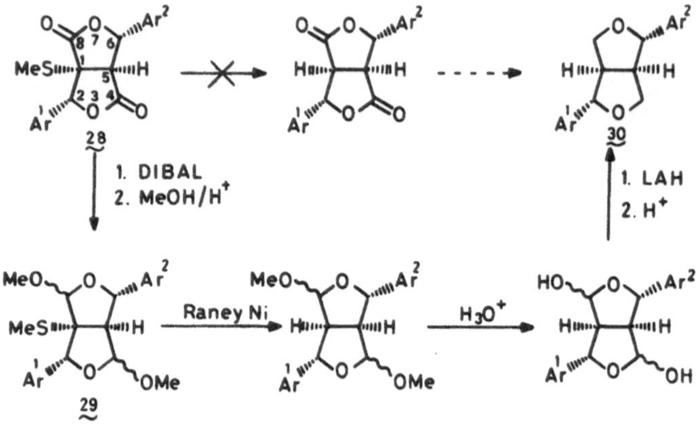

Scheme 17

Scheme 18 shows the solution of this problem and also that of the stereochemistry of intermediates. We proved that the dilactone had both aryl groups equatorial by interchanging the order of introduction of the piperonyl and veratryl groups. After equilibration with ca 0.2% HCl/MeOH the two series did not give the same compound, as expected, but removal of the methylthio group in each case, followed by mild acid equilibration gave the same all equatorial dimethoxyacetal, 31. The conditions of equilibration are chosen so that they are not sufficiently acid to affect the aryl groups. We then sought a method for reducing the acetals directly to the tetrahydrofurans without ring opening. This was accomplished in a non-aqueous acid reduction, with a mixture of triethylsilane and trifluoroborane etherate. However, in these strong acid conditions the aryl groups equilibrated and the product was a mixture of readily separable diequatorial methyl piperitol and the two epi-isomers, methyl xanthoxylol and fargesin, separable only by preparative HPLC. In view of the equilibration, the same reagent was reacted with the methylthio precursor acetals. This produced only one main isomer, 32, with a yield of 89% and it is the isomer in which specifically the aryl group adjacent to the methylthio group is axial and the other is equatorial. The methylthio group thus completely controls the stereochemistry of the reaction. Removal of the methylthio group gave the corresponding epi-lignan. The whole

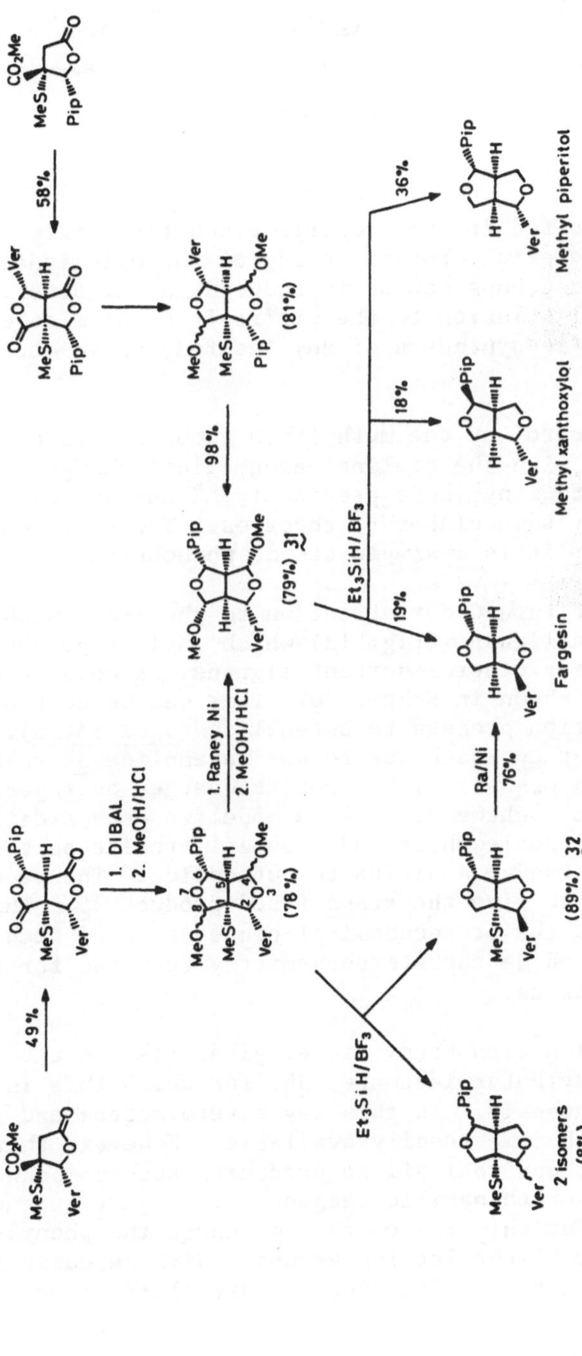

Scheme 18

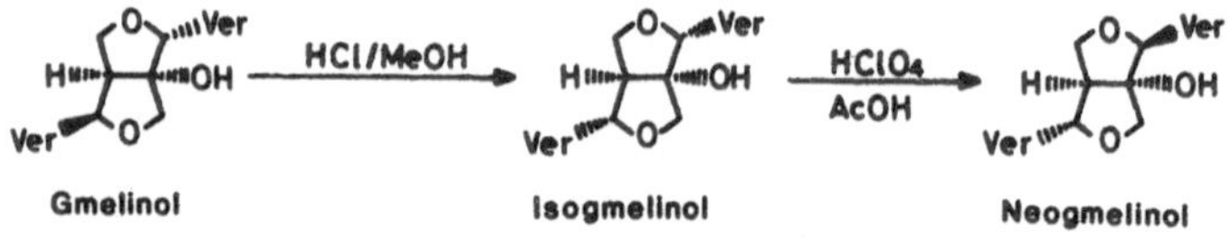

Scheme 19

process resulted in the aromatic group first introduced
becoming the axial aromatic group of the epi-lignan.  As
the aromatic groups can be introduced in any order, this
is a general solution to the difficult problem of the
stereospecific synthesis of any desired, unsymmetrical epi-
lignan.

The control by the methylthio group has an analogy
(Scheme 19) with the gmelinol-neogmelinol change which was
characterised many years previously,[54] and at that time
explained by some elaborate theories.  These are quite
unnecessary; it is a simple steric phenomenon.

We next turned our attention to the 1-arylnaphthalene
and tetralin lignans (Fig. 13) which include podophyllotoxin and many other important lignans.  A model biomimetic
sequence is shown in Scheme 20.  This can be combined with
tandem addition process to butenolide (vide infra).  More
generally our approach was to use butenolide 34 as an
example of a $C_4$-unit with opposite charges on adjacent
carbon atoms (Scheme 21).  As a model we used a dithane
anion as the nucleophile and a benzyl bromide as the electrophile in tandem addition to butenolide.  The reaction
worked well to give the trans-fused product 35.  This seems
to be always the stereochemical consequence of such tandem
additions, and is the stereochemistry required for the
podophyllotoxins.

Reduction with Raney nickel gives rise to the natural
trans-dibenzylbutyrolactones, 36, for which this is the
simplest synthesis.  In this way enterolactone and its
derivatives become readily available.  However, that was
a diversion; our real aim in producing such compounds was
to see whether thiophilic reagents could give ring closure,
and it was for this reason that we chose the phenylthio
groups as excellent leaving groups.  With mercuric salts
(Scheme 22) not only did ring closure-elimination occur,

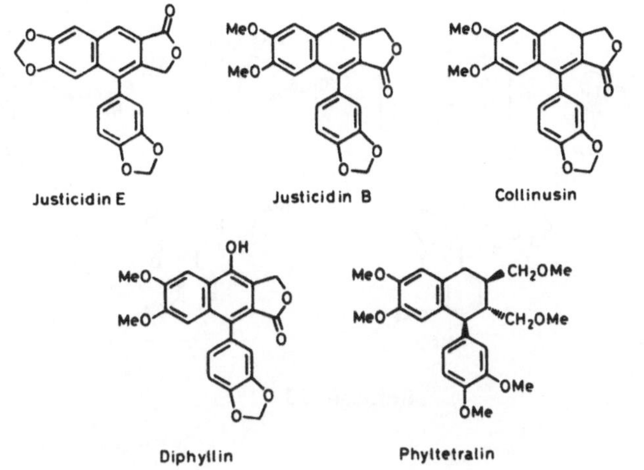

Fig. 13. Physiologically active, structure related lignins.

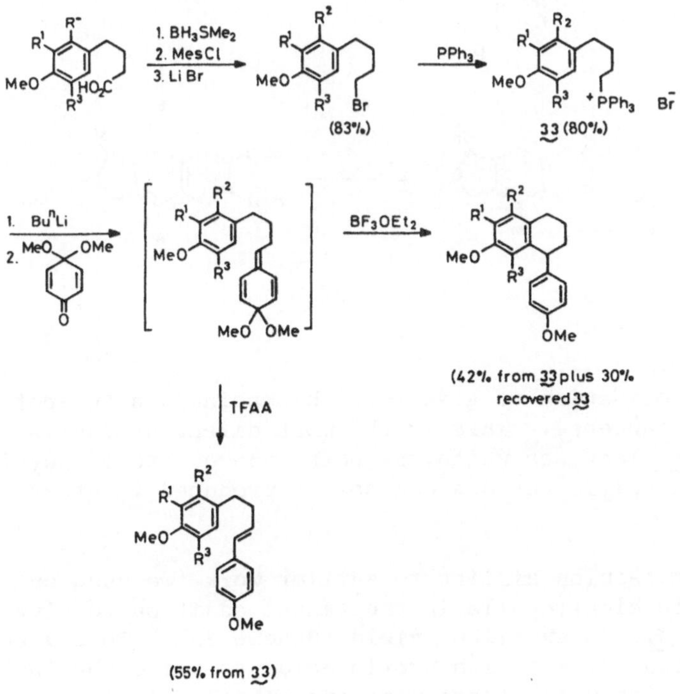

Scheme 20

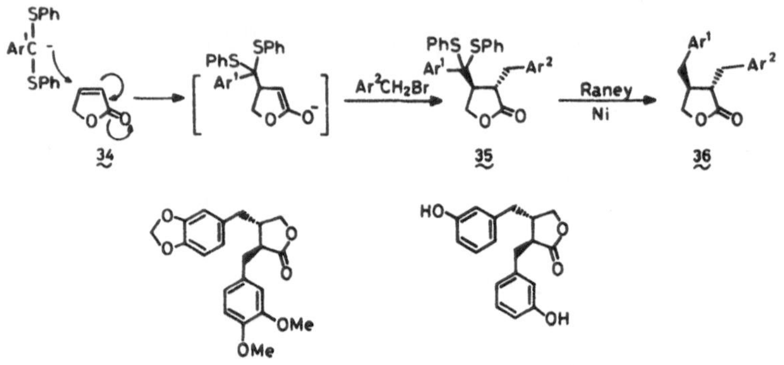

Scheme 21

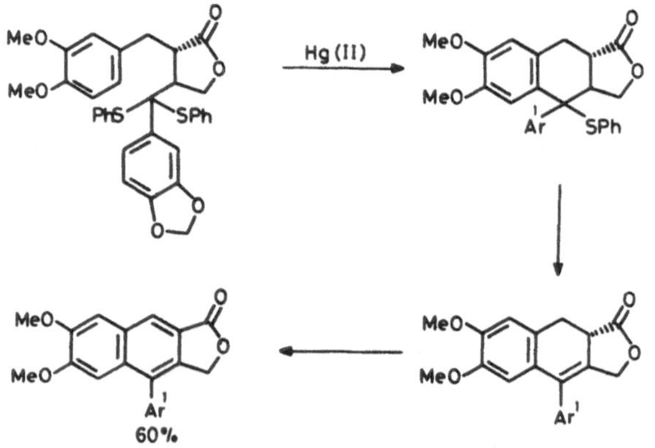

Scheme 22

but also oxidation to give retrochinensin in a two-pot reaction sequence. This is the most direct synthesis of such <u>retro</u>-lactones which, as noted above, are of physiological significance and can now be produced in great variety.

In a fashion similar to earlier work, we used an aldehyde as an electrophile in the tandem addition to give one alcohol, <u>37</u>, in excellent yield (Scheme 23). This alcohol in turn could react with protic acids to yield the 'normal' lactones, for which there were excellent analogies, or with thiophilic reagents, for which we had provided the analogies,

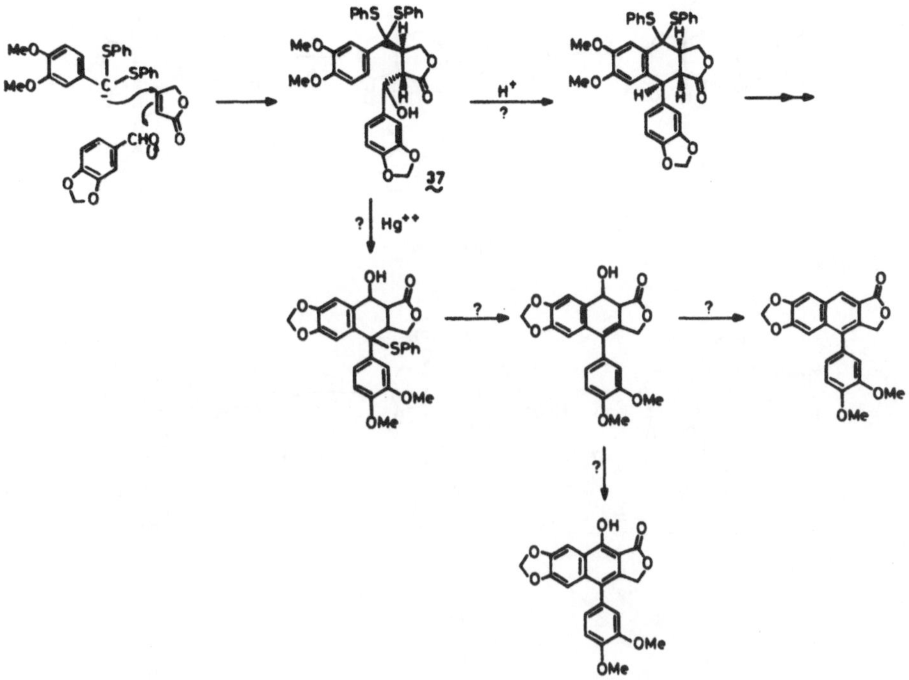

Scheme 23

to yield the <u>retro</u>-series. Both reactions would give
4-functionalised lignans.

To check the protic ring closure (Scheme 24) we
removed the phenylthio groups from <u>37</u> and closed the ring
quantitatively. Note the ring closure gives the <u>trans</u>-aryl
group at C-4. This is ψ-equatorial, not as required for
podophyllotoxin. This result is similar to that of
Ziegler.[55]

We then tried the analogous reaction with the alcohol
containing the two phenylthio groups (Scheme 25). Surprisingly it was the phenylthio group that was the preferred
leaving group, pointing to its lability when a <u>para</u>-alkoxy
group is present. This led to an excellent, short synthesis
of the <u>retro</u>-naphthalene lactones. Stannic chloride gave
the same result plus a most interesting pair of rearrangement products <u>38</u>. These may arise by dehydration, bridging

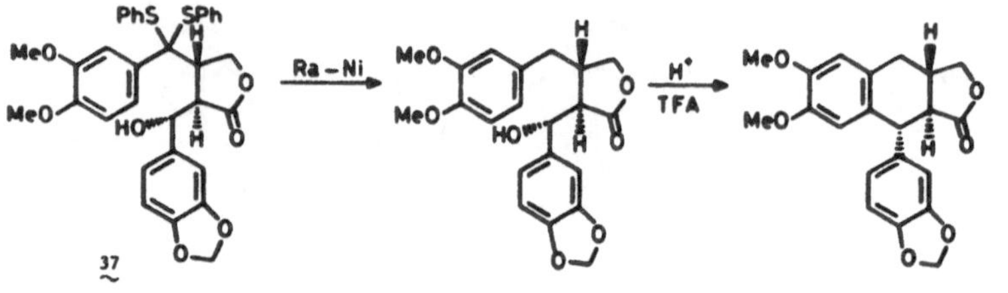

Scheme 24

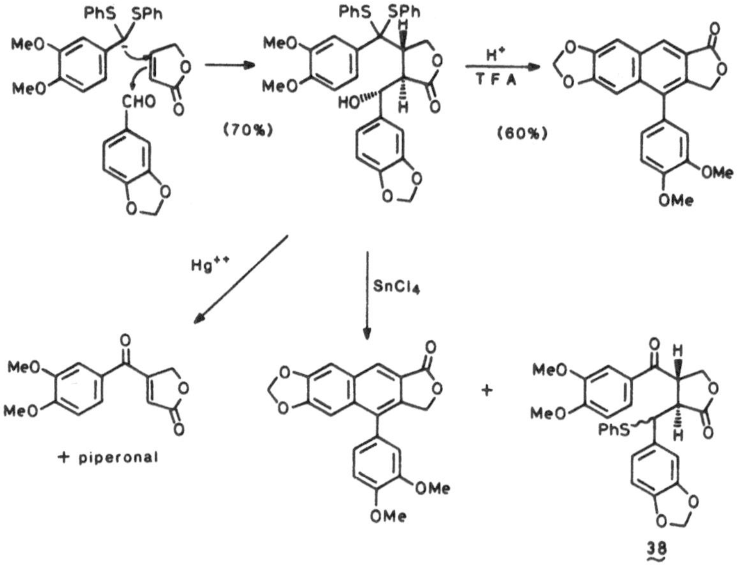

Scheme 25

by the phenylthio group and hydration. The trans-ring
junction is retained. Mercuric salts led to complete
breakdown of the molecule. Thus our carefully chosen model
reactions did not go as expected in the fully functionalised
substrate. The rearrangement product 38 interested us
greatly and we found that it was produced quantitatively
from the alcohol by use of perchloric acid in ethyl acetate
(Scheme 26). The mixture could be separated and it was
interesting to note that the threo isomer yielded an elimination product plus the erythro-isomer on treatment with
acid. However, the mixture was converted quantitatively to

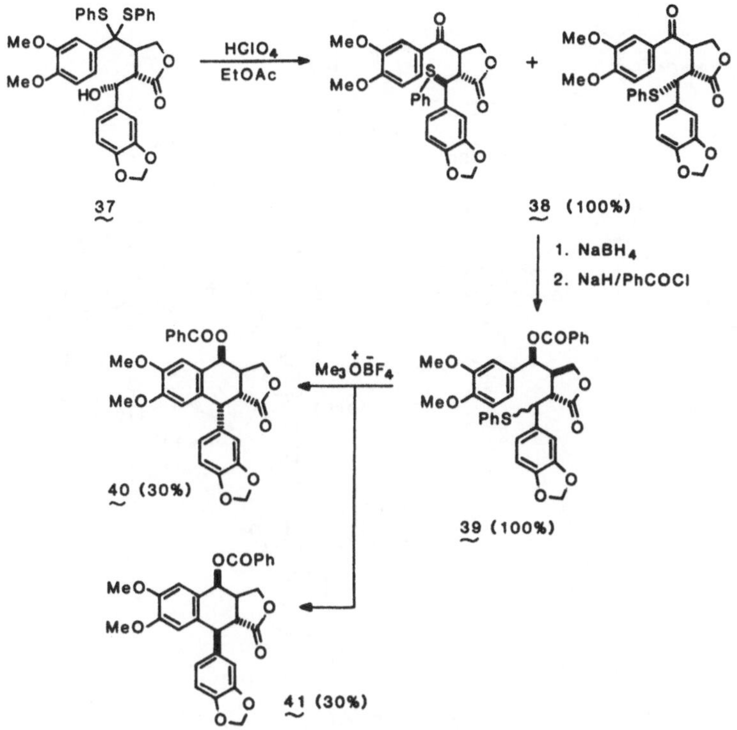

Scheme 26

the corresponding benzoate, 39, and then reacted with
trimethyloxonium tetrafluoroborate to produce corresponding
cyclised products in 60% yield.  Of great interest is the
fact that the 1:1 mixture of phenylthio isomers gave a 1:1
mixture of the 4-cis-aryl compound, 40, and the trans-
isomer, 41.  Whenever similar reactions involving acid
catalysed benzylic alcohol displacements occur, we find
that the trans-isomer results from the $S_N1$ displacement
which presumably occurs.  Here however the displacements
seem to be uniquely $S_N2$.

We did not follow up this observation in the series
shown by using the separate isomers; instead we examined a
different series in which only one phenylthio group was
present in order to obtain more controlled reactions.  In
practise this proved an excellent route to both the retro-
and the normal lactones (Scheme 27) in high yields.

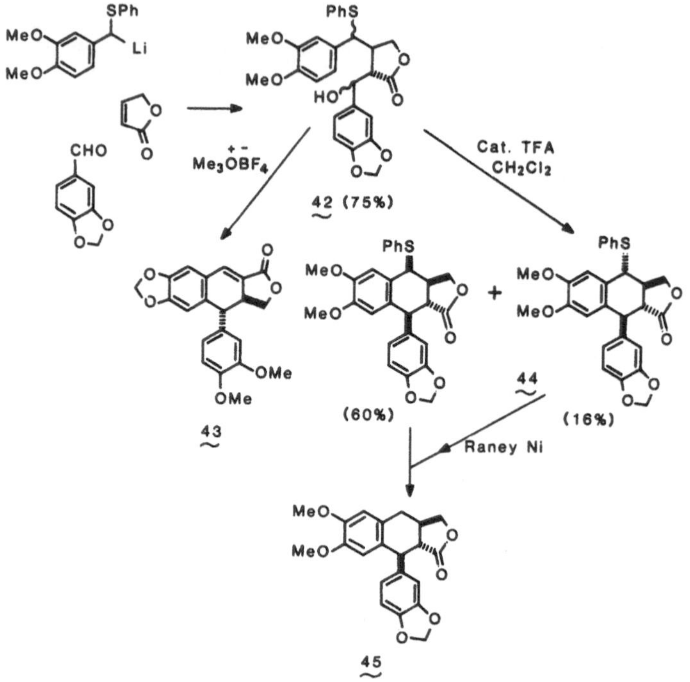

Scheme 27

The tandem additions proceeded as usual to give 42, and methylation of the mixture of stereoisomers gave the unsaturated retro-lactone, 43, an excellent precursor for retro-lactones at a variety of oxidation levels. Gentle acid treatment yielded the 'normal' lactones, 44, shown to be stereoisomers at C-4, by desulphurisation to trans-dihydrocollinusin, 45.

We then turned to the equivalent podophyllotoxin series (Scheme 28). Tandem addition proceeded without problem, to give 46, which on desulphurisation to 47, followed by acid treatment yielded deoxyisopodophyllotoxin 48 in high yields. It was noticeable however that neat trifluoracetic acid was required for the cyclisation rather than the catalytic amounts previously used. We attribute this to the 4'-methoxy group being sterically congested and out of plane with the aromatic ring. When the usual acid conditions were applied to 46, the phenylthio group was

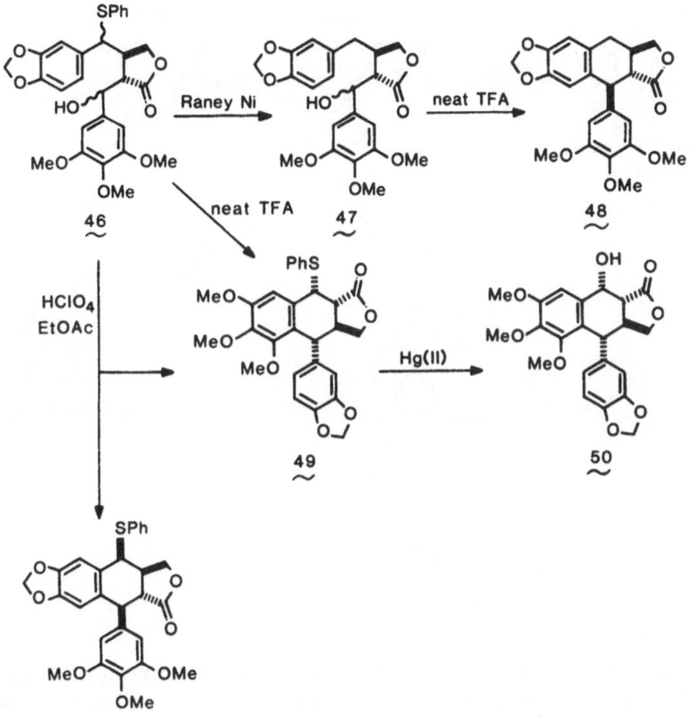

Scheme 28

once more the favored leaving group and a smooth rearrangement occurred to give the retro-lactone 49 in high yield.
Heating of 49 with mercuric trifluoroacetate yielded the
podophyllotoxin analogue, 50, in acceptable yield (65%).
Further studies showed that perchloric acid treatment of
46 gave the retro-product 49, and also the desired compound
51. At this stage we decided that, for the podophyllotoxin
series with the trimethoxybenzyl alcohol group, it was
better to use an alkylthio group as the anion stabilising
functionality.

Accordingly the series of reactions shown in Scheme 29
was carried out. Reaction of the usual tandem addition
products 52, with perchloric acid yielded the desired
isomer 53 in 47% yield and the retro-lactone 54 in 43%

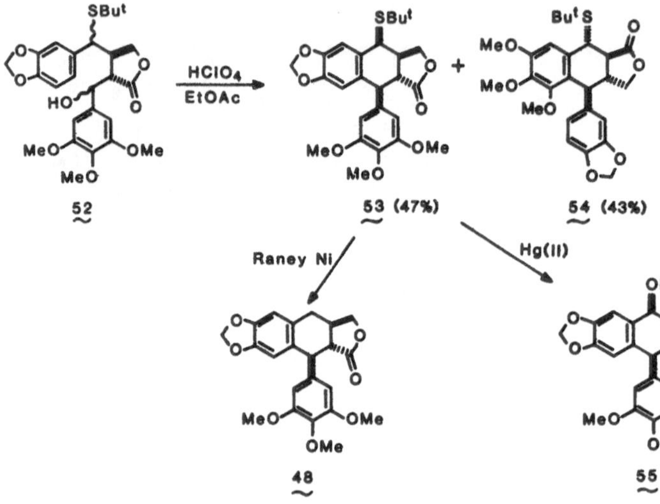

Scheme 29

yield.  The separation was easy as all the products in the
'normal' series are far less soluble than in the retro-
series and simply crystallise out.  Once more the compounds
were converted to 4-deoxyisopodophyllotoxin, 48, and
epiisopodophyllotoxin, 55.  The route is direct, but it
suffers from the problem of retro-cyclisation, which we
should be able to overcome, particularly in the 4'-
demethyl series.  It remains to be seen whether we can
use the dithiane rearrangement process in this series.


ACKNOWLEDGMENTS

      I thank my colleagues Dr. R.S. Ward, Dr. I.T. Kay
(I.C.I. Pesticides), Dr. R. Venkateswarlu, Mr. P. Collins
and particularly Mr. M. Pritchard who is battling bravely
through the aryltetralin and arylnaphthalene series.  The
Royal Society, the S.E.R.C. and I.C.I. are also thanked
for supporting the work financially.

REFERENCES

1. RAO, C.B.S., ed. 1978. Chemistry of Lignans, Andhra
   University Press, Waltair, Andhra Pradesh, India,
   377 pp.
2. GOTTLIEB, O.R. 1978. Neolignans. In C.B.S. Rao,
   ed., op. cit. Reference 1, pp. 277-306.
3. WAGNER, H. 1973. The antihepatotoxic principle of
   Silybum marianum Gaertn. In Recent Flavonoid
   Research. (R. Bognar, F. Kallay, eds.), Akademiai
   Kiado, Budapest, pp. 51-68.
4. ZHUANG, L.-G., O. SELIGMANN, H. WAGNER. 1983.
   Daphneticin, a coumarinolignoid from Daphne
   tangutica. Phytochemistry 22: 617-619.
5. LIN, L.-J., G.A. CORDELL. 1984. Synthesis of
   coumarinolignans through chemical and enzymic
   oxidation. J. Chem. Soc., Chem. Commun., 160.
6. WASSERMAN, H.H., R.K. BRUNNER, J.D. BUYNAK, C.G.
   CARTER, T. OKU, R.P. ROBINSON. 1985. Total
   synthesis of (±)-O-methylorantine. J. Amer. Chem.
   Soc. 107: 519-521.
7. COLE, J.R., R.M. WIEDHOPF. 1978. Distribution of
   lignans. In C.B.S. Rao, ed., op. cit. Reference
   1, pp. 39-64.
8. GOTTLIEB, O.R. 1974. Lignans and neolignans. Rev.
   Latinoamer. Quim. 5: 1-11.
9. STAFFORD, H.A. 1974. The metabolism of aromatic
   compounds. Annu. Rev. Plant Physiol. 25: 459-486.
10. PRYKE, J.A., T. AP REES. 1977. The pentose phosphate
    pathway as a source of NADPH for lignin synthesis.
    Phytochemistry 16: 557-560.
11. ADLER, E. 1977. Lignin chemistry - past, present and
    future. Wood Sci. Technol. 11: 169-218.
12. BIRCH, A.J., A.J. LIEPA. 1977. Biosynthesis of
    lignans. In C.B.S. Rao, ed., op. cit. Reference 1,
    pp. 307-325.
13. ERDTMAN, H. 1933. Dehydrierungen in der Coniferyl-
    reihe. II. Dehydrodi-isoeugenol. Liebigs Annu.
    Chem. 503: 283-294.
14. KURODA, H., M. SHIMADA, T. HIGUCHI. 1975. Purifica-
    tion and properties of O-methyltransferase involved
    in the biosynthesis of gymnosperm lignin.
    Phytochemistry 14: 1759-1763.
15. STÖCKIGT, J., R.L. MANSELL, G.G. GROSS, M.H. ZENK.
    1973. Enzymatic reduction of p-coumaric acid via
    p-coumaroyl-CoA to p-coumaryl alcohol by a cell-free

      system from Forsythia sp.  Z. Pflanzenphysiol. 70:
      305–307.

16.  MONTI, D., M. SANCHEZ.  1971.  Biosynthesis of allyl-
      phenols in Ocyum basilicum.  J. Chem. Soc., Chem.
      Commun., 1108–1109.

17.  BIRCH, A.J.  1963.  Biosynthetic pathways.  In
      Chemical Plant Taxonomy.  (T. Swain, ed),
      Academic Press, London, pp. 141–166.

18.  AYRES, D.C.  1969.  Incorporation of L-[U-$^{14}$C]-β-
      phenylalanine into the lignan podophyllotoxin.
      Tetrahedron Lett., 883–886.

19.  AYRES, D.C., A. FARROW, B.G. CARPENTER.  1981.  The
      biogenesis of podophyllotoxin.  J. Chem. Soc.,
      Perkin Trans I, 2134–2136.

20.  STOECKIGT, J., M. KLISCHIES.  1977.  Biosynthesis of
      arctiin and phillyrin.  Holzforschung 31: 41–44.

21.  FUJIMOTO, H., T. HIGUCHI.  1977.  Biosynthesis of
      liriodendrin by Liriodendron tulipfera.  Wood Res.
      62: 1–10.

22.  JACKSON, D.E., P.M. DEWICK.  1984.  Cinnamic acid
      precursors of podophyllotoxin in Podophyllum
      hexandrum.  Phytochemistry 23: 1029–1035.

23.  JACKSON, D.E., P.M. DEWICK.  1984.  Interconversions
      of aryltetralin lignans in Podophyllum hexandrum.
      Phytochemistry 23: 1037–1042.

24.  TIEN, M., T.K. KIRK.  1983.  Lignin-degrading enzyme
      from the hymenomycete Phanerochaete chrysosporium
      Burds.  Science 221: 661–663.
      TIEN, M., T.K. KIRK.  1984.  Lignin-degrading enzyme
      from Phanerochaete chrysosporium: purification,
      characterization and catalytic properties of a
      unique $H_2O_2$-requiring oxygenase.  Proc. Natl. Acad.
      Sci. 81: 2280–2284.

25.  GOLD, M.H., M. KUWAHARA, A.A. CHIU, J.K. GLENN.  1984.
      Purification and characterization of an extracel-
      lular $H_2O_2$-requiring diarylpropane oxygenase from
      the white rot basidiomycete, Phanerochaete
      chrysosporium.  Arch. Biochem. Biophys. 234:
      353–362.
      KUWAHARA, M., J.K. GLENN, M.A. MORGAN, M.H. GOLD.
      1984.  Separation and characterization of two
      extracellular $H_2O_2$-dependent oxidases from ligni-
      nolytic cultures of Phanerochaete chrysosporium.
      FEBS Lett. 169: 247–250.

26.  KAMAYA, Y., T. HIGUCHI.  1984.  Metabolism of non-
      phenolic diarylpropane lignin substructure model

compound by Coriolus versicolor. FEMS Microbiol. Lett. 22: 89-92.

KAMAYA, Y., T. HIGUCHI. 1984. Metabolism of 3,4-dimethoxycinnamyl alcohol and derivatives by Coriolus versicolor. FEMS Microbiol. Lett. 24: 225-229.

27. GOLD, M.H., M.B. MAYFIELD, T.M. CHENG, K. KRISNANGKURA, M. SHIMADA, A. ENOKI, J.K. GLENN. 1982. A Phanerochaete chrysosporium mutant defective in lignin degradation as well as several other secondary metabolic functions. Arch. Microbiol. 132: 115-122.

28. ANDER, P., K.E. ERIKSSON. 1976. The importance of phenol oxidase activity in lignin degradation by the white-rot fungus Sporotrichum pulverulentum. Arch. Microbiol. 109: 1-8.

29. KOENIGS, J.W. 1972. Production of extracellular hydrogen peroxide and peroxidase by wood-rotting fungi. Phytopathology 62: 100-110.

30. HARKIN, J.M., J.R. OBST. 1973. Syringaldazine, an effective reagent for detecting laccase and peroxidase in fungi. Experientia 29: 381-387.

31. SCHOEMAKER, H.E., P.J. HARVEY, R.M. BOWEN, J.M. PALMER. 1985. On the mechanism of enzymatic lignin breakdown. FEBS Lett. 183: 7-12.

HARVEY, P.J., H.E. SCHOEMAKER, R.M. BOWEN, J.M. PALMER. 1985. Single-electron transfer processes and the reaction mechanism of enzymatic degradation of lignin. FEBS Lett. 183: 13-16.

32. CAMAIONI, D.M., J.A. FRANZ. 1984. Carbon-hydrogen vs. carbon-carbon bond cleavage of 1,2-diarylethane radical cations in acetonitrile-water. J. Org. Chem. 49: 1607-1613.

33. SETCHELL, K.D.R., A.M. LAWSON, F.L. MITCHELL, H. ADLERCREUTZ, D.N. KIRK, M. AXELSON. 1980. Lignans in man and in animal species. Nature 287: 740-742.

SETCHELL, K.D.R., A.M. LAWSON, E. CONWAY, N.F. TAYLOR, D.N. KIRK, G. COOLEY, R.D. FARRANT, S. WYNN, M. AXELSON. 1981. The definite identification of the lignans trans-2,3-bis(3-hydroxybenzyl)-γ-butyrolactone and 2,3-bis(3-hydroxybenzyl)butane-1,4-diol in human and animal urine. Biochem. J. 197: 447-458.

AXELSON, M., K.D.R. SETCHELL. 1981. The excretion of lignans in rats - evidence for an intestinal bacterial source for this new group of compounds.

FEBS Lett. 123: 337-342.

34. STITCH, S.R., J.K. TOUMBA, M.B. GROEW, C.W. FUNKE, J.
    LEEMHUIS, J. VINK, G.F. WOODS. 1980. Excretion,
    isolation and structure of a new phenolic constit-
    uent of female urine. Nature 287: 738-740.

35. AXELSON, M., J. SJOVALL, B.E. GUSTAFSSON, K.D.R.
    SETCHELL. 1982. Origin of lignans in mammals
    and identification of a precursor from plants.
    Nature 298: 659-660.

36. MacRAE, W.D., G.H.N. TOWERS. 1984. Biological
    activities of lignans. Phytochemistry 23: 1207-
    1220.

37. ISSELL, B.F., F.M. MUGGIA, S.K. CARTER, eds. 1984.
    Etoposide (VP-16): Current Status and New Develop-
    ments, Academic Press, New York, 355 pp.

38. JARDINE, I. 1980. Podophyllotoxins. In Anticancer
    Agents Based on Natural Product Models. (J.M.
    Cassady, J.D. Douros, eds.), Med. Chem., Vol. 16,
    Academic Press, New York, pp. 319-351.

39. von KROGH, G. 1983. Condylomhea acuminata 1983:
    up-dated review. Seminars in Dermatology 2: 109-
    129.

40. SOFOWORA, E.A., W.A. ISAAC-SODEYE, L.O. OGUNKOYA.
    1975. Isolation and characterization of an anti-
    sickling agent from Fagara zanthoxyloides root.
    Lloydia 38: 169-171.

41. JIN, J., Z. YU, M. ZHONG. 1982. Antiarrythmic
    principle of Justicia procumbens. Yaoxae Tongbao
    17: 365.

42. PATEL, N.G., C-L.J. WANG. 1984. U.S. Patent No.
    84,318,401, Dec.

43. OHTA, K., K. MUNAKATA. 1970. Justicidin C and D, the
    1-methoxy-2,3-naphthalide lignans, isolated from
    Justica procumbens. Tetrahedron Lett., 923-925.
    WADA, K., K. MUNAKATA. 1970. (-)-Parabenzylactone,
    a new piperolignanolide isolated from Parabenzoin
    trilobium Nakai. Tetrahedron Lett., 2017-2019.

44. MUNAKATA, K. U.S. Patent No. 3,704,247.

45. SCHMIDT, J.M., G.R. PETTIT. 1978. Presence of
    oncornavirus-like particles in the P388 murine
    leukemic cell line. Experientia 34: 659-660.

46. GHOSAL, S., S. BANERJEE, A.W. FRAHM. 1979. Prosta-
    lidins A, B, C and retrochinensin, a new anti-
    depressant: 4-aryl-2,3-naphthalide lignans from
    Justica prostata. Chem. Ind. (London), 854-855.

47.  MAY, G., G. WILLUHN.  1978.  Antivirale Wirkung
     wassriger Pflanzenextrakte in Gewebekulturen.
     Arzneim. Forsch./Drug Res. 28: 1-7.
     BEDOWS, E., G.M. HATFIELD.  1982.  An investigation
     of the antiviral activity of Podophyllum peltatum.
     J. Nat. Prod. 45: 725-729.
48.  LOIKE, J.D., S.B. HORWITZ.  1976.  Effect of VP-16-
     213 on the intracellular degradation of DNA in
     HeLa cells.  Biochemistry 15: 5443-5448.
49.  WOZNIAK, A.J., W.E. ROSS.  1983.  DNA damage as a
     basis for 4'-demethylepipodophyllotoxin-9-(4,6-0-
     ethylidene-β-D-glucopyranoside) (Etoposide)
     cytotoxicity.  Cancer Res. 43: 120-124.
50.  WARD, R.S.  1982.  The synthesis of lignans and
     neolignans.  Chem. Soc. Rev. 11, 75-125.
51.  PELTER, A., R.S. WARD, P. COLLINS, R. VENKATESWARLU,
     I.T. KAY.  1983.  A general synthesis of 2,6-
     diaryl-3,7-dioxabicyclo[3.3.0]octane lignans.
     Tetrahedron Lett. 24: 523-525.
     PELTER, A., R.S. WARD, P. COLLINS, R. VENKATESWARLU,
     I.T. KAY.  1985.  A general synthesis of 2,6-
     diaryl-3,7-dioxabicyclo[3.3.0]octane lignans
     applicable to unsymmetrically substituted compounds.
     J. Chem. Soc., Perkin Trans 1, 587-594.
     POHMAKOTR, M., V. REUTRAKUL, T. PHONGRADIT, A.
     CHANSRI.  1982.  Dianion of diethyl succinate:
     reactions with alkylating agents and carbonyl
     compounds.  Chem. Lett., 687-690.
     MAHALANABIS, K.K., M. MUMTAZ, V. SNIECKUS.  1982.
     Dimetalated tertiary succinamides: Synthesis of
     several classes of lignans including the mammalian
     urinary lignans enterolactone and enterodiol.
     Tetrahedron Lett. 23: 3975.
52.  PELTER, A., R.S. WARD.  1973.  Substituted furofurans.
     In C.B.S. Rao, ed., op. cit. Reference 1, pp. 227-
     275.
53.  ANJANEYULU, A.S.R., K.J. RAO, V.K. RAO, L.R. ROW, C.
     SUBRAHMANYAM, A. PELTER, R.S. WARD.  1975.  The
     structures of lignans from Gmelina arborea Linn.
     Tetrahedron 31: 1277-1285.
     ANJANEYULU, A.S.R., A.M. RAO, V.K. RAO, L.R. ROW, A.
     PELTER, R.S. WARD.  1975.  The structure of
     Gummadiol - a lignan Hemi-acetal.  Tetrahedron Lett.,
     1803-1806.
     ANJANEYULU, A.S.R., P.A. RAMAIAH, L.R. ROW, A. PELTER,
     R.S. WARD.  1975.  The structure of Wodeshiol -

the first of a new series of lignans. Tetrahedron
Lett., 2961-2964.

ANJANEYULU, A.S.R., A.M. RAO, V.K. RAO, L.R. ROW, A.
PELTER, R.S. WARD. 1975. The isolation and
structure of 6"-bromo-isoarboreol - the first
bromine-containing lignan. Tetrahedron Lett.,
4697-4700.

PELTER, A., R.S. WARD. E.V. RAO, K.V. SASTRY. 1976.
Revised structures for pluviatilol, methyl pluviatilol and xanthoxylol. Tetrahedron 32: 2783-2788.

ANJANEYULU, A.S.R., A.M. RAO, V.K. RAO, L.R. ROW, A.
PELTER, R.S. WARD. 1977. Novel hydroxy lignans
from the heartwood of Gmelina arborea. Tetrahedron
33: 133-143.

PELTER, A., R.S. WARD, C. NISHINO. 1977. Revised
structures for epiaschantin and epimagnolin.
Tetrahedron Lett., 4137-4140.

PELTER, A., R.S. WARD, D.J. WATSON, P. MURRAY-RUST,
J. MURRAY-RUST. 1978. On the question of
distinguishing between 2,6- and 2,4-diaryl-3,7-
dioxabicyclo[3.3.0]octanes. Tetrahedron Lett.,
1509-1512.

PELTER, A., R.S. WARD, D.J. WATSON, P. COLLINS, I.T.
KAY. 1979. The unambiguous synthesis of lignans
of the 2,6-diaryl-3,7-dioxabicyclo-[3.3.0]octane
series. The synthesis of eudesmin and 4,8-
dihydroxysesamin. Tetrahedron Lett., 2275-2278.

ROW, L.R., R. VENTKATESWARLU, A. PELTER, R.S. WARD.
1980. Acid catalysed rearrangements of arboreol:
a biomimetic synthesis of gmelanone. Tetrahedron
Lett. 21: 2919-2922.

WARD, R.S., A. PELTER, I.R. JACK. P. SATYANARAYANA,
B.V. GOPALA-RAO, P. SUBRAHMANYAM. 1981.
Reactions of paulownin, gmelinol and gummadiol with
2,3-dichloro-5,6-dicyanobenzoquinone. Tetrahedron
Lett. 22: 4111-4114.

PELTER, A., R.S. WARD, D.J. WATSON, P. COLLINS, I.T.
KAY. 1982. Synthesis of 2,6-diaryl-4,8-dihydroxy-
3,7-dioxabicyclo[3.3.0]octanes. J. Chem. Soc.,
Perkin Trans 1, 175-181.

54. BIRCH, A.J., P. MACDONALD, A. PELTER. 1967. A
revised structure for neogmelinol: determinations
of configurations in tetrahydrofuranoid lignans.
J. Chem. Soc. (C), 1968-1972.

55. ZIEGLER, F.E., J.A. SCHWARTZ. 1978. Synthetic studies on lignan lactones: aryl dithiane route to (±)-podorhizol and (±)-isopodophyllotoxone and approaches to the stegane skeleton. J. Org. Chem. 43: 985-991.

Chapter Nine

BIOSYNTHESIS OF ISO-CHORISMATE-DERIVED QUINONES

ECKHARD LEISTNER

Institut für Pharmazeutische Biologie
Nussallee 6
53 Bonn 1
West Germany

INTRODUCTION

    Recent reviews on quinones have concentrated on
different aspects of their biology and chemistry.  The
biosynthesis, ecology and toxicology of quinones are
treated[1] in a book edited by Higuchi.  An article[2] on the
biosynthesis of chorismate-derived quinones refers to work
on quinone-producing plant cell cultures.  It is demonstrated that there are quite a few quinones which are
produced in cell cultures in rather large amounts.  This
is amazing for two reasons:  firstly, secondary plant
products are quite often produced in rather small amounts
in cell cultures and secondly, quinones originate from very
different biosynthetic precursors such as acetate or
phenylalanine, tyrosine and/or mevalonic acid.  Thus the
observation that quinones are produced in rather large
amounts in cell cultures is unlikely to be due to a single
regulatory phenomenon.  The different biosynthetic pathways
leading to quinones are outlined in another review.[3]  Work

on the biosynthesis of vitamin K is most advanced when
compared to the investigation of the biosynthesis of other
quinones.[4]

    Until recently it was believed that chorismic acid is
the branch point compound that links the shikimate pathway
to the vitamin K biosynthetic pathway.[1-4]  Recent results
show that this is not true (vide infra).[5]  Since vitamin K
was assumed to be derived from chorismic acid,[6] it was
concluded that plant quinones which, like vitamin K, are
synthesized in plants via the shikimate pathway[3] are also
derived from chorismic acid.  Since chorismic acid is not
the immediate precursor of the benzene ring of vitamin K,[5]
it probably is also not the immediate precursor of plant
quinones such as alizarin, lucidin, or juglone.[3]

    There is another interesting aspect to the biosyn-
thesis of vitamin K:  in plants and microorganisms, but not
in animals, an aromatic ring may be formed via the shikimate pathway.  Different metabolites of the shikimate
pathway may undergo aromatization; these include shikimate,
chorismate, arogenate or iso-chorismate.  In the case of
vitamin K iso-chorismic acid (I) undergoes aromatization
by a hitherto unknown and unusual reaction sequence.  Likewise, the second aromatization process involved in the
biosynthesis of the naphthalene ring system of vitamin K
is also unusual.  Thus only one mole of coenzyme A is
required for each mole of vitamin K formed although two
moles of coenzyme A would be expected.[7]


BIOSYNTHESIS OF VITAMIN K$_2$ (MENAQUINONES)

The Initial Aromatization

    The formation of o-succinylbenzoic acid (OSB) (II) is
the key reaction in the biosynthesis of vitamin K$_2$.[4]  This
reaction links the shikimic acid pathway to the vitamin K
biosynthetic pathway (Fig. 1).  It has been assumed that
chorismic acid is the branch point for vitamin K biosynthesis.[6]  This assumption was based on the finding that
mutants of Escherichia coli produce vitamin K although they
seemed to be blocked between chorismic acid and iso-chorismic acid (I).  Indeed, work in this laboratory confirmed[6,8]
that chorismic acid was converted to a product (2,3-
dihydroxybenzoic acid) of iso-chorismic acid (I) in cell

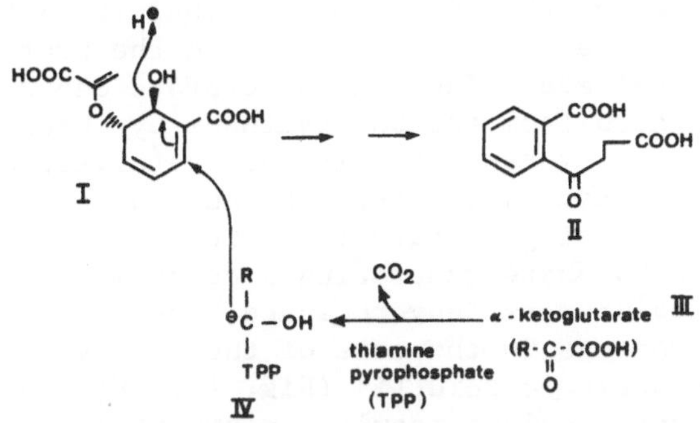

Fig. 1.  The initial aromatization.  (See note added in proof).

free extracts of wild type E. coli $K_{12}$ but not in extracts of the mutant (E. coli AN 154).  These data seem to suggest that in the mutant E. coli AN 154 the conversion of chorismic acid to iso-chorismic acid (I) is blocked and – since vitamin K is produced in good yield in the mutant – that chorismic acid is the metabolite introduced directly into the vitamin K biosynthetic pathway.  Some authors have pointed out, however, that from a mechanistic point of view iso-chorismic acid (I) would be a more favorable substrate for the reaction initiating vitamin K biosynthesis.[9,10] Indeed, a specially designed experiment showed that the mutant E. coli AN 154 is not blocked between chorismic acid and iso-chorismic acid; simultaneous incubation of a radio-actively labelled sample of chorismic acid with unlabelled iso-chorismic acid (I) as carrier in a crude protein extract of the mutant revealed that the reisolated iso-chorismic acid (I) was radioactive.[11]  The inability of the mutant to synthesize products of iso-chorismic acid (I) is therefore not due to a block between chorismic and iso-chorismic acid (I) but must be assigned to another reason.

In agreement with the assumption[4,6] that chorismic acid is the branch point, it was postulated that a cell free system catalyzed the conversion of chorismic acid and α-ketoglutaric acid (III) to o-succinylbenzoic acid[12] (II). Later it became evident, however, that the chorismic acid sample employed in these experiments[12] contained trace amounts of iso-chorismic acid[5] (I).  Indeed, iso-chorismic acid (I) is converted to o-succinylbenzoic acid (II) in the presence of α-ketoglutaric acid (III) in almost 90% yield.[5]

A prerequisite for the high yield is that the α-ketoglutarate dehydrogenase is precipitated from the protein extract
by protamine sulfate. The α-ketoglutarate dehydrogenase can
also be separated from the OSB synthase by chromatofocussing.
Thus both enzyme systems are distinct catalytic entities.[5]
This is an important observation because both enzyme systems
decarboxylate α-ketoglutarate (III) in the presence of
thiamine pyrophosphate generating a carbanion. In the case
of the α-ketoglutarate dehydrogenase, the carbanion reduces
lipoic acid whereas in the case of the OSB synthase, it
attacks iso-chorismic acid (I) (Fig. 1). Elimination of OH⁻
and the pyruvate residue results in aromatization, the first
aromatic product being o-succinylbenzoic acid (II) (Fig. 1).
An intermediate in this aromatization has been trapped[12].
(See note added in proof). The enzyme decarboxylating α-ketoglutarate (III) in the presence of thiamine pyrophosphate is a
subunit of the OSB-synthase. The subunit can be separated
from the holoenzyme by chromatography on DEAE sephadex. The
decarboxylating enzyme has a molecular weight of 66,500
whereas the intact OSB synthase has a molecular weight of
190,000.[11] The separation of the decarboxylating enzyme
from the OSB synthase has made it possible to directly
isolate the product of the decarboxylation, i.e. the
thiamine pyrophosphate adduct of succinic semialdehyde (IV).
The adduct desintegrates into succinic semialdehyde and
thiamine pyrophosphate. Both products were identified,[11]
the aldehyde after conversion to its hydrazone and the
thiamine pyrophosphate by its ability to stimulate the
growth of a thiamine auxotrophic strain of Lactobacillus
viridescens.

Quantitative determination of both the aldehyde and the
thiamine pyrophosphate suggests, as expected, a molar ratio
of one for the two components in the labile adduct. The
reaction mechanism for the initial aromatization can therefore be drawn as depicted in Figure 1.

## The Second Aromatization

The enzymic conversion of o-succinylbenzoic acid (II)
to 1,4-dihydroxy-2-naphthoic acid (DHNA) (V) is a reaction
dependent on ATP, $Mg^{2+}$ and coenzyme A.[4] The product DHNA
is a precursor of vitamin $K_2$. One may assume that both
carboxyl groups in OSB (II) are activated during the conversion to DHNA (V). Activation of the "aliphatic" group would
provide for the acidity of a proton on the methylene group

adjacent to the carboxyl carbon. Nucleophilic attack of the resulting carbanion on the "aromatic" carboxyl carbon would result in ring closure and elimination of the "activating unit" at the "aromatic" carboxyl group. Aromatization would then be the final step in the formation of DHNA (V). Since the activated intermediate between OSB (II) and DHNA (V) had been isolated,[7] its characterization was possible. At least two questions had to be answered. First, are both carboxyl groups in OSB activated? Second, if only one carboxyl group is activated, which one is activated? It had been suggested,[4] and experimental evidence seemed to indicate, that the "aromatic" carboxyl group is activated.[13] But this turned out to be wrong.[14]

There are at least three possible biochemical mechanisms for the activation of carboxyl groups; a phosphoric acid anhydride, an adenylate, or a thiolester may be formed. Since radioactivity was not incorporated from either $\gamma-^{32}P$ ATP or [2-$^3$H adenin]-ATP, into the activated intermediate, a coenzyme A ester was likely to be present.[7] Indeed, the activated intermediate was shown to contain coenzyme A and OSB (II) in the ratio 1:1. Thus, hydrolysis of the activated intermediate and quantitative determination of OSB (II) and coenzyme A by HPLC showed that the activated OSB (II) was a monoester of coenzyme A. In addition, quantitative determination of OSB (II) by its known specific radioactivity and of coenzyme A by means of a coupled enzyme assay gave an identical result.[7] Finally, when the o-succinylbenzoate:coenzyme A ligase preparation was incubated with $^{14}$C-OSB and $^3$H-coenzyme A, the $^3$H/$^{14}$C ratio of the resulting, activated OSB also indicated a ratio of OSB (II):coenzyme A of 1.[7]

The remaining question concerned which carboxyl group is activated. It had been proposed,[4] and experimental evidence seemed to suggest,[14] that the "aromatic" carboxyl group is activated. On the other hand, Leete had shown that OSB (II) is a precursor of certain alkaloids in orchids, the biosynthesis of which is likely to proceed through reduction of the "aliphatic" carboxyl group. Since reduction of a carboxyl group requires previous activation with ATP and coenzyme A, activation of the "aliphatic" carboxyl group of o-succinylbenzoic acid (II) was indicated. The question was eventually answered by synthesis of both coenzyme A esters of OSB via their monoimidazolides.[14] The "aromatic" coenzyme A ester (VII, Fig. 3) was not

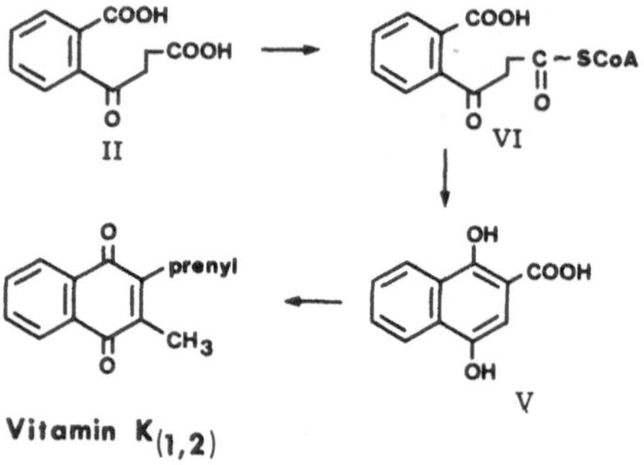

Fig. 2.  The second aromatization.

converted by the naphthoatsynthase to DHNA (V), but the
"aliphatic" coenzyme A ester (VI) was converted to DHNA
(V) in a 50 to 60% yield without any cofactor requirement.[14]
The bis-coenzyme A ester, in which both carboxyl groups are
activated, was converted to DHNA (V) only in extremely low
yield and possibly only after hydrolysis of one of the
coenzyme A ester bonds.  We had to take into account,
however, that combinations of the three coenzyme A esters
["aliphatic" ester (VI), "aromatic ester (VII), and di-
coenzyme A ester] would participate in DHNA (V) biosynthesis.  Enzymic synthesis of DHNA (V) was only observed
when the incubation mixture contained the aliphatic
coenzyme A ester (VI).  Addition of either the "aromatic"
(VII) or di-coenzyme A ester or both did not increase DHNA
(V) synthesis.

The second aromatization process therefore proceeds as
shown in Figure 2.  These findings are at variance with
previous assumptions[4] and with experimental evidence which
seemed to indicate that the aromatic carboxyl group is
activated.[13]  In previous experiments we attempted[13] to
determine the site of activation of o-succinylbenzoic acid
(II) in a different way:  the enzymically formed OSB-
coenzyme A ester was isolated and methylated with diazomethane (Fig. 3).  Mild alkaline hydrolysis yielded the
"aliphatic" methylester of o-succinylbenzoic acid (X).
From this it was concluded that the coenzyme A residue
occupies the "aromatic" rather than the "aliphatic"
carboxyl group.  However, this is an erroneous conclusion.

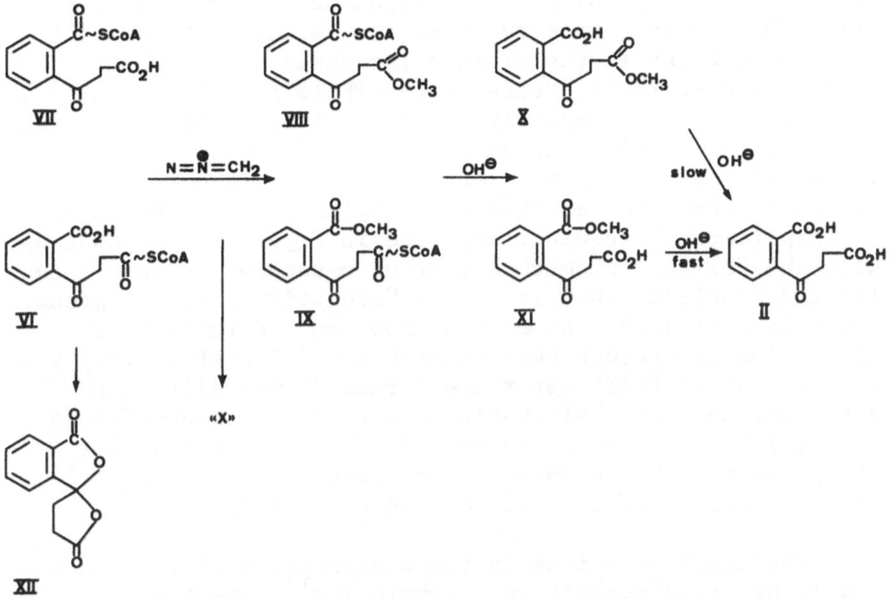

Fig. 3. Experimental approach which led to the erroneous
conclusion[13] that the "aromatic" rather than the "aliphatic"
carboxyl group of o-succinylbenzoic acid is activated
during vitamin $K_2$ biosynthesis.

The reason for this error is shown in Figure 3. When
the coenzyme A ligase preparation[7] (which is a non-purified
enzyme extract) is incubated with coenzyme A, ATP, $Mg^{2+}$
and o-succinylbenzoic acid at pH 7.9, a mixture of the
"aliphatic" (VI), "aromatic" (VII) and di-coenzyme A ester
is formed. This mixture can be separated by HPLC.    In
contrast, if the same experiment is carried out at the
lower pH of 6.5, only the "aliphatic" ester (VI) is formed
and no "aromatic" (VII) or di-coenzyme A ester is detect-
able under these conditions. After incubation at pH 7.9
the mixture of mono-coenzyme A esters was purified by
paper chromatography.[13]  While the "aromatic" coenzyme A
ester (VII) is rather stable, the "aliphatic" ester (VI)
has a half life of 15 minutes at neutral pH and room
temperature.[7]  The product of decomposition is the spiro-
dilactone of o-succinylbenzoic acid (XII) (Fig. 3).   After
methylation with diazomethane a mixture of the methylated
coenzyme A esters (VIII, IX) is formed including an unknown

product ("X", Fig. 3) of the "aliphatic" coenzyme A ester
(VI). Due to this fact the relative amount of "aromatic"
coenzyme A ester in the mixture increased drastically.
When this mixture is treated with alkali for short time[13]
the thioester bonds hydrolyze first. Thus methylesters (X,
XI) of o-succinylbenzoic acid are formed. Moreover, we
found that the "aromatic" methylester (XI) is more labile
in alkali than the "aliphatic" methylester (X). Hence in
the experimental sequence depicted in Figure 3, one ends up
with the "aliphatic" methylester of o-succinylbenzoic acid
(X) and concludes that it is the "aromatic" carboxyl group
which is activated. As we now know, only part of this
conclusion is correct because both the "aliphatic" (VI) and
the "aromatic" (VII) ester are formed enzymically at pH
7.9. But only the "aliphatic" ester (VI) is converted to
DHNA (V) by naphthoatsynthase. It should be noted that
biosynthetic schemes shown in previous publications[1,7,13]
are in error; a more recent one is correct.[14]

The final reactions in the biosynthesis of vitamin K
involving prenylation[16] and methylation[17] have been
described.


BIOSYNTHESIS OF VITAMIN $K_1$ (PHYLLOQUINONE)

Phylloquinone is a constituent of higher plants. Its
structure is similar to vitamin $K_2$ in that it also contains
the menadione nucleus to which, however, a phytyl residue
(one unsaturated and three saturated isoprene units) is
attached (Fig. 4). Phylloquinone occurs in chloroplasts of

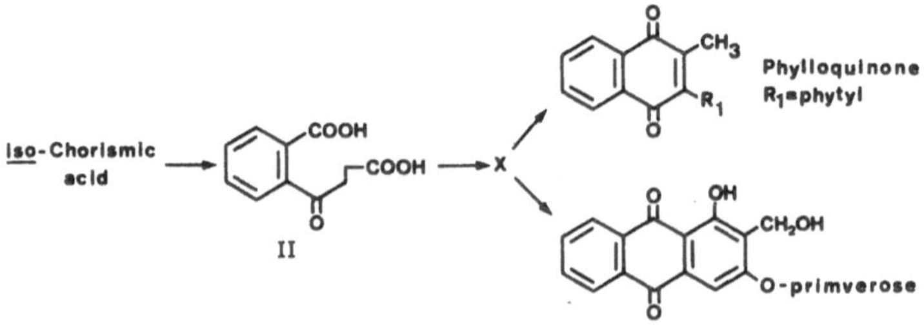

Fig. 4. Branched biosynthetic pathways leading to phylloquinone (Vitamin $K_1$) and anthraquinones (e.g. lucidinprim-
veroside).

all green plants.  Its biosynthesis also proceeds from o-succinylbenzoic acid (II).  The final steps in phylloquinone biosynthesis have been localized in subplastidic fractions.  Thus the prenylation of DHNA takes place in the envelope of the chloroplast whereas the final methylation reaction occurs in isolated thylakoids provided protein of the stroma is added.[18]  Stromal protein alone is unable to catalyze this reaction.  It is therefore assumed that the stromal protein contains a cofactor essential for the prenylation reaction.  Although chromoplasts of Capsicum do not contain vitamin $K_1$, enzymes involved in its biosynthesis seem to be present.  Addition of DHNA (V) and (S)-adenosyl-L-methionine to chromoplasts resulted in formation of phylloquinone.[19]

## BIOSYNTHESIS OF QUINONOID NATURAL PRODUCTS IN DIFFERENT TAXA OF HIGHER PLANTS

### Morinda, Rubia and Galium

Phylloquinone is also a constituent of a green photosynthetic cell suspension culture of Morinda lucida Benth. (Rubiaceae).[20]  The amounts of lipoquinones (i.e. phylloquinone, plastoquinone, α-tocopherol, ubiquinone) isolated from the cultured cells are similar to those isolated from the intact leaf.[20]  This is also true of the chlorophyll content of leaf and culture (Table I).

This suspension culture of Morinda cells is being studied because the biosynthesis of anthraquinones and its glycosides can be triggered in the culture.  Upon transfer of cells into darkness and a medium containing sucrose the chlorophyll decreases and the lipoquinones (including phylloquinone) disappear from the cells (Table 1).  Simultaneously, anthraquinone pigments are formed (Fig. 4) and the cultured cells turn yellow and eventually red.  Thus, in the suspension culture photoautotrophy correlates with lipoquinone synthesis while heterotrophy correlates with anthraquinone synthesis.  This reflects the situation in the intact plants where lipoquinones are associated with chloroplast whereas anthraquinones occur in the roots.

The fact that phylloquinone and anthraquinone biosynthesis share a common biosynthetic pathway in the initial steps raises two questions:  how is the metabolic shift

Table 1. Alternative formation of phylloquinone and anthraquinones in a photoautotrophic and heterotrophic cell suspension culture of Morinda lucida.

|  | Chlorophyll a + b $\mu g \cdot g^{-1}$ d.wt. | Phylloquinone $\mu g \cdot g^{-1}$ d.wt. | Anthraquinones $\mu mol \cdot g^{-1}$ fr.wt. |
|---|---|---|---|
| Intact leaf | 2212 | 34.1 | 0.0 |
| Photoauto-trophic culture (light, no sucrose) | 2981 | 9.7 | 0.1 |
| Hetero-trophic culture (darkness, sucrose) | 1550 | 0.0 | 6.6 |

The heterotrophic cultures were generated by transfer of photoautotrophic cells into darkness and growth in sucrose. Data of the heterotrophic culture were taken 5 days after transfer.

triggered in the culture, and which regulatory phenomena are involved in phylloquinone and anthraquinone biosynthesis. Various observations indicate that anthraquinone biosynthesis can be repressed, induced or increased in cell suspension cultures.[2] Thus one plant effector (i.e. 2,4-dichlorophenoxyacetic acid) may support growth of a cell culture of Morinda citrifolia but repress anthraquinone synthesis completely whereas another hormone (e.g. naphthalene acetic acid) may support both growth and anthraquinone formation.[21] Cells producing anthraquinones have a different shape when compared to those cells which are devoid of any anthraquinones. Thus it appears that morphology of cells and pigment production are closely related. A systematic evaluation of the influence of different structural analogs of 2,4-dichlorophenoxyacetic

acid on anthraquinone synthesis showed that 4-iodophenoxy-
acetic acid stimulated anthraquinone synthesis more than the
corresponding 4-bromo- or 4-chlorophenoxyacetic acid.
Highest yields of pigments were obtained in the presence of
2,3-dimethyl- and 2-bromophenoxyacetic acids.[22]  Another
factor that influences anthraquinone formation in cell
cultures is the nitrogen source.[21,23]  Thus L-tryptophan
as well as its precursors (indole, anthranilic acid) and
degradation product (kynurenic acid) suppress anthraquinone
formation.  Suppression of pigment formation was most effective, however, with 2-aminobenzothiazole, a nonphysiological
substance.  Precursors of the anthraquinones such as
shikimic acid or o-succinylbenzoic acid (II) did not
alleviate the inhibitory effect.  It was concluded that
the late steps in anthraquinone biosynthesis were blocked
and that a primary metabolite like tryptophan may completely
inhibit secondary product formation.[23]

  The influence of various sugars on anthraquinone
synthesis has also been investigated.[21]  Among the sugars
tested sucrose plays an important role in the cell cultures
of M. citrifolia.  Sucrose is most efficient in supporting
growth and pigment production.[21]  This is also evident from
our experiments with the photoautotrophic cells of M.
lucida.  Reversion of phylloquinone to anthraquinone formation (Table I) is contingent upon the presence of sucrose.
The sugar cannot be replaced by mannitol.[20]  Thus sucrose
plays an important nutritive or regulatory role in secondary
product formation in these cell cultures.  There is another
observation which is interesting with respect to the regulation of anthraquinone synthesis in cell cultures.  Anthraquinones are not only produced in Morinda but also in cell
cultures of Galium and Rubia.[24,25]  Plants of Galium mollugo
also contain iridoids while the cell suspension cultures
are devoid of these metabolites.[24]  Iridoids are biosynthetically derived from mevalonic acid and sugars which, of
course, also contribute to anthraquinone biosynthesis in
addition to o-succinylbenzoic acid (II).[1-3]  The amount of
anthraquinones in the Galium cell cultures can be stimulated
by the addition of o-succinylbenzoic acid (II) alone.  This
shows that mevalonic acid and sugars occur in the cultured
cells at a non-limiting level.  In spite of this, iridoids
which are derived from these metabolites are not produced.
Thus, the presence in a cell culture of appropriate precursors does not necessarily result in the synthesis of
secondary products.[24]  These and the other regulatory

phenomena are unexplained.  It may be that a deeper insight
into the enzymology of anthraquinone formation in these
cultures will provide for a better understanding of the
observations listed above.

One of the open questions in anthraquinone biosynthesis
is the nature of the branching compound X (Fig. 4).  In
vitamin $K_2$ biosynthesis 1,4-dihydroxy-2-naphthoic acid (V)
is involved (vide supra).  It may also be an intermediate in
phylloquinone[19] and anthraquinone biosynthesis.[1-3]  Feeding
experiments established that anthraquinones in Rubia,[1-3]
Morinda[26] and Galium[27,28,29] are derived from shikimic acid,
α-ketoglutaric acid and mevalonic acid with o-succinylbenzoic acid (II)[27,28,29] being a key intermediate.  Incorporation of 4-(2'-[$^{13}$C]carboxyphenyl)-4-oxobutanoic acid (i.e.
labelled o-succinylbenzoic acid) (II) into lucidinprimvero-
side, the major pigment in G. mollugo cell cultures, was
accomplished in a chemostat under conditions of limiting
phosphate.  The specific incorporation was determined by
mass spectrometry and shown to be as high as 83.9%.  The
label in the anthraquinone lucidin was located by NMR spectroscopy.  It is assumed that the biosynthetic sequence
proceeds as shown in Figure 5.

It is noteworthy that prenylation of DHNA (V) occurs
at carbon 3 rather than carbon 2 as occurs in vitamin K
biosynthesis (vide supra).  This is not only evident from
the mode of incorporation of labelled o-succinylbenzoic
acid (II) into lucidin (XIII) but also from the finding that
2-methoxycarbonyl-3-prenyl-1,4-naphthoquinone (XIV)[30,31] and
the diglycoside of its hydroquinone[29] (XV) are natural
products occurring both in the intact plant[30,31] and in cell
suspension culture[29] of G. mollugo.

Enzyme studies on the biosynthesis of anthraquinones
are greatly hampered by the fact that cultured cells of G.
mollugo contain rather large amounts of anthraquinones which
on homogenization inactivate enzymes.  A systematic study of
different methods aimed at the removal of anthraquinones and
isolation of active enzyme preparations from G. mollugo
cells[32] has led to the detection of o-succinylbenzoic acid:
coenzyme A ligase.[33]  The activated o-succinylbenzoic acid
[viz. the coenzyme A thiolester (VI), Fig. 5] was converted
to DHNA (V) by an enzyme extract from Mycobacterium phlei
indicating that the bacterial coenzyme A ester and the ester
from G. mollugo are identical.

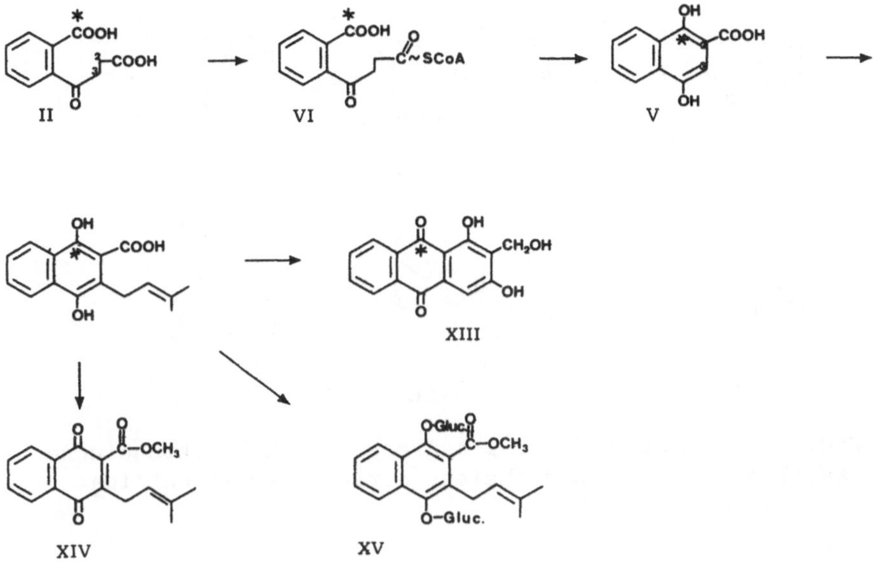

Fig. 5.   Biosynthesis of lucidin (XIII) and related naphthalene derivatives in Rubiaceae (Rubia, Morinda, Galium).
The asterisk indicates $^{13}$C labelled C-atoms.


## Catalpa

While the biosynthesis of vitamin $K_1$ (phylloquinone)
may proceed through 1,4-dihydroxy-2-naphthoic acid (DHNA)
(V),[18,19] an alternative route has been suggested[34] (Fig.
6).   Vitamin K with only one prenyl unit (XVI) has been
isolated from callus tissue of Catalpa ovata.   This naphthoquinone is associated with 1-hydroxy-2-methylanthraqui-
none (XVII) and o-succinylbenzoic acid (II) is a precursor
of these compounds.   During their biosynthesis 2-prenyl-
carboxyoxotetralone (XVIII) and catalponone (XIX) are
formed.   This was established by trapping labelled XVIII
and XIX after administration of radioactive o-succinylbenzoic acid (II) to C. ovata tissue cultures.   The
chirality of XVIII and XIX was shown to be 2S (XVIII) and
2R (XIX) respectively.   Formation of catalponone (XIX) does
not proceed through DHNA (V).[35]   The biosynthesis of anthraquinone in the tissue culture of C. ovata may therefore be
different from the cultures of Rubia, Morinda and Galium in
that carboxyoxotetralone (XX) is prenylated instead of

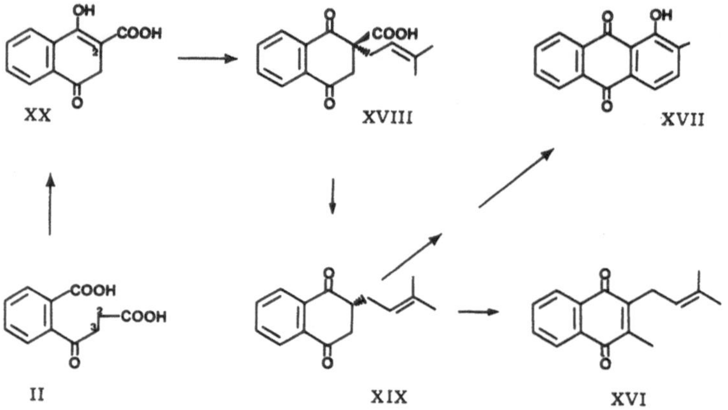

Fig. 6. Biosynthesis of 1-hydroxy-2-methylanthraquinone (XVII) and related naphthalene derivatives in Catalpa (Bignoniaceae).

DHNA (V) and that prenylation takes place at the C-atom corresponding to the C-2 atom of o-succinylbenzoic acid rather than C-3 atom (compare Fig. 5).

Streptocarpus

The genus Streptocarpus belongs to the family Gesneri-aceae.  Streptocarpus dunni contains several pigments in its roots and leaves, the major one of which is 1-hydroxy-2-hydroxymethylanthraquinone (XXI, Fig. 7).  The pigment was shown to be derived from o-succinylbenzoic (II) and mevalonic acid.[36]  Streptocarpus plants and cell cultures are a rich source of both anthraquinones and naphthaqui-nones[37] and investigation of their structure and biosyn-thesis has shown that a considerable variation in structure and biosynthesis is brought about by different prenylation reactions.[38,39]  Potential precursors labelled with $^{13}C$ and deuterium were administered to a plant cell culture of S. dunni.  Unusually high incorporation rates for o-succinyl-benzoic acid (II) ranged between 7 and 20%.[38,39]  The o-succinylbenzoic acid (II) was labelled in the "aromatic" carboxyl carbon.  The mode of incorporation of o-succinyl-benzoic acid into anthraquinones (e.g. 1-hydroxy-2-hydroxymethylanthraquinone, XXI) and into naphthoquinones (e.g. α-dunnione, XXII) were elucidated using NMR techniques.

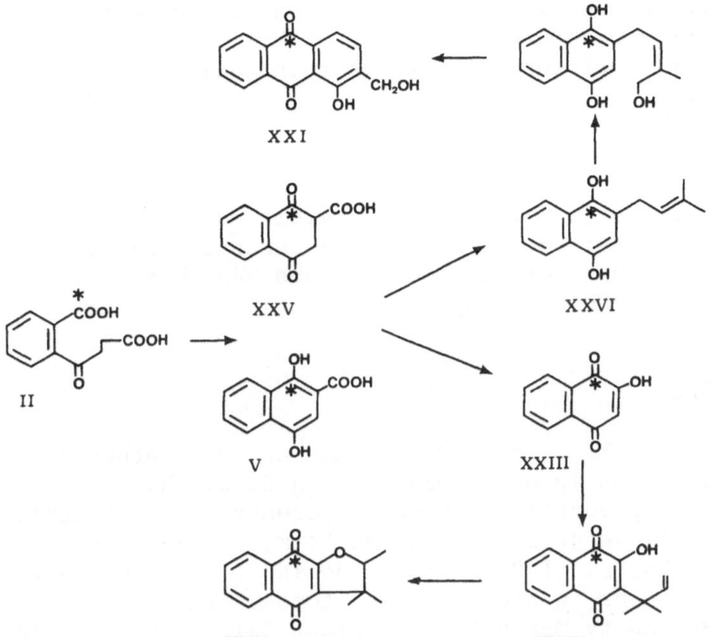

Fig. 7.  Biosynthesis of 1-hydroxy-2-hydroxymethylanthra-
quinone (XXI) and related naphthalene derivatives in
Streptocarpus (Gesneriaeceae).  The asterisk indicates [13]C
labelled C-atoms.

During the biosynthesis of these anthraquinones prenylation
occurs in a different manner than observed with Rubia,
Galium and Morinda (vide supra).  The C-2 position (rather
than C-3) of the hypothetical naphthalene derivative was
prenylated (compare the Catalpa experiments) during the
reactions leading to anthraquinones.  The pathway to naphthoquinones, however, proceeds via lawsone (2-hydroxy-
1,4-naphthoquinone, XXIII) which forms an o-prenylether
which, after conversion to 2-hydroxy-3-(1,1-dimethylallyl)-
1,4-naphthoquinone (XXIV), is eventually converted to
α-dunnione (XXII).

    At present it is unclear whether 2-carboxy-2,3-
dihydro-1,4-naphthoquinone (XXV) or 1,4-dihydroxy-2-
naphthoic acid (V) or both are intermediates in the
biosynthetic pathway depicted in Figure 7.  Another interesting aspect of the proposed pathway[39] (Fig. 7) is that

# 258 ECKHARD LEISTNER

oxidation of one of the methyl groups of 2-dimethylallyl-1,4-dihydroxynaphthohydroquinone (XXVI) is assumed to occur prior to ring closure leading to anthraquinones.

## ACKNOWLEDGMENTS

The author's work reported herein was supported by grants from Deutsche Forschungsgemeinschaft, Minister für Wissenschaft und Forschung des Landes NRW (West Germany), and Fonds der Chemischen Industrie.

## REFERENCES

1. LEISTNER, E. 1985. Occurrence and biosynthesis of quinones in woody plants. In Biosynthesis and Biodegradation of Wood Components. (T. Higuchi, ed.), Academic Press, New York, pp. 273-290.
2. LEISTNER, E. 1985. Biosynthesis of chorismate-derived quinones in plant cell cultures. In Primary and Secondary Metabolism of Plant Cell Cultures. (K.-H. Neumann, W. Barz, E. Reinhard, eds.), Springer, Berlin, Heidelberg, pp. 215-224.
3. LEISTNER, E. 1981. Biosynthesis of plant quinones. In The Biochemistry of Plants. (E.E. Conn, ed.), Vol. 7, Academic Press, New York, pp. 403-423.
4. BENTLEY, R., R. MEGANATHAN. 1981. Biosynthesis of vitamin K (menaquinone) in bacteria. Microbiol. Rev. 46: 241-280.
5. WEISCHE, A., E. LEISTNER. 1985. Cell-free synthesis of o-succinylbenzoic acid from iso-chorismic acid, the key reaction in vitamin $K_2$ (menaquinone) biosynthesis. Tetrahedron Lett. 26: 1487-1490.
6. YOUNG, I.G. 1975. Biosynthesis of bacterial menaquinones: menaquinone mutants of Escherichia coli. Biochemistry 14: 399-406.
7. HEIDE, L., S. ARENDT, E. LEISTNER. 1982. Enzymatic synthesis, characterization, and metabolism of the coenzyme A ester of o-succinylbenzoic acid, an intermediate in menaquinone (vitamin $K_2$) biosynthesis. J. Biol. Chem. 257: 7396-7400.
8. YOUNG, I.G., T.J. BATTERHAM, F. GIBSON. 1969. The isolation, identification and properties of iso-chorismic acid, an intermediate in the biosynthesis of 2,3-dihydroxybenzoic acid. Biochim. Biophys. Acta 177: 389-400.

9.  HASLAM, E.  1974.  The Shikimate Pathway, Butter-
    worths, London, 316 pp.
10. DANSETTE, P., R. AZERAD.  1970.  A new intermediate in
    naphthoquinone and menaquinone biosynthesis.
    Biochem. Biophys. Res. Commun. 40: 1090-1095.
11. WEISCHE, A., W. LEISTNER, unpublished.
12. MEGANATHAN, R., R. BENTLEY.  1983.  Thiamine pyrophos-
    phate requirement for o-succinylbenzoic acid
    synthesis in Escherichia coli and evidence for an
    intermediate.  J. Bacteriol. 153: 739-746.
13. KOLKMANN, R., G. KNAUEL, S. ARENDT, E. LEISTNER.
    1982.  Site of activation of o-succinylbenzoic acid
    during its conversion to menaquinones (vitamin $K_2$).
    FEBS Lett. 137: 53-56.
14. KOLKMANN, R., E. LEISTNER.  1985.  Synthesis and
    revised structure of the o-succinylbenzoic acid
    coenzyme A ester, an intermediate in menaquinone
    biosynthesis.  Tetrahedron Lett. 26: 1703-1704.
15. LEETE, E., G.B. BODEM.  1976.  Biosynthesis of
    shihunine in Dendrobium pierardii.  J. Am. Chem.
    Soc. 98: 6321-6325.
16. SAITO, Y., K. OGURA.  1981.  Biosynthesis of menaqui-
    nones.  Enzymatic prenylation of 1,4-dihydroxy-2-
    naphthoate by Micrococcus luteus membrane fractions.
    J. Biochem. (Tokyo) 89: 1445-1452.
17. SAMUEL, O., R. AZERAD.  1972.  C-méthylation des
    desméthylménaquinones: II. Spécificité du système
    enzymatique de méthylation de Mycobacterium phlei
    vis-à-vis du substrat quinonique.  Biochimie 54:
    305-317.
18. KAIPING, A., J. SOLL, G. SCHULTZ.  1984.  Site of
    methylation of 2-phythyl-1,4-naphthoquinol in
    phylloquinone (vitamin $K_1$) synthesis in spinach
    chloroplasts.  Phytochemistry 23: 89-91.
19. GAUDILLIERE, J.-P., A. d'HARLINGUE, B. CAMARA, R.
    MONEGER.  1984.  Prenylation and methylation
    reactions in phylloquinone (vitamin $K_1$) synthesis
    in Capsicum annuum plastids.  Plant Cell Rep. 3:
    240-242.
20. IGBAVBOA, U., H.-J. SIEWEKE, E. LEISTNER, J. RÖWER,
    W. HÜSEMANN, W. BARZ.  1985.  Alternative formation
    of anthraquinones and lipoquinones in heterotrophic
    and photoautotrophic cell suspension of Morinda
    lucida Benth.  Planta, submitted.
21. ZENK, M.H., H. EL-SHAGI, U. SCHULTE.  1975.  Anthra-
    quinone production by cell suspension cultures of

_Morinda_ citrifolia. Planta medica, Supplement, 79-101.

22. ZENK, M.H., U. SCHULTE, H. EL-SHAGI. 1984. Regulation of anthraquinone formation by phenoxyacetic acids in _Morinda_ cell cultures. Naturwissenschaften 71: 266.

23. EL-SHAGI, H., U. SCHULTE, M.H. ZENK. 1984. Specific inhibition of anthraquinone formation by amino compounds in _Morinda_ cell cultures. Naturwissenschaften 71: 267.

24. BAUCH, H.-J., E. LEISTNER. 1978. Aromatic metabolites in cell suspension cultures of _Galium mollugo_ L. Planta medica 33: 105.

25. SCHULTE, U., H. EL-SHAGI, M.H. ZENK. 1984. Optimization of 19 Rubiaceae species in cell cultures for the production of anthraquinones. Plant Cell Rep. 3: 51.

26. LEISTNER, E. 1975. Isolierung, Identifizierung und Biosynthese von Anthrachinonen in Zellsuspensionskulturen von _Morinda_ citrifolia. Plant medica, Supplement, 214-224.

27. BAUCH, H.-J., E. LEISTNER. 1978. Attempts to demonstrate incorporation of labelled precursors into aromatic metabolites in cell suspension cultures of _Galium mollugo_ L. Planta medica 33: 124-127.

28. INOUE, K., Y. SHIOBARA, H. NAYESHIRO, H. INOUYE, G. WILSON, M.H. ZENK. 1979. Site of prenylation in anthraquinone biosynthesis in cell suspension of _Galium mollugo_. J. Chem. Soc., Chem. Commun., 957-959.

29. INOUE, K., Y. SHIOBARA, H. NAYESHIRO, H. INOUYE, G. WILSON, M.H. ZENK. 1984. Biosynthesis of anthraquinones and related compounds in _Galium mollugo_ cell suspension cultures. Phytochemistry 23: 307-311.

30. HEIDE, L., E. LEISTNER. 1981. 2-Methoxycarbonyl-3-prenyl-1,4-naphthoquinone, a metabolite related to the biosynthesis of mollugin and anthraquinones in _Galium mollugo_ L. J. Chem. Soc., Chem. Commun., 334-336.

31. HEIDE, L., E. LEISTNER. 1982. Versuche zur Synthese natürlich vorkommender prenylierter Naphthalinderivate. Nachweis eines neuen Prenylchinonderivates in _Galium mollugo_. Z. Naturforsch. 37C: 354-362.

32. HEIDE, L., E. LEISTNER. 1983. Enzyme activities in extracts of anthraquinone-containing cells of Galium mollugo. Phytochemistry 22: 659-662.

33. HEIDE, L., R. KOLKMANN, S. ARENDT, E. LEISTNER. 1982. Enzymic synthesis of o-succinylbenzoyl-CoA in cell-free extracts of anthraquinone producing Galium mollugo L. cell suspension cultures. Plant Cell Rep. 1: 180-182.

34. INOUE, K., S. UEDA, Y. SHIOBARA, J. KIMURA, H. INOUYE. 1981. Quinones and related compounds in higher plants. Part 11. Role of 2-carboxy-2,3-dihydro-1,4-naphthoquinone and 2-carboxy-2-(3-methyl-but-2-enyl)-2,3-dihydro-1,4-naphthoquinone in the biosynthesis of naphthoquinone congeners of Catalpa ovata callus tissue. J. Chem. Soc., Perkin Trans. I, 1246-1258.

35. UEDA, S., K. INOUE, T. HAYASHI, H. INOUYE. 1975. Zur Biosynthese des Catalponols und artverwandter Stoffe. Tetrahedron Lett., 2399-2401.

36. STÖCKIGT, J., U. SROCKA, M.H. ZENK. 1973. Structure and biosynthesis of a new anthraquinone from Streptocarpus dunnii. Phytochemistry 12: 2389-2391.

37. INOUE, K., S. UEDA, H. NAYESHIRO, H. INOUYE. 1983. Quinones of Streptocarpus dunnii. Phytochemistry 22: 737-741.

38. INOUE, K., S. UEDA, H. NAYESHIRO, H. INOUYE. 1982. Biosynthesis of quinones of Streptocarpus dunnii cell cultures. J. Chem. Soc., Chem. Commun., 993-994.

39. INOUE, K., S. UEDA, H. NAYESHIRO, N. MORITONE, H. INOUYE. 1984. Biosynthesis of naphthoquinones and anthraquinones in Streptocarpus dunnii cell cultures. Phytochemistry 23: 313-318.

Note added in proof: A non-aromatic intermediate has been identified. (See G.T. Emmons, I.M. Campbell, R. Bentley. 1985. Vitamin K (menoquinone) biosynthesis in bacteria: Purification and probable structure of an intermediate prior to o-succinylbenzoate. Biochem. Biophys. Res. Commun. 131: 956-985.) Its structure shows that pyruvate elimination preceeds dehydration during OSB biosynthesis.

Chapter Ten

NATURALLY OCCURRING QUINONES AS BIOREDUCTIVE ALKYLATING
AGENTS

HAROLD W. MOORE AND J. OLLE KARLSSON

Department of Chemistry
University of California
Irvine, California  92717

INTRODUCTION

Bioreductive alkylation is the term used to describe
the effect of those compounds which express their mode of
biological action as alkylating agents, but do so subsequent
to their reduction in vivo.[1]  That is, they are pro-drugs
which are activated by a bioreduction.  Quinones are a class
of compounds ideally suited to function as the reducible
moiety of bioreductive alkylating agents since their facile
reduction in vivo and in vitro to the corresponding hydro-
quinones is a well known and extensively studied reaction.[2]
If the quinone is further substituted with a side-chain
bearing a leaving group X at the 1-position of the substituent, then quinonemethide formation can result by an
elimination of HX from the hydroquinone.[2,3]  The reactive
quinonemethide is suggested as the discrete alkylating
agent and functions as such by a Michael addition of a
biologically important nucleophile (Nu⁻:DNA, protein,
carbohydrate, etc.) to the enone of the methide.  This

postulate is represented by the sequence of reactions
outlined in Scheme 1, i.e., 1 → 2 → 3 → 4.  Further
comments concerning the details of this proposed mechanism
of action follow.  1) Although a two-electron reduction is
represented, a one-electron process can also be envisaged.
That is, rather than the hydroquinone 2, a semiquinone
radical-anion could be formed initially.  This could then
proceed to the radical form of 3 and subsequently to a
radical anion of 4.  2) The facility of quinonemethide
formation, and thus biological activity, may be enhanced
as the half-wave reduction potential ($E_{1/2}$) of the quinone
decreases.  That is, electron-releasing substituents
attached to the quinone nucleus will retard the reduction
step.  However, once reduced, the electron-rich hydroquinone will be activated towards elimination of HX and thus
formation of quinonemethide.  This would be particularly
true if the electron-releasing substituents are ortho and/or
para to the side-chain bearing the leaving group.  In such
cases, the electron-releasing substituents would facilitate
loss of the leaving group X and thus enhance quinonemethide
formation.  3) The ease of quinonemethide formation should
also be affected by the character of the leaving group, X.
Specifically, good leaving groups (weak bases) in the sense
of the $SN_1$ or $SN_2$ reaction would also enhance quinonemethide
formation from the hydroquinone 2.  4) It is also noted
that the generalized structure 1 is an oversimplification
since the quinone nucleus need not be limited to the
benzoquinones.  Indeed, quinonemethides can be envisaged
as arising by a reductive elimination from appropriately

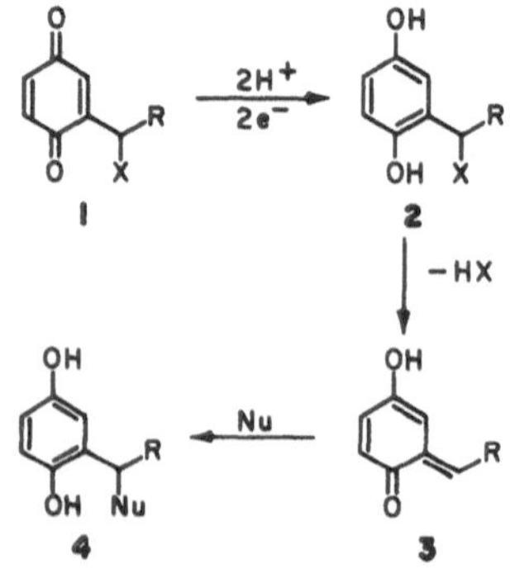

Scheme 1

substituted benzoquinones, naphthoquinones, anthraquinones, anthracyclines, and many other related systems. Additionally, the leaving group need not necessarily be located at the position adjacent to the quinone nucleus. Rather, it could be bonded to a number of possible vinylogous sites in compounds more complex than 1 and still give rise to quinonemethides or related reactive intermediates.

Not all of the above proposed structural requirements have been probed experimentally. However, Sartorelli and co-workers at Yale have elegantly shown that a number of benzoquinones, naphthoquinones, and anthraquinones, meeting the general structure 1, do express antineoplastic activity.[1,4] Furthermore, the biological activity is indeed generally enhanced as the half-wave reduction potential of the quinones decreases and as the leaving group ability increases. These workers have further suggested that quinones of structure 1 might be logical chemotherapeutic agents to attack hypoxic tumor cells, i.e., those cells remote from blood vessels such as are found at the core of solid tumors.[5] The oxygen deficiency of hypoxic cells suggests them to be potentially good candidates for selective cancer chemotherapy by bioreductive alkylating agents since they are believed to provide a more efficient reducing environment than do oxygen-rich tumor or normal cells. Thus, their ability to reduce the quinone 1 to the "activated" hydroquinone 2 may be enhanced. Indeed, some progress in this direction has been reported.[6] For example, mitomycin C, as well as some 9,10-anthraquinones which are substituted so as to allow quinonemethide formation, have recently been shown to possess significantly enhanced toxicity towards hypoxic tumor cells as compared to analogous oxygen-rich cells.[7]

As noted above, the generation of electrophilic quinonemethides by reductive activation is documented and many of the precursor quinones express biological activity. In addition to this bioreductive activation, one can also envisage quinonemethides as being generated in a proton transfer equilibration as outlined in Scheme 2. For such a process to be particularly facile, it may be necessary for the group R to be an acidifying substituent ($COCH_3$, $CO_2CH_3$, CN, etc.) and/or an acidic hydroxyl group to be appropriately situated to function as an internal acid catalyst. Thus, hydroxyquinones of structural type 5 and 6 would be reasonable candidates. The potential biological

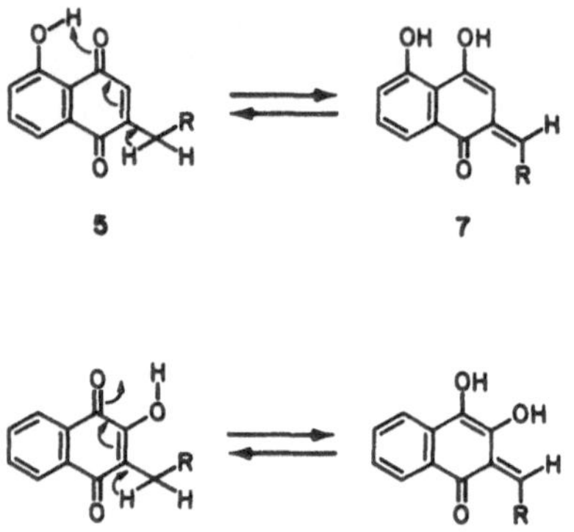

Scheme 2

activity of such compounds might then be due to an equilibrium concentration of the respective quinonemethides 7 and
8.

## NATURALLY OCCURRING QUINONES AS POTENTIAL PRECURSORS TO QUINONEMETHIDES

The above discussion is focussed upon the conversion
of quinones to quinonemethides by processes involving
either reductive elimination or proton transfer.  It is
also possible that quinonemethides may function as alkylating agents and thus those quinones which are amenable to
the above transformations may express biological activity
as antineoplastic agents.  With these ideas in mind, a
survey of naturally occurring quinones was accomplished to
determine those examples which can be catalogued according
to the above structural requirements.[8]  Approximately 300
compounds were found and a few selected examples are
discussed below.

### Selected Natural Products

The term, bioreductive alkylation, was coined by Lin,
Cosby, Shansky, and Sartorelli in 1972[1] to explain the
antineoplastic activity of some synthetic quinones of
general structure 1.  However, the genesis of the concept
stems from the earlier report of Iyer and Szybalski[7] who

suggested that the natural product, mitomycin C, expressed
its antineoplastic activity as a cross-linking agent for
DNA, and that this alkylation occurred only subsequent to
reduction in vivo.  Indeed, they demonstrated the cross-
linking of calf thymus DNA by mitomycin C in vitro, and
this occurred only in the presence of a reducing agent.  A
possible molecular mechanism of action of mitomycin C 9
which falls within the framework of bioreductive alkylation
is outlined in Scheme 3.  The salient features of this
proposed mechanism follow:  1) mitomycin C is reduced to
the hydroquinone which then eliminates $CH_3OH$ to give the
indole 10;  2) opening of the aziridine ring by an elimination reaction would give the quinonemethide 11;  3)
nucleophilic addition of DNA to the quinonemethide would
give the monoalkylated adduct 12;  4) intramolecular $SN_2$
displacement of the carbamate would result in the cross-
linked adduct 13.  Two interesting model experiments which
lend support for this proposed mechanism have recently been
reported.  Keller et al. have observed the first 1-
substituted mitocene to be generated from mitomycin C under
reductive conditions.[9]  Specifically, an adduct analogous
to 13 was obtained when a mixture of mitomycin C and

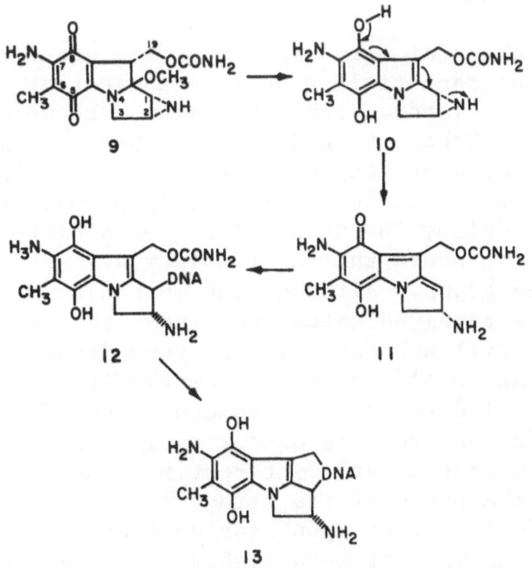

Scheme 3

potassium ethylxanthate was treated with the reducing agent
sodium dithionite.  This product most likely arises from
nucleophilic attack of the xanthate ion at position 1 of the
quinonemethide 11.  A related result was reported by
Hashimoto, Shudo, and Okamoto who generated the quinonemethide by reduction of mitomycin C with $H_2$/Pd/C and trapped
it in situ with 5'-guanylic acid to give 14.[10]  These

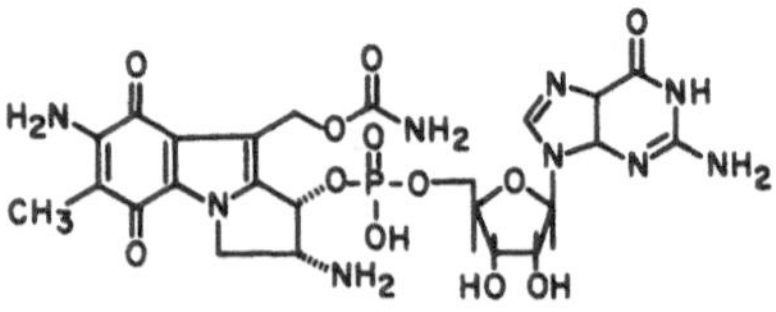

14

investigators have recently provided evidence that mitomycin
C covalently binds to the guanine base of DNA both in vitro
and in vivo.[11]  For example, reductive activation of mitomycin C in the presence of calf thymus DNA followed by
enzymatic degradation to the drug-bound nucleotide was
accomplished.  The structure of the adduct was shown to
involve covalent binding at position-1 of mitomycin to
positions 7 and 8 of the guanine base.  These two adducts
were also detected as being formed in vivo.

A major class of naturally occurring quinones which
show marked antibacterial as well as anticancer activity
are the anthracyclines.  Perhaps the best known member is
adriamycin (15, Scheme 4, R=OH) which shows significant
activity against a variety of neoplasms including leukemia,
lung cancer, breast cancer, sarcoma, lymphoma, and neuro-
blastoma.[12]  Adriamycin can be viewed as a bioreductive
alkylating agent where the sugar moiety at C-7 is considered
as the leaving group.  Although the mechanism of action of
adriamycin has received extensive study, the detailed mode
of action is still unknown.  It has been demonstrated to
intercalate into DNA[13] and most recently it has been shown
to covalently bind to the macromolecule when the reaction is
carried out in the presence of a reducing agent.[14]  This
last result is particularly interesting since such would be
predicted on the basis of its structure and the concept of
quinonemethide formation upon bioreduction.  The site of
covalent binding was not established.  However, position 7
would be expected according to the mechanism outlined in
Scheme 4.

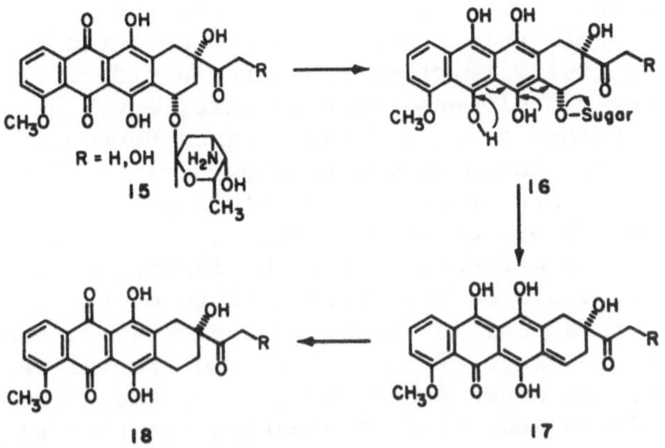

Scheme 4

Reduction of adriamycin to the hydroquinone 16, followed by elimination of the sugar moiety, would give the quinonemethide 17, which could function as the alkylating agent. It is of interest to point out that dithionite reduction of daunomycin (15, R=H) gives a nearly quantitative yield of the 7-deoxy derivative 18.[15] This transformation most certainly involves the quinonemethide 17. In fact this intermediate has now been detected when daunomycin was treated with the reducing agent 2,5,5-trimethyl-2-oxomorpholin-3-yl.[16] Protonation of 17 by solvent $CH_3OH$ was shown to give 18. Interestingly, quinonemethide 17 has been demonstrated thus far to show only nucleophilic properties.[17] However, that obtained by analogous reductive elimination from 11-deoxydaunomycin has been demonstrated to function as an electrophile,[18] and that from alklavinone expresses reactivity as both a nucleophile and an electrophile.[19]

A large number of anthracycline antibiotics are now known, and the list grows yearly. All of these compounds possess structural features analogous to those of adriamycin and are thus amenable to quinonemethide formation upon bioreduction. Some of the more important members of this series from a biological activity point of view are aklavin, steffimycin, baumycin $A_1$, $A_2$, $B_1$, $B_2$, $C_1$, $C_2$, nogalamycin, musettamycin, rhodirubin A,B, marcellomycin, carminomycin, aclacinomycin A,B, isoquinocycline B, pyrromycin, cinerubin A,B, and rudolphomycin.

A few additional, general comments regarding the natural products survey which suggest potentially important structure-activity relationships are in order. For example, C-glycosides of quinones are very rare, e.g., aquayamycin, P 1894B, carminic acid, and kidamycin. However, all of those which are known appear to show biological activity. Other examples are hedamycin, neopluramycin, and pluramycin A, all of which are closely related to kidamycin. In addition, it is noteworthy that the leaving group in these C-glycosides would be the ether linkage of the sugar; this is located in a position strictly analogous to the O-glycosidic linkage of most of the anthracyclines, e.g., adriamycin. As noted above, members of the anthracyclines readily undergo reductive elimination reaction to quinone-methide intermediates. Thus, an analogous reductive elimination of the C-glycosidic quinones would be a most

## I. Benzoquinones

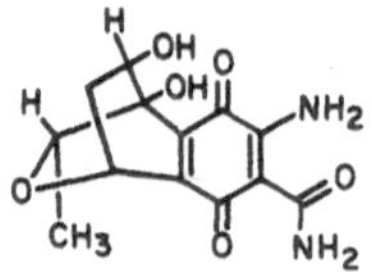

U58,431
Antibiotic

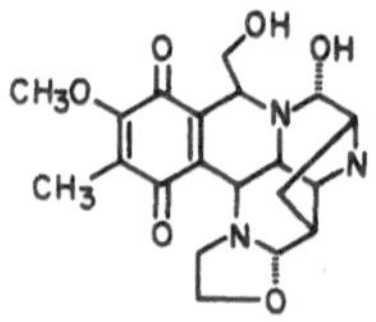

Naphthyridinomycin
Antibiotic

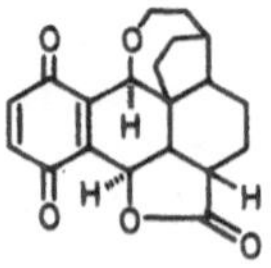

Pleurotin
Antibiotic

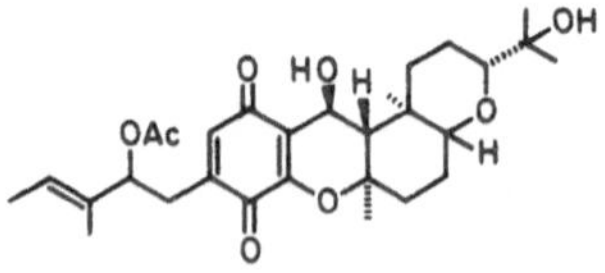

Stemphone
Antibiotic

reasonable possibility. However, unlike the anthracyclines, this should be a favorably reversible reaction since it would be intramolecular in character. No study has appeared whose objectives are the synthesis and biological evaluation of C-glycosidic quinones. This could be a worthwhile endeavor.

## 2. Naphthoquinones

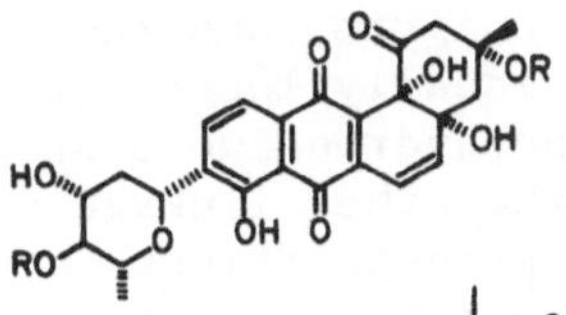

R = H    **Aquayamycin**
         Antibiotic

R =      **P 1894B**
         Proline Hydrase Inhibitor

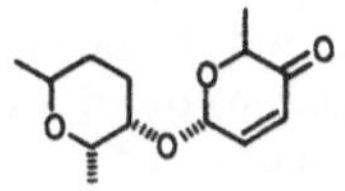

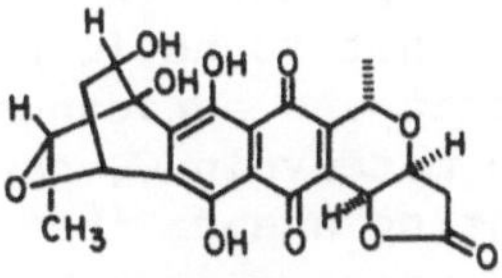

**Granaticin**
Antitumor – Antibiotic

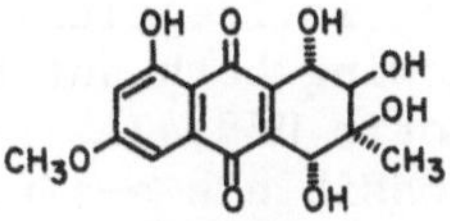

**Altersalanol A**
Phytotoxin

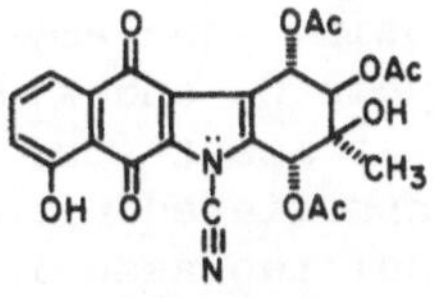

**Kinamycin C**
Antibiotic

## 3. Anthaquinones

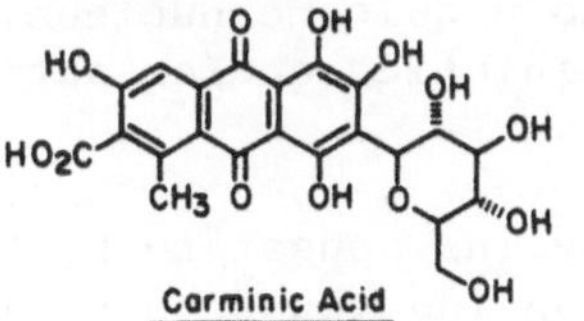

**Carminic Acid**
Antitumor

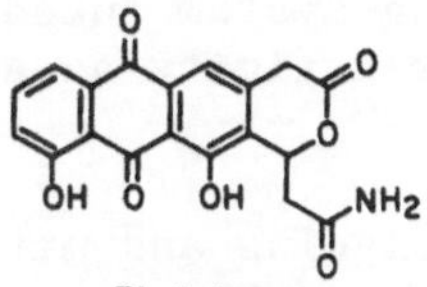

**Ekatetrone**
Antitumor – Antibiotic

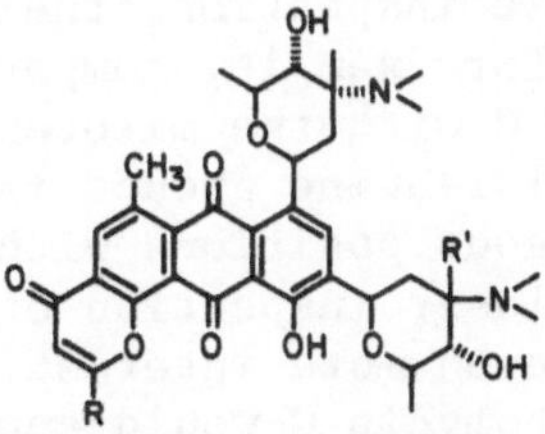

R 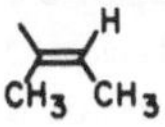 =      R' = H    **Kidamycin**
                   Antitumor – Antibiotic

Antibiotic U58,431 (sarubicin A) is one of only four known naturally occurring primary amino-1,4-quinones. The others are mitomycin C, streptonigrin, and rhodoquinone. The amino group in U58,431 is ideally located to facilitate quinonemethide formation at the hydroquinone stage by assisting in the cleavage of the ether linkage of the bicyclic ring situated in the _para_-position. A mitomycin C derivative having the amino and methyl groups on the quinone ring interchanged would provide an analogous situation; this would be a most interesting compound to compare with mitomycin C with regard to its ability to crosslink DNA and its biological action. Also note that the bicyclic ring of U58,431 is identical to that found in granaticin, a compound showing both antibiotic and anticancer properties. Antibiotic U58,431, like mitomycin C, can also be viewed as a possible bis-alkylating agent; that is, a bis-quinonemethide can be viewed as arising from the above-mentioned ether cleavage followed by dehydration. Other potential bis-alkylating agents listed here would be naphthyridinomycin, pleurotin, aquayamycin, P 1894B, altersalanol A, kinamycin C, and kidamycin. Granaticin is unique in that it has at least four possible alkylating sites, i.e., the two associated with the bicyclo[2.2.2]ring as well as an additional two associated with the γ-lactone and pyrane rings. Thus, conceivably it could produce the equivalent of a tetraquinonemethide under reductive conditions. It is also interesting to note that granaticin is one of approximately 25 natural quinones having the pyrane-γ-lactone ring system fused to a quinone nucleus, and all of these show activity as antibiotics and/or antifungal agents.

The kinamycins and mitomycins appear to be the only examples of naturally occurring indolequinones. Both meet the structural requirements for bioreductive alkylating agents, and upon close inspection, their similarity is rather remarkable. For example, comparison of the hydroquinone of kinamycin C with the mitocene hydroquinone 10 reveals the potential leaving groups in both compounds to be in strictly analogous positions with respect to the indole nucleus. A closer inspection of these natural products suggest an even more interesting structural similarity. That is, mitomycin C would appear to be "activated" towards alkylation by reduction and loss of $CH_3OH$ to give 10 which functions as the ultimate precursor to the quinone-methide 11. The driving force for methide formation would

be enhanced by the release of the strain energy of the
aziridine ring.  This elimination would be further assisted
by the C-8 hydroxy group as well as the availability of the
non-bonding electron pair on the indole nitrogen.  Note
that this electron pair would not assist bond cleavage of
the aziridine ring at any quinonoid stage since it would be
vinylogously conjugated to the quinone carbonyl group.  The
electron pair of the pyrrole nitrogen in kinamycin would be
even less available for assisted ionization than that of
mitomycin since it is in conjugation with both the quinone
carbonyl and the cyano substituent.  Thus, both natural
products would be expected to be reasonably stable in their
quinone forms.  However, kinamycin, like mitomycin, may
unleash its alkylating sites by a sequence of reduction and
elimination steps.  For example, reduction to the hydroquinone followed by loss of HCN and proton transfer would
result in the decyanated quinone form of kinamycin.  A
second reduction step would then provide an indole having
the electron pair on the indole nitrogen appropriately
situated to assist ionization as well as the electron pair
on one of the hydroquinone hydroxy groups.  This sequence
of reactions would result in 19 which would function as a
potent alkylating agent.  A less esoteric possibility
worth considering is that the activity of kinamycin is
simply due to HCN which is released upon bioreduction.  In
any regards, it is again noteworthy to point out that no
systematic study of the chemistry or biological properties
of N-cyano indolequinones has appeared.

19

Finally, it is noted that both kinamycin C and altersalanol A contain an identical, polyoxygenated, cyclohexenyl
ring which bears the potential leaving groups for quinonemethide formation.  An analogous structural feature is also
found in bostrycin, altersalanol B, and kinamycin A, B,
and D.

SYNTHETIC APPROACHES TO BIOREDUCTIVE ALKYLATING AGENTS

Synthesis and Chemistry of (2-Alkynylethenyl)ketenes - A
New Quinone Synthesis

Our experimental efforts have focussed upon developing
new and efficient synthetic routes to quinones possessing
structural features necessary for bioreductive alkylating
agents.  One such study has resulted in the discovery of a
new quinone synthesis.[20]  Interestingly, this involves the
previously unknown (2-alkynylethenyl)ketenes, specifically
those of general structure 21 (Scheme 5).  Such compounds
are accessible from the corresponding alkynylcyclobutenones
20.  The ketenes thus generated undergo ring closure to the
unique zwitterions 22 and 23 and these, in turn, proceed to
products 24 and 25 via a trimethylsilyl transfer.

The results of this study are outlined in Table 1 and
Scheme 5.  The 4-alkynyl-2,3-dimethoxy-4-trimethylsiloxy-
cyclobutenones 20a-e were thermolyzed in refluxing p-xylene
(135°C) to give both five- and six-membered ring products.
The ring closure pathway is influenced by the substituent
R; electron withdrawing groups favor the formation of cyclo-
pentenediones 25 and electron releasing groups favor the
quinones 24.  The structures of these products are based
upon spectral and chemical properties.  For example, the
quinones 24b-e all show a positive leucomethylene blue test
and can be easily reduced to their corresponding hydroqui-
nones.

The proposed mechanism for the conversion of 20 to 24
and 25 in refluxing p-xylene is provided in Scheme 5.  The
cyclobutenone 20 would be anticipated to be in equilibrium
with the (2-alkynylethenyl)ketenes 21.  The ketene can then
undergo ring closure to yield 22 and/or 23.[21,22]  As is
indicated by the product distributions, the mode of ring
closure (21 to 22 vs. 21 to 23) is dictated by the electronic influence of the alkyne substituent R.  Transfer of
the trimethylsilyl group to the negative site then gives
the products 24 and 25.  The zwitterionic intermediates 22
and 23 seem to be the best way to explain the results
obtained.  However, recent results in our laboratories have
indicated that they might also be regarded as being diradical in nature (see below for another case of zwitterion-
diradical dualism).  It is interesting to note that the
quinone 24e possesses features for potential bioreductive

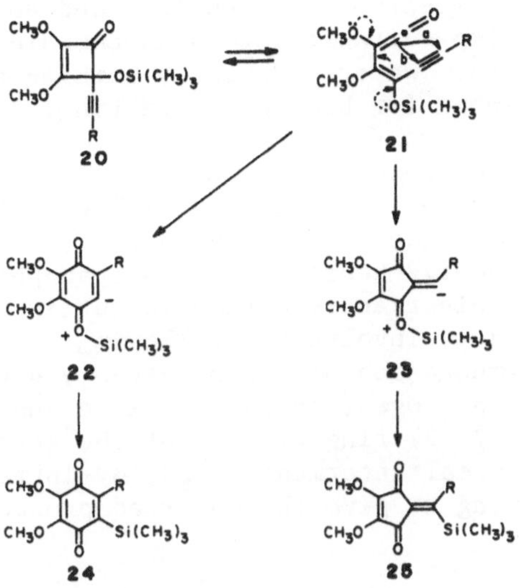

Scheme 5

Table 1

| | | Isolated Yields (%) | |
| --- | --- | --- | --- |
| R | | 24 | 25 |
| a) | $-CO_2C_2H_5$ | -- | $33^a$ |
| b) | $-C_6H_5$ | $13^b$ | 52 |
| c) | $-\underline{n}-C_4H_9$ | 75 | -- |
| d) | $-CH_2C_6H_5$ | 74 | -- |
| e) | $-CH_2OSi(CH_3)_3$ | 80 | -- |

[a] A yield of 43% was obtained starting from a chromatographed (silica gel) sample of the cyclobutenone 1a. However, the yield of 24a was only 21% after chromatography due to hydrolysis.

[b] 24b and 25b could not be separated by chromatography – instead the mixture was reduced with sodium dithionite. The hydroquinone of 24b was then separated from 25b by chromatography and later reoxidized to 24b with $Ag_2O$. The yield of 24b is based on the isolated yield of its hydroquinone.

action, i.e., an electron-rich nucleus and an oxygen function in a position suitable for quinonemethide formation
after elimination. We are currently working toward deriva-
tizing this quinone for biological testing.

Efforts to explore other routes to (2-alkynylethenyl)
ketenes have resulted in further exciting developments.
Specifically, a unique, thermally induced, rearrangement of
4-alkynyl-3-azido-1,2-benzoquinone to cyanophenols has been
discovered.[23] This transformation is outlined in Scheme 6
and is envisaged to involve the following steps:[24]
1) thermal fragmentation of the previously unknown azidoquinone 26 to dinitrogen, carbon monoxide and (2-alkynyl-
ethenyl)ketene 27; 2) ring closure of the ketene to the
dipolar or diradical intermediate 28; 3) intra- or intermolecular trapping to give the observed products.

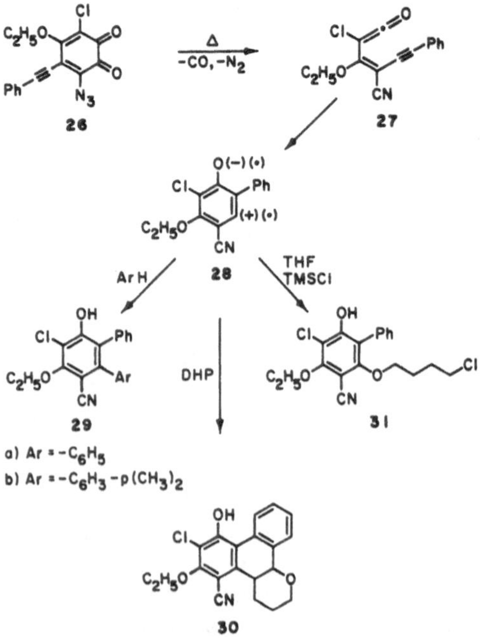

Scheme 6

Thermolysis of 26 in benzene or p-xylene gave, respectively, 29a (84%) and 29b (56%).[26,27] A particularly interesting product was realized when 26 was decomposed in refluxing cyclohexane containing an excess (10 eq) of dihydropyran (DHP). Here, the annulated product 30 was obtained in 37% yield. Finally, thermolysis of 26 in refluxing cyclohexane containing THF and TMSCl gave the highly functionalized phenol 31 in 72% isolated yield.[28] This last product can be seen to arise from 28 via silylation of the phenoxide oxygen and trapping of the aryl cation by THF. The resulting oxonium ion is then opened by attack of Cl⁻, and desilylation in work-up gives the observed product 31.

In contrast to the above examples involving intermolecular condensations of the cyclized intermediate 28, thermolysis of the azidoquinone 32 in benzene resulted in the intramolecular trapping of 33 to give 34 (33%).

Direct evidence for the intermediacy of (2-alkynyl-ethenyl)ketenes in these reactions was obtained from a direct trapping experiment. That is, refluxing a solution of 26 in hexane containing 10 equivalents of dicyclohexyl-carbodiimide gave a 44% yield of 35, the anticipated product from a DCC/ketene cycloaddition. Evidence for the electronic state of the intermediates 28 and 33 formed upon ring closure of these ketenes is less definitive. As noted above, a diradical or zwitterion can be envisaged, and the products, 29, 30, 31 and 34 are best viewed as arising from the dipolar species. However, direct evidence for a diradical intermediate was obtained for a closely related example. Specifically, hexynylcyanoketene[25] was generated in

refluxing benzene in the presence of diphenylacetylene.  It
was anticipated that the resulting cyclobutenone 36 (Scheme
7) would undergo electrocyclic ring opening to the ketene
37 and that this would proceed to products via the diradical
or dipolar intermediate 38.  If this intermediate is zwitterionic, incorporation of the solvent (phenylation) would
be anticipated.  On the other hand, a diradical intermediate would be expected to result in products arising from
hydrogen atom abstraction from the butyl side chain.  The
radical process was observed since the only products
detected were 39, 40, 41, and 42, formed in the ratio of
1:1:0.2:0.3 respectively, with a combined yield of 54%.

## Novel Synthetic Routes to Heterocyclic Quinones

An interesting route to heterocyclic quinones from
azidoquinones was recently discovered in our laboratory.[26]
As formally outlined in Scheme 8, the method combines a
number of reactions which alone have received little attention.  Together, they constitute one mechanistic sequence
which rationalizes the formation of the observed products

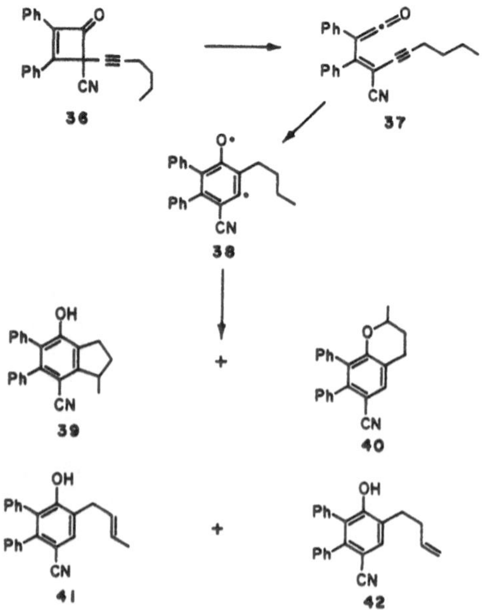

Scheme 7

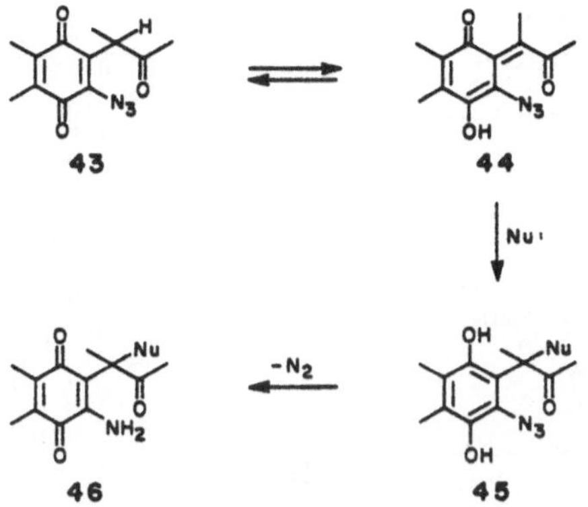

Scheme 8

and provides insight to a variety of synthetic targets.  The
generalized sequence of steps include:  1) equilibration of
an appropriately substituted azidoquinone, 43, to the
quinonemethide 44, and its trapping in situ by Michael
addition to give the unstable azidohydroquinone 45;[27]
2) spontaneous disproportionation of the azidohydroquinone
to the aminoquinone, 46;[28] and 3) in some cases subsequent
ring closure or rearrangements induced by nucleophilic
attack of the resulting amino group on a proximate electrophilic site.

Schemes 9, 10 and 11 illustrate examples of the above
reaction sequence.  Refluxing an aqueous THF solution of
the azidoquinone, 47, for 1.5 hours resulted in the formation of the aminoquinone, 48a (79%) (Scheme 9).  When the
reflux time was extended to 5 h the indolequinone, 49a,
formed by ring closure was isolated in 74% yield.  In an
analogous fashion, 48b (77%) was obtained when 46 was
decomposed in methanol.  However, 49b was not obtained by
extending the reflux time (15 h), but was formed upon reduction of 48b with $Na_2S_2O_4$ followed by subsequent oxidation
of the hydroquinone in air.  Decomposition of 47 in acetic
acid (90°C) gave 48c (65%).

A number of unusual transformations were observed
when the azidoquinones were treated with enolates of acidic
ketones.  For example, when 51 was slowly added to a THF
solution of dimedone containing a catalytic amount of

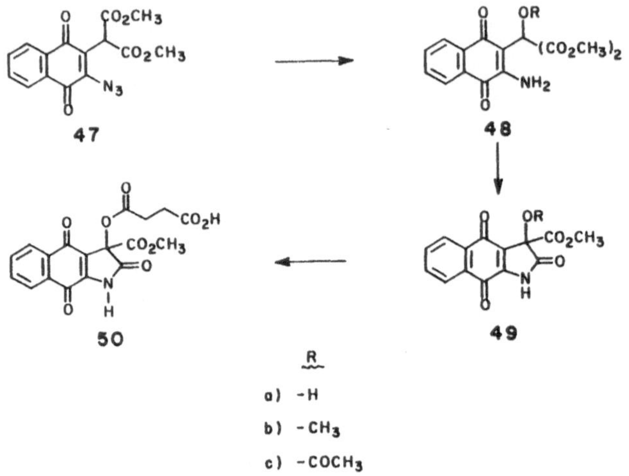

Scheme 9

sodium hydride, 52 was isolated in 68% yield (Scheme 10).
Surprisingly, when the related azidoquinone, 53, was
subjected to the above reaction conditions, a different
reaction pathway was followed.  Here a 64% yield of the
indolequinone 54 was realized.

The same type of unusual reaction was observed in
the last transformation described here, i.e., observed
when the azidoquinone 55 was treated with dimedone and NaH
and THF.  In this case, the tetracyclic indolequinone, 59,
was realized in 63% yield (Scheme 11).

A mechanistic rationale for the formation of 59
involves the conversion of 55 to 59 which is in analogy to
the aminoquinone formation outlined in Scheme 9.  Subsequent transformation of 56 to 59 is an example of "criss-
cross annulation".[29]

Finally, it is noted that the generalized reaction in
Scheme 9 provides a facile entry to a variety of quinones
which meet those structural requirements outlined for
bioreductive alkylation agents.  For example, succinylation
of 49a gives 50, a water soluble (Na salt) quinone which
undergoes an immediate conversion to the desuccinoylated
quinone upon treatment with sodium dithionite.  This

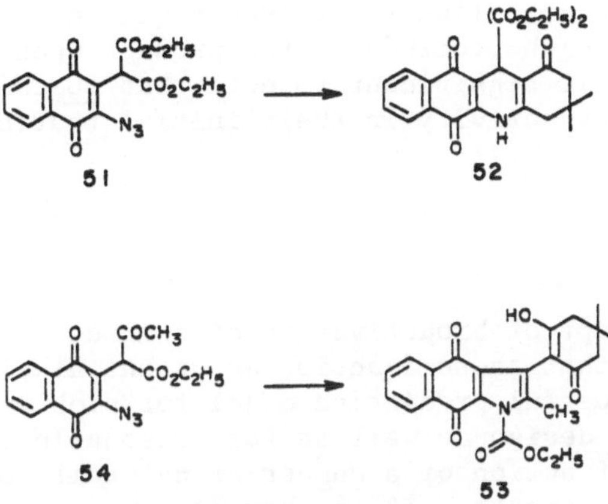

Scheme 10

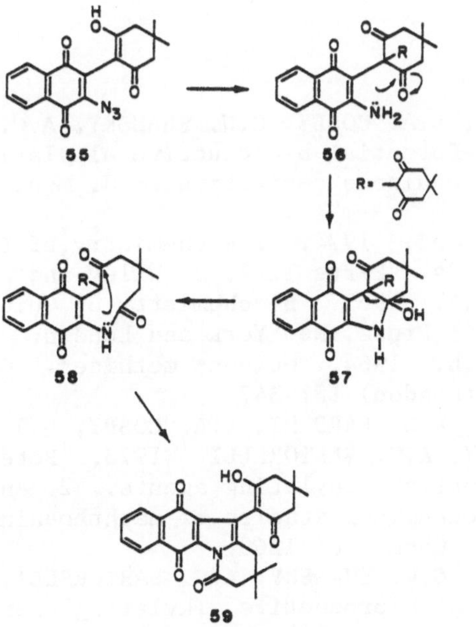

Scheme 11

transformation most likely involves a quinonemethide
intermediate which proceeds to the product upon proton
transfer.  It is significant to note that 50 and 49a both
show anticancer activity in their initial testing in the
P-388 system.


CONCLUSION

The concept of bioactivation of quinones to quinonemethides and that these function as potent alkylating agents
provides a powerful predictive model for problems of
rational drug design as well as for reasonable explanation
of the mode of action of a number of naturally occurring
antineoplastic agents.  It is, however, very speculative
and its implications are likely to be less far reaching
than this article implies.  Nevertheless, the idea does
provide the foundation for research problems in a variety
of areas.  These include synthetic organic chemistry, medi-
cinal chemistry, toxicology, biochemistry and pharmacology.
It is anticipated that many new active compounds will
result from the simple idea of quinone-quinonemethide
interconversions.


REFERENCES

1.  LIN, A.J., L.A. COSBY, C.W. SHANSKY, A.C. SARTORELLI.
        1972.  Potential bioreductive alkylating agents.
        1. Benzoquinone derivatives.  J. Med. Chem. 15:
        1247.
2.  PATAI, S., ed.  1974.  The Chemistry of the Quinonoid
        Compounds.  Parts 1, 2, J. Wiley and Sons, Inc.
    MORTON, R.A.  1965.  Biochemistry of Quinones.
        Academic Press, New York and London.
3.  TURNER, A.B.  1964.  Quinone methides.  Quart. Rev.
        Chem. (London) 18: 347.
4.  LIN, A.J., R.S. PARDINI, L.A. COSBY, B.J. LILLIS, C.W.
        SHANSKY, A.C. SARTORELLI.  1973.  Potential
        bioreductive alkylating agents.  2. Antitumor effects
        and biochemical studies of naphthoquinone derivatives.
        J. Med. Chem. 16: 1268.
    LIN, A.J., C.W. SHANSKY, A.C. SARTORELLI.  1974.
        Potential bioreductive alkylating agents.  3. Synthe-
        sis and antineoplastic activity of acetoxymethyl and
        corresponding ethyl carbamate derivatives of benzo-
        quinones.  J. Med. Chem. 17: 558.

LIN, A.J., B.J. LILLIS, A.C. SARTORELLI. 1975. Potential bioreductive alkylating agents. 5. Antineoplastic activity of quinoline-5,8-dione, naphthazarine and naphthoquinones. J. Med. Chem. 18: 917.

LIN, A.J., A.C. SARTORELLI. 1976. Potential bioreductive alkylating agents. 7. Antitumor effects of phenyl-substituted 2-chloromethyl-3-phenyl-1,4-naphthoquinones. J. Med. Chem. 19: 1336.

LIN, A.J., A.C. SARTORELLI. 1973. 2,3-Dimethyl-5,6-bis(methylene)-1,4-benzoquinone. The active intermediate of bioreductive alkylating agents. J. Org. Chem. 38: 813.

5. KENNEDY, K.A., B.A. TEICHER, S. ROCKWELL, A.C. SARTORELLI. 1980. The hypoxic tumor cells: a target for selective cancer chemotherapy. Biochem. Pharmacol. 29: 1

6. KENNEDY, K.A., S. ROCKWELL, A.C. SARTORELLI. 1980. Preferential activation of mitomycin C to cytotoxic metabolites by hypoxic tumor cells. Cancer Res. 40: 2356.

7. IYER, V.N., W. SZYBALSKI. 1964. Mitomycins and porfiromycin-mechanism of activation and cross-linking of deoxyribonucleic acid. Science 145: 55.

8. MOORE, H.W., R. CZERNIAK. 1981. Naturally occurring quinones as potential bioreductive alkylating agents. Med. Res. Rev. 1: 249.

9. KELLER, P.J., J.F. KOZLOWSKI, U. HORNEMANN. 1979. Formation of 1-ethylxanthyl-2,7-diaminomitosen and 1,10-diethylxanthyl-2,8-diaminodecarbamoylmitosene in aqueous solution upon reduction-reoxidation of mitomycin C in the presence of potassium ethylxanthate. J. Am. Chem. Soc. 101: 7121.

10. HASHIMOTO, Y., K. SHUDO, T. OKAMOTO. 1980. Acylation of 5-guanylic acid by reductively activated mitomycin C. Chem. Pharm. Bull. 28: 1961.

11. HASHIMOTO, Y., K. SHUDO, T. OKAMOTO. 1983. Modification of deoxyribonucleic acid with reductivity activated mitomycin C. Chem. Pharm. Bull. 31: 861.

12. PERRY, S. 1974. Summary and general discussion. Cancer, Chemother. Rep. Part 1 58: 117.

13. PIGRAM, W.J., W. FULLER, L.D. HAMILTON. 1972. Stereochemistry of intercalation: intercalation of daunomycin with DNA. Nature, New Biol. 235: 17.

14. SINHA, B.K., C.F. CHIGNELL. 1979. Binding mode of chemically activated semiquinone free radicals from quinone anticancer agents to DNA. Chem.-Biol. Interact. 28: 301.

15.  SMITH, T.H., A.N. FUJIWARA, D.W. HENRY, W.W. LEE.
     1976. Synthetic approaches to adriamycin. Degra-
     dation of daunorubicin of nonasymmetric tetracycline
     retone and refunctionalization of a ring to adria-
     mycine. J. Am. Chem. Soc. 98: 1969.
16.  KLEYER, D.L., T.H. KOCH. 1983. Spectroscopic obser-
     vation of the tautomer of 7-deoxydaunomycinone from
     elimination of daunosamine from daunomycin hydro-
     quinone. J. Am. Chem. Soc. 105: 2504.
17.  KLEYER, D.L., T.H. KOCH. 1983. Electrophilic trap-
     ping of the tautomer of 7-deoxydaunomycinone. A
     possible mechanism for covalent binding of dauno-
     mycin to DNA. J. Am. Chem. Soc. 105: 5154.
18.  RAMAKRISHNAN, K., J. FISHER. 1983. Nucleophilic
     trapping of 7,11-di-deoxyanthracyclinone quinone
     methide. J. Am. Chem. Soc. 105: 7187.
19.  KLEYER, D.L., G. GAUDIANO, T.H. KOCH. 1984. Spectro-
     scopic and kinetic evidence for the tautomer of
     7-deoxyaklavinone as an intermediate in the reduc-
     tive coupling of aclacinomycin A. J. Am. Chem.
     Soc. 106: 1105.
20.  KARLSSON, J.O., N.V. NGUYEN, L.D. FOLAND, H.W. MOORE.
     1985. (2-Alkynylethenyl)ketenes. A new benzoqui-
     none synthesis. J. Am. Chem. Soc. (in press).
21.  MARVELL, E.N. 1980. Thermal Electrocyclic Reactions.
     Academic Press, New York, pp. 124-190.
     JACKSON, D.A., M. REY, A.S. DREIDING. 1983. Prepara-
     tion of 2-vinylcyclobutanones and their conversion
     to cyclopentenones. Tetrahedron Lett., 4817.
     BERGE, J.M., M. REY, A.S. DREIDING. 1982. Addition
     of vinylketenes to enamines. A method for the
     preparation of 6,6-dialkylcyclohexa-2,4-dienones
     and 4,4-dialkylcyclobutenones. Helv. Chim. Acta
     65: 2230.
     DANHEISER, R.L., S.K. GEE, H. SARD. 1982. A [4+4]
     annulation approach to eight-membered carbocyclic
     compounds. J. Am. Chem. Soc. 104: 7670.
     HUSTON, R., M. REY, A.S. DREIDING. 1982. Vinylketenes
     as synthons for bicyclo[4.2.1] nonadienones. Helv.
     Chim. Acta 65: 451.
     DÖTZ, K.H., B. TRENKLE, U. SCHUBERT. 1981. Addition
     to ynamines to vinylketenes. Angew. Chem. 93: 296.
     DANHEISER, R.L., H. SARD. 1980. (Trimethylsilyl)
     vinylketene. A stable vinylketene and reactive
     enophile in [4_2]cycloadditions. J. Org. Chem.
     45: 4810.

22. DANHEISER, R.L., H. GEE. 1984. A regiocontrolled annulation approach to highly substituted aromatic compounds. J. Org. Chem. 49: 1674.

23. NGUYEN, N.V., K. CHOW, J.O. KARLSSON, H.W. MOORE. 1985. Chemistry of azidoquinones. Conversion of 3-azido-5-alkynyl-1,2-benzoquinones to cyanophenols via (2-alkynylethenyl)ketenes. J. Am. Chem. Soc. (submitted for publication).

24. MOORE, H.W. 1979. Zwittazido cleavage. Acc. Chem. Res. 12: 125.

25. NGUYEN, N.V., H.W. MOORE. 1984. In situ generation and reactions of hexynylcyanoketene. J. Chem. Soc., Chem. Commun., 1066.

26. HAMDAN, A.J., H.W. MOORE. 1985. A novel synthetic route to heterocyclic quinones. J. Org. Chem. (in press).

27. SMITH, L.I., E.W. KAISER. 1940. The reaction between quinones and metallic enolates. XI. Duroquinone and enolates of cyanoacetic ester and β-diketones. J. Am. Chem. Soc. 62: 138.
    JURD, L. 1978. Quinones and quinone methides III. A novel side-chain amination reaction of Z-(1-phenylethyl)-1,4-benzoquinones. Aust. J. Chem. 31: 347.

28. MOORE, H.W., H.R. SHELDEN. 1968. Rearrangements of azidoquinones. Reaction of thymoquinone and 2,5-dimethyl-1,4-benzoquinone with sodium azide in trichloroacetic acid. J. Org. Chem. 33: 4019.

29. ODA, K., T. OHMUMA, Y. BAN. 1984. A facile removal of the arenesulfonyl group by electrochemical reduction of sulfonamides in a new cooperative system of anthracene and ascorbic acid: the control of criss-cross annulation. J. Org. Chem. 49: 953.

Chapter Eleven

BIOCHEMISTRY OF PLANT COUMARINS

STEWART A. BROWN

Department of Chemistry
Trent University
Peterborough, Ontario
Canada K9J 7B8

INTRODUCTION

Among natural products deriving at least in part from
the shikimate pathway, the coumarins are certainly among
the most numerous and are probably the most varied in
structure. With over 800 known in the early 1980s[1] and
numerous new ones being identified each year, mostly from
plants, the total is now closing in on a thousand. An
indication of the structural variety is given by the
compounds depicted in Figure 1, which is by no means
exhaustive. The central feature of coumarins is the
2H-benzopyran-2-one nucleus which, unsubstituted, is
the compound known simply as coumarin (1, Fig. 1).

In the more complex structures, such as the furano-
and pyranocoumarins, phenylcoumarins, and coumarin ethers,
there are numerous instances in which other derivations,
primarily from acetate, are known or presumed to be
involved. In the present context the elaboration of
these non-shikimate-derived residues is of secondary

287

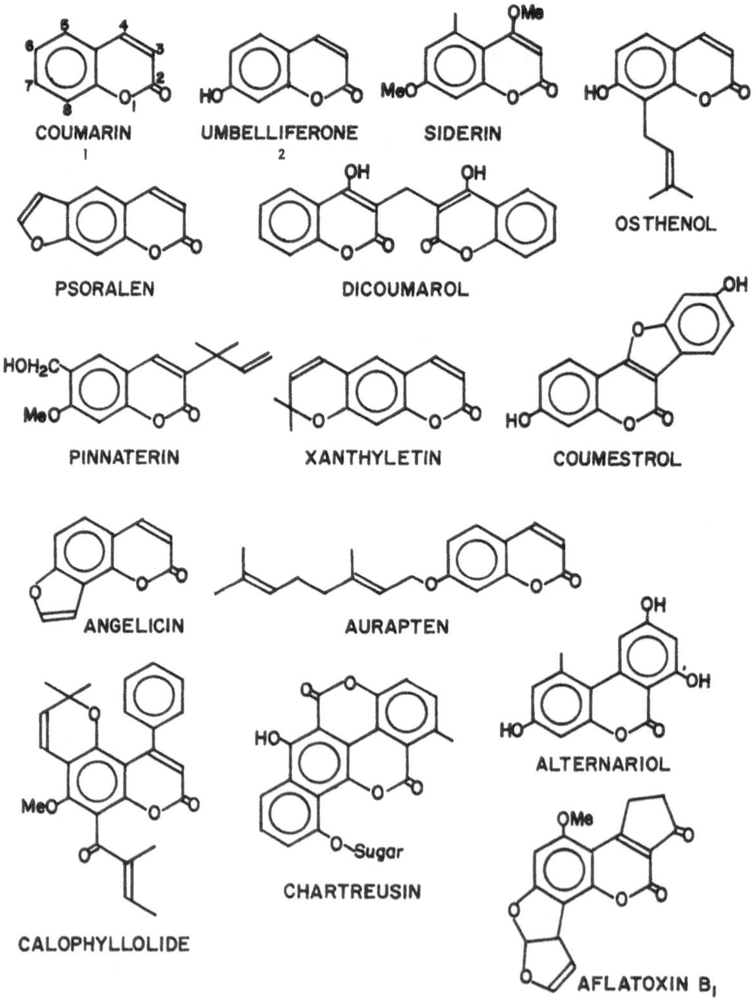

Fig. 1.   Examples of naturally occurring coumarins.

interest, and will not be emphasized in this discussion.
However, it is worthy of note that some coumarins, such as
the fungal aflatoxins,[2] alternariol[3] and 5-methylcoumarins[4]
(the last category also occurring in plants), have no components of shikimate origin, being entirely acetogenins.[1] The
phenylcoumarins such as coumestrol, of plant origin, are
derived in part from acetate and in part from shikimate.[1]
Coumestrol and other 3-phenylcoumarins are isoflavonoids
from a biosynthetic standpoint, and will not be covered in
this review.


BIOSYNTHESIS

## Coumarin and Umbelliferone

    Direct evidence for the formation of coumarin from
shikimate came from the radiotracer experiments of Kosuge
and Conn[5] in 1959, and they and others have also demonstrated cinnamic acid and a number of its ring-oxygenated
derivatives, well known to originate from shikimate, to be
precursors of various coumarins. The committed step in the
biosynthesis of coumarins is hydroxylation of the benzene
ring ortho to the side-chain of a cinnamic acid. In a few
species the substrate is cinnamic acid itself, leading to
the formation of coumarin, but for most coumarins, which
bear 7-oxygenation, it is known or assumed that p-coumaric
(4'-hydroxycinnamic) acid is the substrate, with discrete
enzymes involved in the two cases. The product in the
latter instance is 7-hydroxycoumarin (umbelliferone) (2,
Fig. 1), and thus it is this compound rather than coumarin
which, from the biosynthetic standpoint, is the parent
compound of the vast majority of coumarins.

    Once ortho hydroxylation has been effected the stage
is set for lactone ring formation, via a light-mediated
trans-cis isomerization of the side-chain double bond,[6]
after which lactonization occurs spontaneously. However,
there are instances in which this lactone ring formation
is prevented by glucosylation of the free hydroxyl group,
yielding a trans-o-glucosyloxycinnamic acid (3, Scheme 1).
This trans isomer can then be isomerized to the cis (4,
Scheme 1), so that the intact cell contains a coumarinyl
glucoside, which has been termed "bound coumarin". The
existence of this has long been recognized in plants such
as sweet clover (Melilotus spp.),[7] but it was thought that

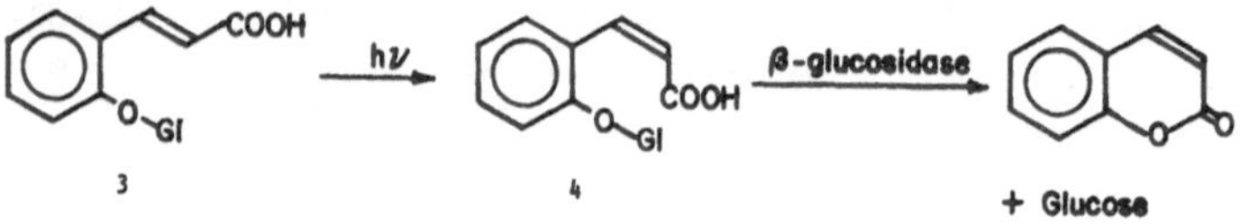

Scheme 1

these cells also contained the free lactone, until 1961 when Haskins and Gorz[8] showed that leaves of M. alba autoclaved before work-up contained only traces of free coumarin. Nevertheless, the distinct odor of coumarin from cut sweet clover shows that the free form is contained in disrupted cells.

The explanation for this phenomenon has been provided in recent years by the work of Conn and his collaborators. Kosuge and Conn[9] had earlier identified in Melilotus a β-glucosidase specific for the cis glucoside which, when the cells are disrupted, is cleaved releasing the cis-ortho-hydroxyacid which immediately lactonizes (Scheme 1). Although compartmentation of enzyme and substrate was postulated at the time to explain the lack of hydrolysis in intact cells, evidence about the nature of this compartmentation was not obtained until 1981. At that time Boudet's group[10] in France devised a procedure for the isolation of vacuoles from plant cells which involved lysis of the plasmalemma during centrifugation of protoplasts through a diethylaminoethyl dextran layer. Conn's group, in collaboration with Boudet's laboratory,[11] then established by the use of this technique that the compartmentation was at the subcellular, not the tissue, level. There was no β-glucosidase activity in these vacuoles, which contained the entire substrate complement - both the cis and trans forms of o-coumaric acid. The site of the glucosidase is less certain, but evidence from these studies suggested that it was either in the intercellular spaces of the intact leaf or loosely attached to the cell wall. In either case the problem of inaccessibility of the substrate to the enzyme until the cells are broken has now been resolved.

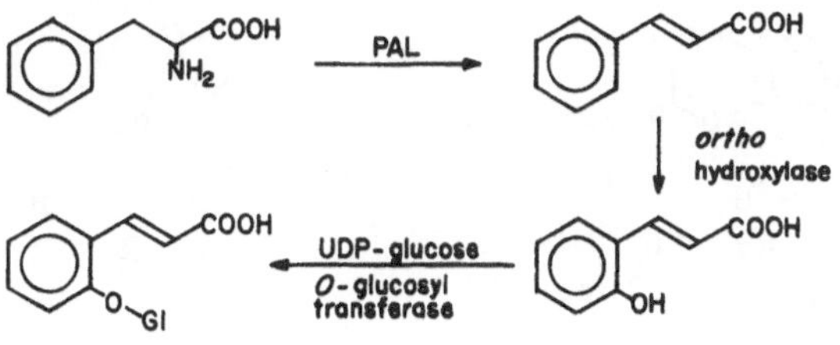

Scheme 2

Parallel studies in Conn's laboratory have also dealt
with the problem of intracellular localization of two other
enzymes involved in coumarin formation in M. alba,
phenylalanine ammonia lyase (PAL) and o-coumaric acid-O-
glucosyltransferase, which mediates the glucosylation of
the ortho hydroxyl group (Scheme 2).[12]  As Gestetner and
Conn[13] had earlier reported localization in the chloroplast
of the ortho hydroxylase catalysing the committed step of
coumarin formation − ortho hydroxylation of cinnamic acid −
it was logical to examine this organelle for the two enzymes
mediating the immediately preceding and following reactions.
Rather surprisingly, no evidence of either enzyme in the
isolated intact chloroplasts was found.  The question then
inevitably arose as to how cinnamic acid and o-coumaric
acid pass in and out of the chloroplast, respectively,
through the permeability barrier of the chloroplast
membranes.  However, Conn and his associates[14] have recently
reexamined the evidence for cinnamic acid ortho hydroxylase
as a chloroplastic enzyme and have been unable to confirm
the earlier findings.  It now appears that experimental
difficulties in purifying the o-coumaric acid from this
reaction may have led to an erroneous inference about the
presence of the enzyme in the chloroplast.  If so, there is
a distinct possibility that none of the reactions of the
pathway from phenylalanine to the bound coumarin in this
species is, in fact, chloroplastic.  However, we should
note that a chloroplastic ortho hydroxylation of trans-
cinnamic acid has been reported in Petunia hydrida.[15]

5  $R_6 = R_7 = H$
6  $R_6 = Me$, $R_7 = H$
7  $R_6 = Me$, $R_7 = glucosyl$

8  $R = H$
9  $R = Me$

10  $R_3 = R_4 = OH$, $R_5 = H$
11  $R_3 = OMe$, $R_4 = OH$, $R_5 = H$
12  $R_3 = R_5 = OMe$, $R_4 = OH$

## Introduction of Further Ring Oxygenation

A number of commonly occurring coumarins and a good many less common ones bear additional oxygenation on the coumarin nucleus. Reference has already been made to umbelliferone, where the 7-hydroxyl group originates from the para hydroxyl group of p-coumaric acid. Other common coumarins have two oxygen functions; aesculetin (5), scopoletin (6), and daphnetin (8), are well known. Early studies[1] provided good evidence that ferulic acid (11) was a precursor of scopoletin in tobacco, and an enzyme capable of ortho-hydroxylating ferulic acid to yield scopoletin has been described from another source.[16] In the light of this and the origin of umbelliferone it was generally assumed that the oxygenation pattern of coumarins is established at the cinnamic acid stage before the coumarin nucleus is elaborated. This assumption broke down in the case of furanocoumarins when a number of studies with tracers in different laboratories[1] strongly indicated that additional ring oxygenation is introduced on the coumarin nucleus to form such compounds as 5- and 8-methoxypsoralen.

Recent tracer studies suggest the same pattern for other simple coumarins as well. Aesculetin was reported[17] in the early 1970s not to be formed from the analogously oxygenated caffeic acid (10) in Cichorium intybus (chicory), in which it occurs primarily as the 7-glucosyloxy derivative,

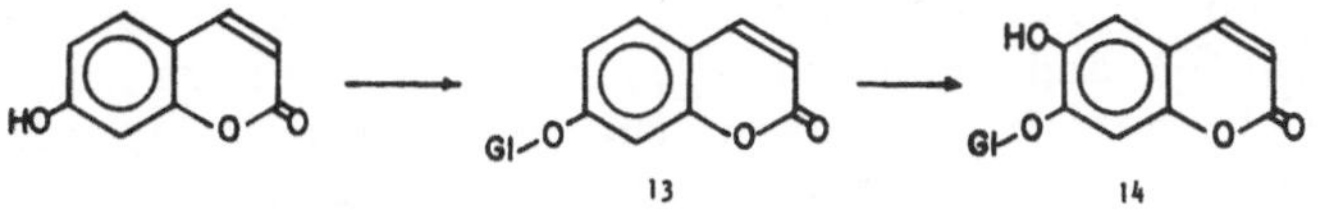

Scheme 3

cichoriin (14, Scheme 3). This finding has now been confirmed and [14]C-labelled umbelliferone has been shown, by comparison, to be a very efficient precursor of aesculetin.[18] These studies also showed incorporation of umbelliferone into its O-glucoside, skimmin, (13, Scheme 3) almost without dilution of the isotope, indicating a negligible endogenous pool. As skimmin has not been reported to occur in C. intybus, but free umbelliferone has,[1] umbelliferone may be glucosylated in a slow, rate-limiting reaction, and the resulting skimmin then metabolized in a more rapid 6-hydroxylation to yield cichoriin (Scheme 3), so that skimmin would not normally accumulate. In this hypothesis free aesculetin would not be an intermediate, but existing evidence by no means rules out its participation in cichoriin formation, and studies on cell-free systems would appear necessary to resolve this question.

Research in Ibrahim's laboratory[19,20] has revealed, in cell cultures of tobacco, a UDP-glucose o-dihydroxycoumarin 7-O-glucosyltransferase mediating glucosylation of aesculetin (5), as well as of daphnetin (8), with strict position specificity. The biosynthetic significance of this finding is unclear, in that neither cichoriin nor daphnin, the corresponding glucosides, have apparently been reported to occur in any species of Nicotiana,[1] and some degree of activity against scopoletin and umbelliferone was also recorded. In a later investigation by French workers,[21] three o-diphenol-O-methyltransferases were identified in tobacco leaves, and all three, in contrast to the cell culture enzyme, methylated both the 6- and 7-hydroxyls of aesculetin, with the apparent involvement of only one active site.

A series of recent papers by Japanese investigators
has reported on factors affecting the formation of
scopoletin (6) and scopolin (7), its O-glucoside, in
tobacco cell cultures.  Among the nutritional factors
studied, only sucrose among carbon sources tested
increased the rate of biosynthesis, although phosphate
and inorganic nitrogen were also effective.[22]  An amino
acid mixture including phenylalanine, as well as malt
extract, stimulated the production of these coumarins.[22]
Several growth factors:  2,4-dichlorophenoxyacetic acid
(2,4-D), indoleacetic acid (IAA) and benzylaminopurine
(BAP) had stimulatory effects during the exponential
growth phase, whereas naphthaleneacetic acid (NAA) and
kinetin, in contrast, caused scopoletin accumulation
during the non-growth phase.[23]  A synergistic effect was
noted with NAA and BAP.  Glucosylation of scopoletin to
scopolin was uniquely stimulated by 2,4-D, which enhances
the activity of UDP-glucose:scopoletin glucosyltransferase
(to which it binds in vivo).  The enhancement resulted
from the activation of a pre-existing enzyme, and not to
synthesis de novo.[24]  The activity of PAL was enhanced by
both BAP and kinetin, but in the latter instance the fact
that the increase was inhibited by actinomycin D and
cycloheximide showed the effect to be on synthesis of the
enzyme.[24,25]  In view of the serious difficulties often
encountered in studying the biosynthesis of coumarins in
organized plants,[26] this research is important and timely
and it is to be hoped that extensions of it to other
coumarins produced in cell cultures will be forthcoming.

In the case of the 7,8-dioxygenated coumarins, there
has been some tentative evidence that here, too, the
additional oxygenation is introduced on umbelliferone,
rather than on p-coumaric acid.  Very recently (S.A.
Brown, unpublished work) I have examined the origin of a
daphnetin (8) glucoside in Daphne mezereum and initial
indications are that the $^{14}$C of umbelliferone is well
incorporated into this glucoside.  If confirmed, this
finding, together with two earlier reports of single
experiments on the origin of hydrangetin (9),[27,28] would
be in accord with a precursor role for umbelliferone in
this context as well.  Since umbelliferone occurs naturally in the two species in question[1] it could then be
regarded as a natural intermediate.

The origin of the other naturally occurring dioxygena-
tion pattern, 5,7, has not been studied in simple coumarins,
but in furanocoumarins, e.g. bergapten (20), oxygenation of
the 5 position follows the elaboration of the furanocoumarin
skeleton.[1]

Coumarins trioxygenated on the benzene ring occur
naturally in three patterns:  5,6,7-, 5,7,8- and 6,7,8-.
The biosynthetic pathways to these are clearly more complex,
not only because of the additional hydroxylation step, but
also because alkyl ether formation must be taken into account
in most cases.  In furanocoumarins either C-6 or C-8 always
bears a side-chain, so that there was no information on the
elaboration of the 6,7,8 pattern from earlier work on this
category of coumarins.  Recently, in collaboration with
D.E.A. Rivett, we have obtained data on the formation of a
simple coumarin with this pattern, isofraxidin (15).[29]  As
its prenyl ether, puberulin (16), this coumarin occurs in
the aerial parts of the South African rutaceous species,
Agathosma puberula.  Again, feeding experiments with
labelled precursors have pointed to umbelliferone as a
precursor.  Scopoletin was used with comparable efficiency,
and the sequence of Scheme 4 was therefore proposed.  This
sequence includes aesculetin, and experimental evidence for
the participation of this intermediate has since been
obtained (S.A. Brown, D.E.A. Rivett and H.J. Thompson,
unpublished work).  In addition to p-coumaric acid we also
tested caffeic (10), ferulic (11), and sinapic (12) acids,
the last-named having the complete oxygenation pattern of
isofraxidin except for the lactone ring oxygen.  All three
were generally poorer precursors of puberulin than was
p-coumaric acid and these results thus contraindicate the
establishment of this coumarin's oxygenation pattern before
lactonization.

15   R = H
16   R = $CH_2-CH = OMe_2$

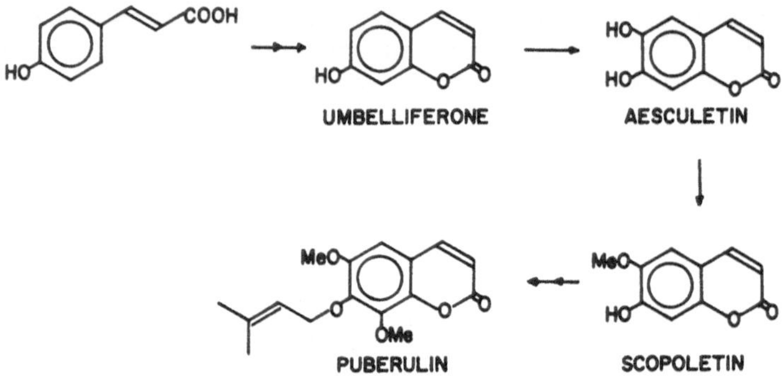

UMBELLIFERONE     AESCULETIN

PUBERULIN     SCOPOLETIN

Scheme 4

The poor utilization of ferulic acid raises a point of interest. As mentioned above, there has long been evidence that scopoletin in tobacco arises from ferulic acid. However, this evidently cannot be the case in Agathosma, since scopoletin is well utilized to form puberulin, but ferulic acid is not. It may be that scopoletin, like the other di- and trioxygenated coumarins thus far studied, originates via umbelliferone in most species, and that the picture in tobacco is exceptional.

The route from scopoletin to puberulin is being investigated, and preliminary experiments (S.A. Brown et al., unpublished work) have resulted in good utilization of isofraxidin in puberulin synthesis. Support for these data in confirmatory studies would point to O-prenylation as the final step in the pathway.

Furanocoumarins

Information on the origin of angular furanocoumarins has not been plentiful, but much more is known about the more widely distributed linear isomers. In brief, the committed step in the pathway is the condensation of mevalonate-derived dimethylallyl pyrophosphate (DMAPP) at position 6 of umbelliferone to yield 7-dimethylsuberosin (Scheme 5). A transferase mediating this step was identified and characterized in Ruta graveolens[30,31] and evidence was obtained that it is a chloroplast enzyme.[31] Cyclization

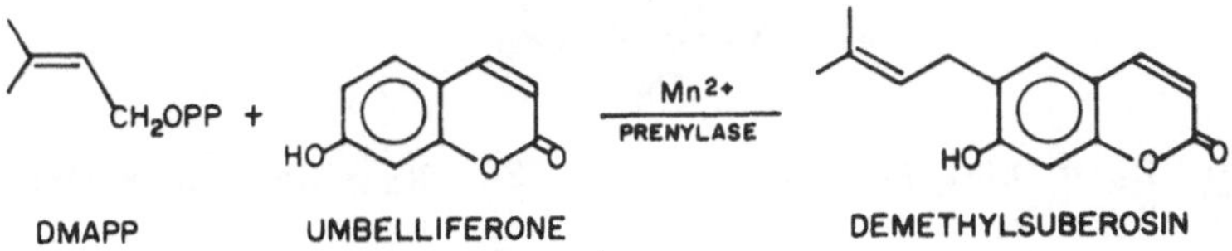

Scheme 5

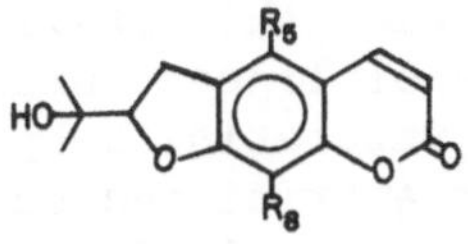

<u>17</u>  R$_5$ = R$_8$ = H
<u>18</u>  R$_5$ = H, R$_8$ = OMe
<u>19</u>  R$_5$ = OMe, R$_8$ = H

of the product of this reaction can then produce marmesin (17), and loss of the hydroxyisopropyl side-chain and desaturation of the ring would then produce the furan structure. Nothing is currently known about the mechanism of this transformation, but recent evidence suggests a two-step process. Matern and Wendorff[32] have reported that microsomes from fungal elicitor-induced parsley cells can produce psoralen from [2-$^{14}$C]marmesin, and can, under certain induction conditions, accumulate an intermediate in this conversion. This approach therefore shows promise of giving us a better understanding of this key step in furanocoumarin elaboration. There is ample evidence that psoralen, the first furanocoumarin formed, can then be further anabolized to oxygenated furanocoumarins.[1] Work by Caporale, Dall'Acqua and their associates also established an alternative pathway in R. graveolens involving rutaretin (18), a compound occurring in this species, en route to xanthotoxin. This sequence, however, seems not to be of general occurrence.[1]

| | | | |
|---|---|---|---|
| 20 | $R_5$ = OMe, $R_8$ = H | 24 | $R_5$ = OH, $R_8$ = OMe |
| 21 | $R_5$ = H, $R_8$ = OMe | 25 | $R_5$ = OMe, $R_8$ = OH |
| 22 | $R_5$ = $R_8$ = OMe | 26 | $R_5$ = OH, $R_8$ = H |
| 23 | $R_5$ = $R_8$ = OH | 27 | $R_5$ = H, $R_8$ = OH |

In more recent studies this Italian group has examined further certain aspects of linear furanocoumarin biosynthesis. The isomer of rutaretin, 5-hydroxymarmesin (19), was shown by the use of a tritiated substrate to serve as a precursor of bergapten (20) in R. graveolens, and also in Ficus carica.[33] 5-Hydroxymarmesin, although a recognized natural product, has apparently not been reported from these two species, and for this reason its status as a natural intermediate, in contrast to rutaretin, remains uncertain.

These workers have also obtained further evidence about the origin of the dioxygenated psoralen, isopimpinellin (22). Something has been known about the origin of this coumarin since the mid-1970s, when work on R. graveolens cell cultures with tracers and enzyme preparations indicated the participation of bergapten (20) and xanthotoxin (21).[34] The sequences shown in Scheme 6 exist in this species, with the route via xanthotoxin predominating. A possible intermediate which we did not test, owing to stability problems, was 5,8-dihydroxypsoralen (23), a hydroquinone that oxidizes spontaneously under even mildly alkaline conditions. However, after stabilizing this compound with a trace of sulfur dioxide, Innocenti et al.[35] found the tritiated substrate to be incorporated into isopimpinellin in R. graveolens shoots with isotope dilutions under 20. These values are less than those previously recorded for bergapten and xanthotoxin, and it now seems not unlikely that the route via the hydroquinone is the major one. If so, the intermediate may well have a transient existence in enzyme-bound form. In any event, attempts to trap it after administration of labelled marmesin (17) failed, as had earlier efforts with [14C]umbelliferone (2) to trap the proposed hydroxymethylpsoralens.[34]

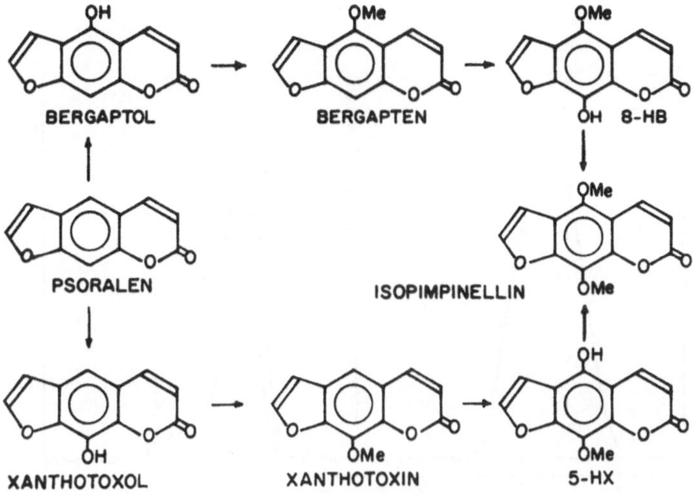

Scheme 6

Investigations with O-methyltransferases have shed some further light on the biosynthesis of isopimpinellin, as well as bergapten and xanthotoxin. The previous studies with cell-free systems on the origin of isopimpinellin[34] had revealed the presence in R. graveolens of an enzyme system able to methylate the hydroxyl groups of both 5-hydroxyxanthotoxin (24) and 8-hydroxybergapten (25) to yield isopimpinellin. Moreover, evidence from mixed substrate experiments indicated that more than one enzyme was participating. In a subsequent investigation[36] these findings were confirmed both for the above substrates and for bergaptol (26) and xanthotoxol (27), which were less effective substrates. Conclusive evidence for the presence of at least two discrete transferases came from studies of the differing effects of divalent cations on the O-methylations at the 5 and 8 positions, different pH optima, distinct differences in stability, and marked changes in ratios during partial purification.

The nature of these two O-methyltransferases has been further elucidated by affinity chromatography. In the initial investigations separations were achieved on columns with an S-adenosyl-L-homocysteine (SAH) ligand coupled to AH-sepharose, and a protein fraction active in the

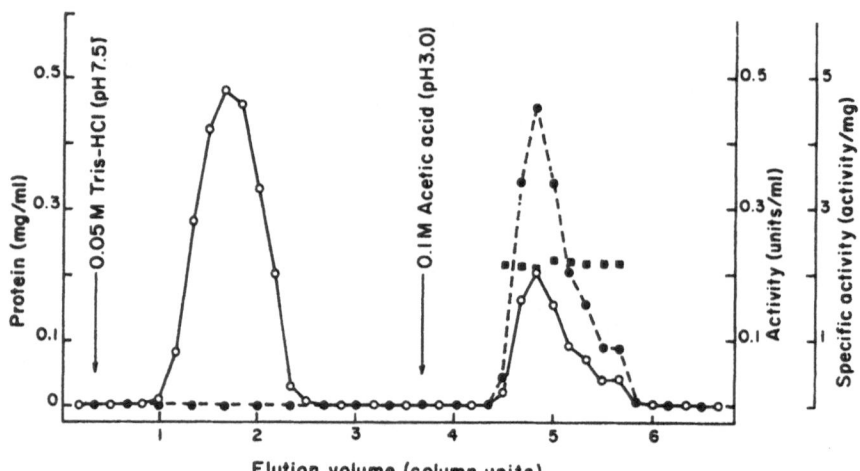

Fig. 2.  Affinity chromatography of O-methyltransferase of
Ruta graveolens on a column of AH-Sepharose 4B coupled with
S-adenosyl-L-homocysteine.  o———o, Protein;  ●———●,
O-methyltransferase activity;  ■ , specific activity.

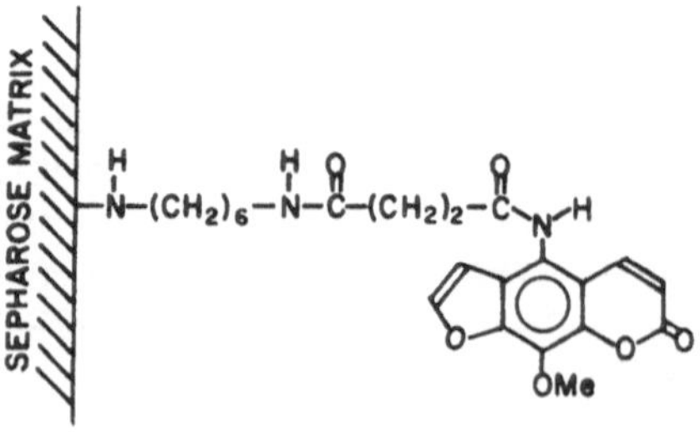

Fig. 3.  Affinity ligand used to separate 5- and 8-hydroxy-
psoralen O-methyltransferases.

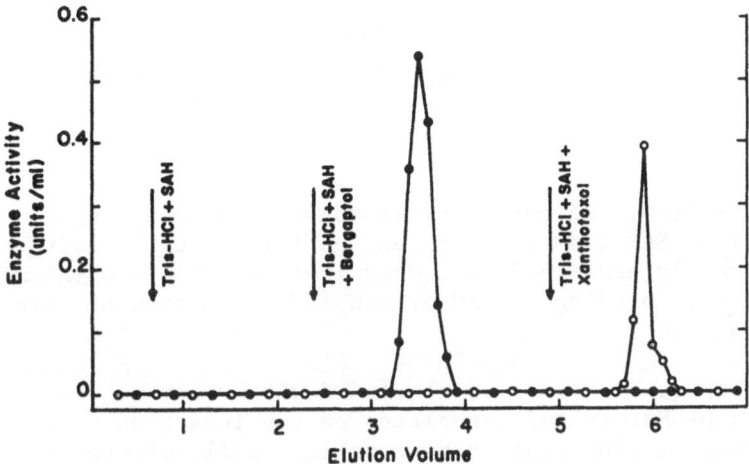

Fig. 4. Separation of the furanocoumarin 5- and 8-O-methyl-
transferases of Ruta graveolens by successive bioelutions
from an AH-Sepharose 4B column linked to 5-(3-carboxypro-
panamido) xanthotoxin. The second protein fraction of
Figure 2 was applied to this column and the irrigant buffer
was changed at the column volume (in column units) shown.
Enzyme activity was measured using as substrate the coumarin
(bergaptol or xanthotoxol) added to the buffer to desorb each
enzyme.

O-methylation of both bergapten and xanthotoxin was
retarded by these columns (Fig. 2).[37]  In subsequent work
5-amino-xanthotoxin was employed as the ligand, linked
through its amino group and a spacer arm to the sepharose
matrix (Fig. 3).  Both transferases were retarded by this
ligand, but they could be desorbed specifically, the 5-O-
methylating enzyme being released by addition of bergaptol
(26) to the irrigant buffer, and the 8-O-methylating enzyme
by xanthotoxol (27).  In each case an electrophoretically
homogeneous enzyme devoid of activity against the other
isomer was obtained (Fig. 4).[38]  Evidence was also presented
for the absence of more than two transferases.

Mechanistic information of considerable interest was
also obtained from these studies with affinity chromatography.[39]  The affinity columns bearing the 5-aminoxanthotoxin
ligand failed to retard both transferases in the absence of
either SAH or S-adenosyl-L-methionine (SAM), the methyl donor

28

of the system, and both were immediately desorbed upon
removal of SAM from the irrigant buffer. These findings are
most readily explained by a compulsory-ordered mechanism in
which prior binding of SAM or SAH induces a binding site for
the coumarin.

This section will be concluded with reference to
studies of Matern and associates on the induction of the
formation of coumarins in Petroselinum cell cultures by
elicitors from fungal pathogens. Elicitors from either
Phytophthora megasperma or Alternaria carthami resulted in
the production by the cultures of linear furanocoumarins, as
well as the dihydropyranocoumarin derivative, graveolone
(28).[40] As furanocoumarins had been observed to inhibit the
germination of fungal pathogen spores, furanocoumarins therefore act as phytoalexins. The elicitors raise for 48-54 h
the levels of several enzymes involved in the formation of
coumarins: PAL, 4'-coumarate:CoA ligase, and dimethylallyl
pyrophosphate:umbelliferone dimethylallyltransferase,[41] this
last enzyme mediating the attachment of the prenyl side-
chain at C-6 of umbelliferone en route to psoralen.


BIOLOGICAL ACTION

This section will review some of the more notable
advances of the past few years in elucidating the biological
action of coumarins in two areas: photobinding to DNA and
plant growth regulation.

Photobinding to DNA

Much attention continues to be given to a phenomenon
whose nature was established fifteen years ago,[42,43] the
cross-linking of the two strands of the DNA double helix by
linear furanocoumarins upon irradiation with long-wave UV.
$C_4$ cyclodiadducts are formed and the photobinding reaction
has proved to be a valuable probe in studies of DNA structure and function.

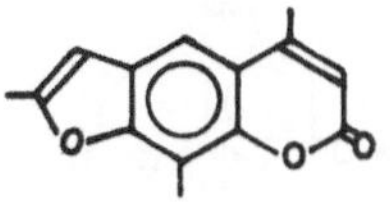

29

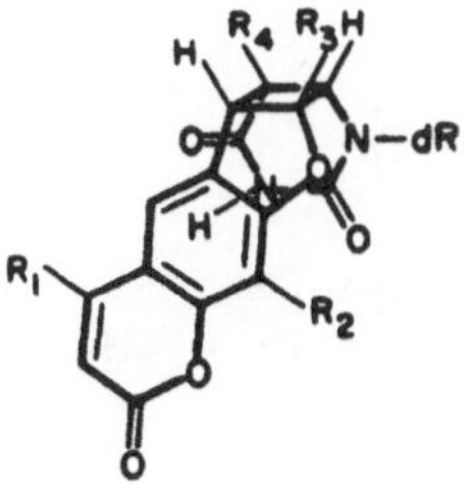

30 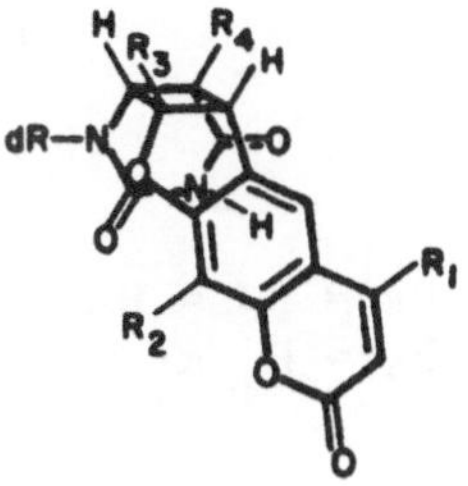                    31

Reprinted with permission from Biochemistry 21: 867. Copy-
right 1982, American Chemical Society.

Several recent publications have dealt with the structure of these radiation-induced adducts with DNA. Rapoport
and collaborators have characterized four mononucleoside
adducts from the photoreaction of DNA with xanthotoxin or
4,5',8-trimethylpsoralen (29), with special attention to
stereochemical considerations.[44] They identified two major
diastereomeric adducts of thymidine with each psoralen
involving the furan ring double bond (30,31). Xanthotoxin
can also yield substantial amounts of monoadducts by reaction of the double bond of the pyrone ring, perhaps because
of a more favorable steric interaction due to the absence
of a 4-methyl substituent. All the adducts have cis-syn
stereochemistry. These same workers have also isolated and
characterized thymidine-psoralen-thymidine photodiadducts
from DNA and four different psoralens.[45] In each case a
single pair of diastereomeric diadducts accounted for over
90% of the total recovered. In another publication from
this laboratory a major species of the photoreaction between
DNA and xanthotoxin was isolated and its crystal structure
determined.[46] Among other things this work indicated that
photoreaction occurs preferentially at both T-A and A-T

32

33  R₅ = R₆ = H          35  R₅ = OMe, R₆ = H
34  R₅ = H, R₆ = OMe     36  R₅ = R₆ = OMe

sequences in natural DNA, and introduces a substantial kink
in the DNA structure.  In an independent study Shim and
Kim[47] have also recovered a pair of diastereomers, monoadducts involving reaction of the xanthotoxin furan ring,
and three pyrone ring monoadducts, after irradiation in
the solid film state.  The 4',5' monoadducts were characterized stereochemically.

As might be expected, such adduct formation is mutagenic, and there is evidence that it can be carcinogenic.
Simple coumarins do not form diadducts, and it has been
generally considered that angular furanocoumarins do not
either, owing to an incorrect geometry precluding intercalation in the helix.  Both angelicin and citropten (32),
which form photomonoadducts involving the pyrone ring, are
less mutagenic in yeast than psoralen and xanthotoxin.[48]
However there has been recent evidence that, under certain
conditions, angelicin (33) can form diadducts to a limited
extent as well.[49,50]  Angelicin is mutagenic in strains of
Escherichia coli deficient in repair functions, although to
a lesser degree than the linear furanocoumarins.[51]  Other
studies on bacteriophage-λ have detected mutagenesis after
irradiation in the presence of angelicin, without any
evidence of cross-linking.[52]

Recent papers dealing with the mechanisms of photo-
addition reactions have concentrated on the question of
singlet oxygen participation.  Detailed consideration of

this question is outside the scope of the present chapter, but the relevant references are appended.[53-60]

Effects of UV irradiation in the presence of furano-coumarins have also been studied with proteins, especially enzymes, and it is now clear that photobinding effects are not confined to nucleic acids. Both linear and angular furanocoumarins bind covalently to proteins in reactions highly dependent on structure.[61] Enzymes are significantly inactivated, as shown in Table 1. The linear isomers appear generally to be most inhibitory, but there is great variation with the structure of both the coumarin and the protein. There are several points of attack on the protein molecule, and in glutamate dehydrogenase and RNase, exposure to UV in the presence of psoralen extensively modified histidine, methionine, tryptophan, phenylalanine and tyrosine residues by photo-oxidation.[62]

## Effects on Plant Growth

Various coumarins have long been known to exert stimu-latory or inhibitory effects on plant growth, and have come to be characterized as plant growth regulators.[67] Research in the past few years has provided additional evidence for such a function of coumarin and scopoletin, which had both been extensively studied previously in this context, and regulatory properties have now been described for several other coumarins.[68] Growth of chinese cabbage seedlings was strongly inhibited by umbelliferone (as well as scopoletin). Although bergapten (20) and sphondin (34) inhibited growth of hypocotyl cuttings of cucumber seedlings, both coumarins accelerated root formation by these cuttings. Vaginidiol (37), isopimpinellin (22), pimpinellin (36), and isoberg-apten (35) also affected growth in these systems. It is becoming apparent that growth regulation by coumarins is a more general phenomenon among this class of compound than

37

Table 1.  Photoinactivation of enzymes by plant psoralens.

| Enzyme | Source | Coumarin* | Reference |
|---|---|---|---|
| Phosphofructokinase | Epidermis | Xanthotoxin | 63 |
| Glutamate dehydrogenase | Bovine liver | Psoralen | 64 |
| 6-Phosphogluconate dehydrogenase | Yeast | Psoralen | 64 |
| Lysozyme | – | Psoralen | 64 |
| DNA polymerase I | E. coli | Xanthotoxin | 65 |
| cAMP phosphodiesterase | Peritoneal macrophages, mouse L1210 cells | Xanthotoxin | 66 |

* The underlined compounds were the most effective among the plant coumarins tested, although in some cases synthetic psoralens caused greater inhibition.

had been realized, and the implications of this need to be more extensively explored.

Evidence continues to accumulate that coumarins interact with other growth regulators. Coumarin ($10^{-3}$M) enhanced IAA oxidase activity in maize seedlings for 12 h after treatment,[69] and, at 3.4 x $10^{-5}$M, in cuttings of Impatiens balsamina.[70] In the former species it strongly inhibited IAA synthetase;[69] thus the effect of coumarin at this concentration is to decrease IAA levels and inhibit seedling growth. However, coumarin at 3.4 x $10^{-4}$M increased endogenous IAA in barley sprouts,[71] but it also somewhat decreased endogenous gibberellin, an effect which persisted into the later growth period. Also in barley amylase synthesis induced by gibberellic acid in seeds was inhibited by coumarin, the effects being gradually overcome by increased gibberellic acid concentrations.[72] In cucumber seedlings coumarin at 1.4 x $10^{-3}$M sharply decreased auxin levels.[73] In seedlings of Amaranthus caudatus the inhibition of light-induced beta-cyanin synthesis by abscisic acid was antagonized by several phenols including coumarin (maximal effect at $10^{-6}$M).[74]

The rates of incorporation by isolated tobacco protoplasts of thymidine, uridine and leucine were reversibly lowered by coumarin, which consistently inhibited cell division.[75]

As measured by incorporation of [U-$^{14}$C]glucose, coumarin was among several compounds which specifically inhibited cellulose formation in cotton fibers developing in unfertilized cotton ovules cultured in vitro.[76] Colvin and Witter[77] have investigated its inhibition of cellulose biosynthesis in Acetobacter xylinum and concluded that it acts as a general cell poison, not specifically interfering with cellulose formation in this species. The general inhibition precedes uridine diphosphoglucose formation. Cell wall regeneration in protoplasts of Marchantia polymorpha is inhibited by coumarin, which apparently inhibits cellulose formation.[78] In contrast, at the relatively high concentration of 1.4 x $10^{-2}$M, coumarin stimulated synthesis of both pectin and α-cellulose, while inhibiting the formation of glycoprotein, hemicellulose, and lignin.[79]

Recent studies have revealed further inhibitory effects of aesculetin (5) and scopoletin (6). In tobacco cultures at 4 x $10^{-4}$M, especially as their glucosides, both coumarins

were inhibitory to 6-phosphogluconate dehydrogenase,
implying a possible function in regulating the pentose
phosphate pathway, which yields the erythrose 4-phosphate
of the shikimate pathway.[80]  IAA oxidase activities of
preparations from Hevea leaves were stimulated by scopoletin,[81] and laccases from two mushrooms, Agaricus bisporus
and Pleurotus ostreatus, were inhibited by aesculetin.[82]

    The questions of inhibition of plant growth and
specific effects at the molecular level were considered at
somewhat greater length in a recent treatise.[67]  One point
which deserves reemphasis is the importance of specifying
the dose when these effects are being assessed.


CONCLUSION

    This chapter has dealt with recent developments in the
biosynthesis of plant coumarins and in selected areas of
their biological action.  In the former field there has been
further exploration of the subcellular localization of intermediates and enzymes of the biosynthetic pathway to coumarin.
Recent findings relating to the elaboration of oxygenation
patterns in simple coumarins are increasingly pointing to
umbelliferone as the precursor of polyoxygenated coumarins
in almost all instances, with additional information now
available on pathways to aesculetin and puberulin.  Important information on factors affecting the syntheses of
coumarins in cell cultures has been forthcoming.  Alternative
pathways to mono- and dimethoxypsoralens have been reported,
and the use of affinity chromatography in studies on the
O-methyltransferases mediating biosynthesis of these
coumarins has yielded extensive data on their nature and
mechanism of action.

    Photobinding of psoralens to DNA continues to be an
active area of study.  New data have been published on the
structures of the radiation-induced adducts.  It is now
known that angular furanocoumarins have limited ability to
form diadducts as well, and the ability of psoralens to
react photochemically with proteins has been recognized and
explored.  The long-accepted ability of coumarins to affect
plant growth continues to attract attention, while the list
of coumarins known to exert such action is lengthening.  The
nature of this action, and of the interaction of coumarins
with other plant growth regulators, is being elucidated in
a number of laboratories.

REFERENCES

1.  MURRAY, R.D.H., J. MÉNDEZ, S.A. BROWN. 1982. The
    Natural Coumarins: Occurrence, Chemistry and
    Biochemistry. John Wiley & Sons Ltd., Chichester.
2.  STEYN, P.S., R. VLEGGAAR, P.L. WESSELS. 1980. The
    biosynthesis of aflatoxin and its congeners. In
    The Biosynthesis of Mycotoxins. (P.S. Steyn, ed.),
    Academic Press, New York, pp. 105-155.
3.  STINSON, E.E., W.B. WISE, A.J. JUREWICZ, P.E. PFEFFER.
    1985. Alternariol: biosynthesis from norlichexan-
    thone investigated by 2-D $^{13}$C NMR. Can. J. Chem.
    (In press).
4.  CHEXAL, K.K., C. FOUWEATHER, J.S.E. HOLKER. 1975. The
    biosynthesis of fungal metabolites. Part VII.
    Production and biosynthesis of 4,7-dimethoxy-5-
    methylcoumarin in Aspergillus variecolor. J. Chem.
    Soc., Perkin Trans. I: 554.
5.  KOSUGE, T., E.E. CONN. 1959. The metabolism of
    aromatic compounds in higher plants. I. Coumarin
    and o-coumaric acid. J. Biol. Chem. 234: 2133-2137.
6.  HASKINS, F.A., L.G. WILLIAMS, H.J. GORZ. 1964. Light-
    induced trans to cis conversion of β-D-glucosyl
    o-hydroxycinnamic acid in Melilotus alba leaves.
    Plant Physiol. 39: 777-781.
7.  ROBERTS, W.L., K.P. LINK. 1937. Determination of
    coumarin and melilotic acid. A rapid micromethod
    for the determination in melilotus seed and green
    tissue. Ind. Eng. Chem., Anal. Ed. 9: 438-441.
8.  HASKINS, F.A., H.J. GORZ. 1961. A reappraisal of the
    relationship between free and found coumarin in
    Melilotus. Crop Sci. 1: 320-323.
9.  KOSUGE, T., E.E. CONN. 1961. Metabolism of aromatic
    compounds in higher plants. III. β-Glucosides of
    o-coumaric, coumarinic, and melilotic acids. J.
    Biol. Chem. 236: 1617-1621.
10. BOUDET, A.M., H. CANUT, G. ALIBERT. 1981. Isolation
    and characterization of vacuoles from Melilotus
    alba mesophyll. Plant Physiol. 68: 1354-1358.
11. OBA, K., E.E. CONN, H. CANUT, A.M. BOUDET. 1981.
    Subcellular localization of 2-(β-D-glucosyloxy)-
    cinnamic acids and their related β-glucosidase in
    leaves of Melilotus alba Desr. Plant Physiol.
    68: 1359-1363.
12. POULTON, J.E., D.E. MCREE, E.E. CONN. 1980. Intra-
    cellular localization of two enzymes involved in

coumarin biosynthesis in <u>Melilotus alba</u>. Plant
Physiol. 65: 171-175.

13. GESTETNER, B., E.E. CONN. 1974. 2-Hydroxylation of
<u>trans</u>-cinnamic acid by chloroplasts from <u>Melilotus
alba</u>. Arch. Biochem. Biophys. 163: 617-624.

14. CONN, E.E. 1984. Compartmentation of secondary
compounds. <u>In</u> Membranes and Compartmentation in
the Regulation of Plant Functions. (A.M. Boudet,
G. Alibert, P.J. Lea, eds.), Annu. Proc. Phytochem.
Soc. Europe, Vol. 24, Oxford University Press,
Oxford, pp. 1-28.

15. RANJEVA, R., G. ALIBERT, A.M. BOUDET. 1977. Metab-
olisme des composés phenoliques chez le <u>Petunia</u>.
V. Utilisation de la phenylalanine par des chloro-
plastes, isolés. Plant Sci. Lett. 10: 225-234.

16. KINDL, H. 1971. Zur Frage der <u>ortho</u>-Hydroxylierung
aromatischer Carbonsäuren in höheren Pflanzen.
Hoppe-Seyler's Z. Physiol. Chem. 352: 78-84.

17. SATÔ, M., M. HASEGAWA. 1972. Biosynthesis of dihy-
droxycoumarins in <u>Daphne odora</u> and <u>Cichorium
intybus</u>. Phytochemistry 11: 657-662.

18. BROWN, S.A. 1985. Biosynthesis of 6,7-dihydroxy-
coumarin in <u>Cichorium intybus</u>. Can. J. Biochem.
Cell Biol. 63: 292-295.

19. IBRAHIM, R.K. 1980. Position specificity of an
<u>o</u>-dihydroxycoumarin glucosyltransferase from
tobacco cell suspension culture. Phytochemistry
11: 2459-2460.

20. IBRAHIM, R.K., B. BOULET. 1980. Purification and
some properties of an <u>o</u>-glucosyltransferase from
tobacco cell suspension culture. Plant Sci. Lett.
18: 177-184.

21. COLLENDAVELLOO, J., M. LEGRAND, P. GEOFFROY, J.
BARTHELEMEY, B. FRITIG. 1981. Purification and
properties of three <u>o</u>-diphenol-O-methyltransferases
of tobacco leaves. Phytochemistry 20: 611-616.

22. OKAZAKI, M., F. HINO, K. NAGASAWA, Y. MIURA. 1982.
Effects of nutritional factors on formation of
scopoletin and scopolin in tobacco tissue cultures.
Agric. Biol. Chem. 46: 601-607.

23. OKAZAKI, M., F. HINO, K. KOMINAMI, Y. MIURA. 1982.
Effects of plant hormones on formation of scopoletin
and scopolin in tissue cultures. Agric. Biol. Chem.
46: 609-614.

24. HINO, F., M. OKAZAKI, Y. MIURA. 1982. Effect of
2,4-dichlorophenoxyacetic acid on glucosylation of

scopoletin to scopolin in tobacco tissue culture.
Plant Physiol. 69: 810-813.

25. HINO, F., M. OKAZAKI, Y. MIURA. 1982. Effects of
    kinetin on formation of scopoletin and scopolin in
    tobacco tissue cultures. Agric. Biol. Chem. 46:
    2195-2202.

26. BROWN, S.A. 1985. Recent advances in the biosynthesis
    of coumarins. Annu. Proc. Phytochem. Soc. Europe
    25: 257-270.

27. BOHM, B.A., R.K. IBRAHIM, G.H.N. TOWERS. 1961. The
    partial elucidation of the structure of a new
    coumarin from Hydrangea macrophylla Ser. Can. J.
    Biochem. Physiol. 39: 1389-1395.

28. BROWN, S.A., G.H.N. TOWERS, D. CHEN. 1964. Biosyn-
    thesis of the coumarins. V. Pathways of umbelli-
    ferone formation. Phytochemistry 3: 469-476.

29. BROWN, S.A., D.E.A. RIVETT, H.J. THOMPSON. 1984.
    Elaboration of the 6,7,8 oxygenation pattern in
    simple coumarins: Biosynthesis of puberulin in
    Agathosma puberula. Z. Naturforsch. 39c: 31-37.

30. ELLIS, B.E., S.A. BROWN. 1974. Isolation of
    dimethylallylpyrophosphate:umbelliferone dimethyl-
    allyltransferase from Ruta graveolens. Can. J.
    Biochem. 52: 734-738.

31. DHILLON, D.S., S.A. BROWN. 1976. Localization,
    purification, and characterization of dimethyl-
    allylpyrophosphate:umbelliferone dimethylallyl-
    transferase from Ruta graveolens. Arch. Biochem.
    Biophys. 177: 74-83.

32. MATERN, U., H. WENDORFF. 1985. Enzymatic synthesis
    of psoralen. Annu. Meeting, Phytochem. Soc. N.A.
    Asilomar, California, p. 22 (abstract).

33. CAPORALE, G., G. INNOCENTI, A. GUIOTTO, P. RODIGHIERO,
    F. DALL'ACQUA. 1981. Biogenesis of linear O-
    alkylfuranocoumarins: a new pathway involving
    5-hydroxymarmesin. Phytochemistry 20: 1283-1287.

34. BROWN, S.A., S. SAMPATHKUMAR. 1977. The biosynthesis
    of isopimpinellin. Can. J. Biochem. 55: 686-692.

35. INNOCENTI, G., F. DALL'ACQUA, G. CAPORALE. 1983. The
    role of 5,8-dihydroxypsoralen in the biosynthesis
    of isopimpinellin. Phytochemistry 22: 2207-2209.

36. THOMPSON, H.J., S.K. SHARMA, S.A. BROWN. 1978.
    O-Methyltransferases of furanocoumarin biosynthesis.
    Arch. Biochem. Biophys. 188: 272-281.

37. SHARMA, S.K., S.A. BROWN. 1978. Affinity chroma-
    tography on immobilized S-adenosyl-L-homocysteine.

Purification of a furanocoumarin O-methyltransferase
from cell cultures of Ruta graveolens L.  J.
Chromatogr. 157: 427-431.

38.  SHARMA, S.K.,  J.M. GARRETT, S.A. BROWN.  1979.
Separation of the S-adenosylmethionine: 5- and
8-hydroxyfuranocoumarin O-methyltransferases of Ruta
graveolens L. by general ligand affinity chromatography. Z. Naturforsch. 34c: 387-391.

39.  SHARMA, S.K., S.A. BROWN.  1979.  Affinity chromatography of Ruta graveolens L. O-methyltransferase.
Studies demonstrating the potential of the technique
in the mechanistic investigation of O-methyltransferases.  Can. J. Biochem. 57: 986-994.

40.  TIETJEN, K.G., D. HUNKLER, U. MATERN.  1983.  Differential response of cultured parsley cells to elicitors
from two non-pathogenic strains of fungi: Identification of induced products as coumarin derivatives.
Eur. J. Biochem. 131: 401-407.

41.  TIETJEN, K.G., U. MATERN.  1983.  Differential response
of cultured parsley cells to elicitors from two
non-pathogenic strains of fungi: Effects on enzyme
activities.  Eur. J. Biochem. 131: 409-413.

42.  DALL'ACQUA, F., S. MARCIANI, G. RODIGHIERO.  1970.
Interstrand cross-linkages occurring in the photoreaction between psoralen and DNA.  FEBS Lett. 9:
121-123.

43.  COLE, R.S.  1970.  Light-induced cross-linking of DNA
in the presence of a furocoumarin (psoralen).
Studies with phage lambda, Escherichia coli, and
mouse leukemia cells.  Biochim. Biophys. Acta 217:
30-39.

44.  KANNE, D., K. STRAUB, H. RAPOPORT, J.E. HEARST.  1982.
The psoralen-DNA photoreaction.  Characterization of
the monoaddition products from 8-methoxypsoralen and
4,5',8-trimethylpsoralen.  Biochemistry 21: 861-871.

45.  KANNE, D., K. STRAUB, J.E. HEARST, H. RAPOPORT.  1982.
Isolation and characterization of pyrimidine-
psoralen-pyrimidine photodiadducts from DNA.  J.
Am. Chem. Soc. 104: 6754-6764.

46.  PECKLER, S., B. GRAVES, D. KANNE, H. RAPOPORT, J.E.
HEARST, S.H. KIM.  1982.  Structure of a psoralen-
thymine monoadduct formed in photoreaction with DNA.
J. Mol. Biol. 162: 157-172.

47.  SHIM, S.C., Y.Z. KIM.  1983.  Photoreaction of
8-methoxypsoralen with thymidine.  Photochem.
Photobiol. 38: 265-271.

48. AVERBECK, D., E. MOUSTACCHI. 1980. Decreased photo-induced mutagenicity of monofunctional as opposed to bifunctional furocoumarins in yeast. Photochem. Photobiol. 31: 475-478.
49. LOWN, J.W., S.-K. SIN. 1978. Photoreaction of psoralen and other furocoumarins with nucleic acids. Bioorg. Chem. 7: 85-95.
50. KITTLER, L., Z. HRADECNA, J. SUEHNEL. 1980. Cross-link formation of phage lambda DNA in situ photochemically induced by the furocoumarin derivative angelicin. Biochim. Biophys. Acta 607: 215-220.
51. VENTURINI, S., M. TAMARO, C. MONTIBRAGADIN, F. BORDIN, F. BACCICCHETTI, F. CARLASSARE. 1980. Comparative mutagenicity of linear and angular furocoumarins in Escherichia coli strains deficient in known repair functions. Chem.-Biol. Interact. 30: 203-207.
52. BELOGUROV, A.A., G.B. ZAVILGELSKI. 1981. Mutagenic effect of furanocoumarin monoadducts and cross-links on bacteriophage lambda. Mutat. Res. 84: 11-15.
53. DE MOL, N.J., G.M.J. BEIJERSBERGEN VAN HENEGOUWEN. 1979. Formation of singlet molecular oxygen by 8-methoxypsoralen. Photochem. Photobiol. 30: 331-335.
54. KITTLER, L., Z. HRADECNA. 1980. Mechanisms of photo-addition of xanthotoxin to thymine and DNA. Indication for exclusion of a singlet oxygen reaction. Stud. Biophys. 78: 51-55.
55. DE MOL, N.J., G.M.J. BEIJERSBERGEN VAN HENEGOUWEN, B. VAN BEELE. 1981. Singlet oxygen formation by sensitization of furocoumarins complexed with, or bound covalently to DNA. Photochem. Photobiol. 34: 661-666.
56. DE MOL, N.J. 1981. Relation between some photobiological properties of furocoumarins and their extent of singlet oxygen production. Photochem. Photobiol. 33: 815-819.
57. DE MOL, N.J. 1981. Role of singlet oxygen in the photosensitizing effect of furocoumarins. Pharm. Weekbl. 116: 641-647, cited in Chem. Abstr. 94: 18103.
58. KRASNOVSKII, A.A., V.L. SUKHORUKOV, A. YA. POTAPENKO. 1983. Photogeneration of singlet oxygen by psoralens. Byull. Eksp. Biol. Med. 95: 59-61, cited in Chem. Abstr. 99: 208220.

59. JOSHI, P.C., M.A. PATHAK. 1983. Production of singlet oxygen and superoxide radicals by psoralens and their biological significance. Biochem. Biophys. Res. Commun. 112: 638-646.

60. ANDERS, A., W. POPPE, C. HERKT-MAETZKY, E.G. NIEMANN, E. HOFER. 1983. Investigations on the mechanism of photodynamic action of different psoralens with DNA. Biophys. Struct. Mech. 10: 11-30, cited in Chem. Abstr. 99: 172028.

61. VERONESE, F.M., O. SCHIAVON, R. BEVILACQUA, F. BORDIN, G. RODIGHIERO. 1981. Drug-protein interaction: The effect of UV-A irradiation of proteins in the presence of psoralens and angelicins. Med. Biol. Environ. 9: 359-364, cited in Chem. Abstr. 96: 30844.

62. SCHIAVON, O., F.M. VERONESE, R. BEVILACQUA, G. RODIGHIERO. 1982. Studies on the mechanism of photoinactivation of enzymes in the presence of furocoumarins. Med. Biol. Environ. 10: 233-236, cited in Chem. Abstr. 98: 67971.

63. MEFFERT, H., W. DIEZEL, N. SOENNICHSEN. 1980. Inactivation and protection of phosphofructokinase by treatment with 8-MOP and UV radiation. Dermatol. Monatsschr. 166: 43-44, cited in Chem. Abstr. 92: 174249.

64. VERONESE, F.M., O. SCHIAVON, R. BEVILACQUA, F. BORDIN, G. RODIGHIERO. 1982. Photoinactivation of enzymes by linear and angular furocoumarins. Photochem. Photobiol. 36: 25-30.

65. GRANGER, M., F. TOULME, C. HÉLÈNE. 1982. Photodynamic inhibition of Escherichia coli DNA polymerase I by 8-methoxypsoralen plus near ultraviolet irradiation. Photochem. Photobiol. 36: 175-180.

66. ROEMER, W., L. KITTLER, G. LOEBER. 1983. Inhibitory effects of furocoumarins plus near UV light (365 nm) on cAMP phosphodiesterase activity. Stud. Biophys. 94: 33-36.

67. BROWN, S.A. 1981. Coumarins. In The Biochemistry of Plants, A Comprehensive Treatise. (E.E. Conn, ed.), Vol. 7, Academic Press, New York, pp. 269-300.

68. SHIMOMURA, H., Y. SASHIDA, H. NAKATA, J. KAWASAKI. 1982. Plant growth regulators from Heracleum lanatum. Phytochemistry 21: 2213-2215.

69. STARIKOVA, V.T. 1982. Effect of coumarin and gibberellin on IAA oxidase and IAA synthetase activities and on hormonal metabolism in corn

seedlings. Fitogorm. Ikh Deistvie Rast. 56-61,
cited in Chem. Abstr. 98: 174732.

70.  DHAWAN, R.S., K.K. NANDA. 1982. Stimulation of
     root formation on Impatiens balsamina L. cuttings
     by coumarin and the associated biochemical changes.
     Biol. Plant 24: 177-182.

71.  SIMONOVA, N.A. 1982. Relationship between growth
     rate of barley sprouts and activity of gibberellin-
     like compounds and auxin. Fitogorm. Ikh Deistvie
     Rast. 118-124, cited in Chem. Abstr. 98: 176299.

72.  CHIA, I.-K., I.-S. KIM. 1980. Effects of coumarin
     as gibberellic acid antagonists (sic) on the
     synthesis of amylase by barley seeds. Hanguk
     Saenghwal Kwahak Yonguwon Nonchong 25: 87-95,
     cited in Chem. Abstr. 94: 186170.

73.  MOROZOVA, E.V., V.V. KATS. 1982. Effect of coumarin
     on the initial stage of cucumber seed germination
     and phytohormone activity. Fitogorm. Ikh
     Deistvie Rast. 74-83, cited in Chem. Abstr.
     98: 174734.

74.  RAY, S.D., K.N. GURUPRASAD, M.M. LALORAYA. 1983.
     Reversal of abscisic acid-inhibited betacyanin
     synthesis by phenolic compounds in Amaranthus
     caudatus seedlings. Physiol. Plant 58: 175-178.

75.  ZELCER, A., E. GALUN. 1980. Culture of newly
     isolated tobacco protoplasts:cell division and
     precursor incorporation following a transient
     exposure to coumarin. Plant Sci. Lett. 18:
     185-190.

76.  MONTEZINOS, D., D.P. DELMER. 1980. Characterization
     of inhibitors of cellulose synthesis in cotton
     fibers. Planta 148: 305-311.

77.  COLVIN, J.R., D.E. WITTER. 1980. On the inhibition
     of cellulose biosynthesis by coumarin. Plant Sci.
     Lett. 18: 33-38.

78.  MORI, K., H. MATSUSHIMA, M. TAKEUCHI. 1982. A study
     of cell wall regeneration of protoplasts in
     Marchantia polymorpha. In Plant Tissue Culture,
     Proceedings of the 5th International Congress of
     Plant Tissue Cell Culture. (A. Fujiwara, A.
     Maruzen, eds.), Tokyo, pp. 45-46, cited in Chem.
     Abstr. 99: 172969.

79.  SUNDARESAN, R.V.S., W.C. KIMMINS. 1981. Effect of
     coumarin on the synthesis and composition of cell
     walls in Phaseolus vulgaris L. Ann. Bot. (London)
     47: 283-285.

80.   AL-QUADAN, F., S.H. WENDER, E.C. SMITH.  1981.  The
        effect of some phenolic compounds on the activity
        of 6-phosphogluconate dehydrogenase from tobacco
        tissue cultures.  Phytochemistry 20: 961-964.
81.   HASHIM, I., L.A. WILSON, K.H. CHEE.  1978.  Regulation
        of indoleacetic acid oxidase activities in Hevea
        leaves by naturally occurring phenolics.  J. Rubber
        Res. Inst. Malays. 26: 105-111, cited in Chem.
        Abstr. 92: 90998.
82.   GIOVANNOZZI-SERMANNI, G., M. LUNA.  1981.  Laccase
        activity of Agaricus bisporus and Pleurotus
        ostreatus.  Mushroom Sci. 11: 485-496, cited in
        Chem. Abstr. 97: 19562.

Chapter Twelve

SOME STRUCTURAL AND STEREOCHEMICAL ASPECTS OF COUMARIN
BIOSYNTHESIS

DAVID L. DREYER

U.S. Department of Agriculture
800 Buchanan Street
Albany, California 94710

INTRODUCTION

Although predictions on the course of chemical progress
can be widely off the mark, the main outlines of coumarin
chemistry seem to be broadly established if one takes as a
criterion, the new coumarin structures reported in the past
decade. With the exception of a series of 3-substituted
4-hydroxy-5-methylcoumarins, probably biosynthesized from
acetate and largely confined to the tribe Vernonieae
(Compositae),[1] most of the structures of new coumarins
reported in this period are predictable variants of established structural patterns. Methods of coumarin isolation
and structure determination are straightforward by modern
standards. The literature contains a sizable data base of
spectroscopic parameters making structure determination a
fairly routine procedure.

Most of the common classes of coumarin systems have
been synthesized although routes to furanocoumarins in high
yield continue to be of interest. Most synthetic effort
in the past decade has centered on isopentenyl substituted
coumarins, due largely to Murray and his students.[2] In
common with many other groups of natural products, emphasis
in coumarin chemistry has shifted away from purely chemical
matters of isolation, structure and synthesis to areas of

317

biosynthesis, ecological significance, chemotaxonomy, and
related biological areas.

    Although coumarins are found in plants from 12 to 14
families, they are most frequently associated with the
otherwise unrelated families Rutaceae and Umbelliferae and
to a lesser extent with the Compositae and Leguminosae.  The
Umbelliferae produce an extensive series of $C_{10}$ and $C_{15}$
terpenoid ethers of umbelliferone (1) as well as a lengthy
series of esters of decursidinol (2), columbianetin (3),
vaginidiol (4) and khellactone (5).  These dihydrofurano-
and dihydropyranocoumarins are not commonly found in plants
of the Rutaceae but occur frequently in Umbellifers.  Linear
furanocoumarins, e.g. psoralen (6), are distributed between
plants of both the Rutaceae and Umbelliferae with a high
frequency of occurrence; however angular furanocoumarins,
e.g. angelicin (7), occur primarily in the Umbellifereae.

UMBELLIFERONE(1)                    DECURSIDINOL (2)

COLUMBIANETIN(3)                    VAGINIDOL(4)

KHELLACTONE(5)

PSORALEN(6)                         ANGELICIN (7)

Even though coumarins are classified as phenolics and
most have a shikimate origin, they largely occur in the
non-polar fractions of plant extracts and seldom contain
free phenolic groups.  Phenolic groups are often masked as
methyl or isopentenyl ethers or are found as parts of
further hetero rings.  Acid used in isolation procedures to
separate alkaloids from neutral constituents can easily
cleave isopentenyl ethers.  Free phenolic groups, when found
in coumarins, are often artifacts arising from such acid-
catalyzed cleavage.

The chemical aspects of coumarin chemistry have
recently been comprehensively summarized.[3]  In view of the
availability of this treatise it appears useful here to
focus on some special topics in coumarin chemistry where
outstanding problems remain in an attempt to identify those
research areas which might benefit by further work.


FORMAL DIELS-ALDER DIMERIC COUMARINS

Certain plants of the Rutaceae produce a number of
different coumarin dimers in which the monomeric units, an
isopentadienyl-substituted coumarin, are joined in a Diels-
Alder fashion.[4-8]  The structural features resulting from
this mode of dimerization are not restricted to coumarins in
the Rutaceae but can also be found in two sets of dimeric
2-quinolones,[9,10] indole alkaloids[11,12] and related isopentenyl substituted compounds.[13-15]  A remarkable feature of
all of these dimers is that they occur as racemates in
spite of the fact that they all have two or more asymmetric
centers.  The racemic nature of these coumarin dimers has
raised the possibility that these substances might arise by
a non-enzymatic Diels-Alder reaction during biosynthesis.

These dimers, 8-12, co-occur in the same species with
related monomeric units which by relatively simple chemical
conversions could easily give rise to the necessary diene
system required by the Diels-Alder reaction.  In several
cases[6,13] the natural dimers have been synthesized from the
anticipated isopentadienyl moiety, demonstrating that such
precursors are viable intermediates.  Examples are the
conversion of mexoticin (13) to toddasin (9),[6] and rosefuran
(14) to a diastereoisomeric mixture of diclausenan A and B
(12).[13]  Other cases have been reported[16] of totally
different products resulting from attempted synthesis.

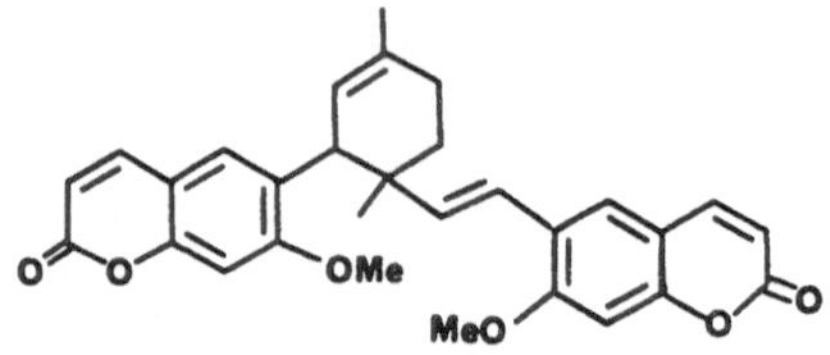

THAMNOSIN (CYCLOBISUBERODIENE) (8)

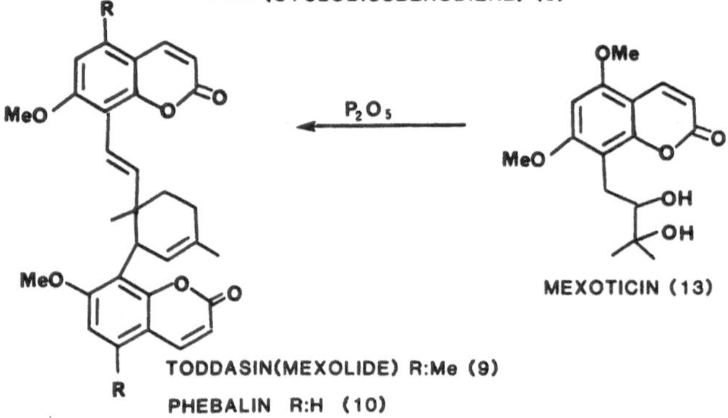

MEXOTICIN (13)

TODDASIN(MEXOLIDE) R:Me (9)
PHEBALIN R:H (10)

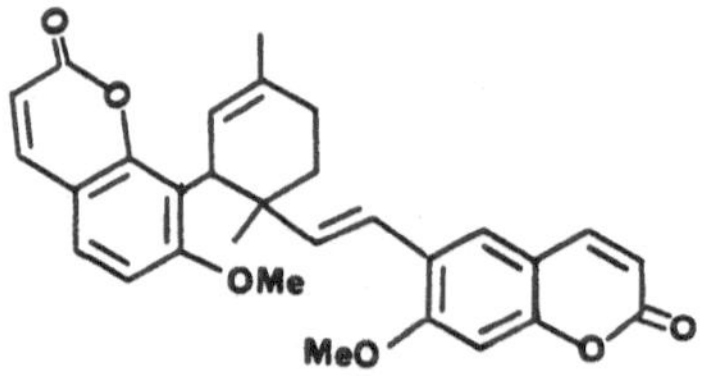

ISOTHAMNNOSIN A (11)

ROSEFURAN (14)

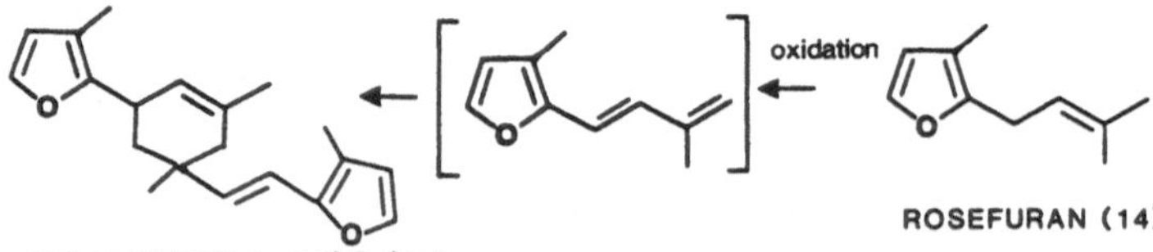

DICLAUSENAN A and B (12)
(steriomeric mixture)

The isopentadienyl system can react with either of the
two double bonds in the dienophile.  Because of the unsymmetrical nature of the dienophile, there are two possible Diels-
Alder adducts for each double bond.  This leads to a total
of four different possible structures (Figs. 1 and 2).  The
formation of coumarin dimers in every case published follows
Figure 1.  Moreover, only one of the two possible adducts
has been reported.  Since each of the different coumarin
adducts contains two asymmetric centers, there are two
diastereoisomers possible and with one possible exception[8]
only one of these has apparently been isolated in each case.
The non-coumarin optically inactive Diels-Alder dimers,

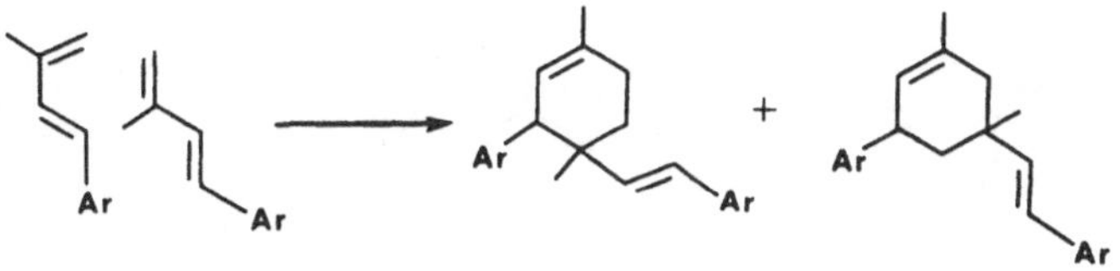

Fig. 1.  Diels-Alder products from reaction with the terminal double bond of the dienophile.

diclausenan A and B,[13] not only occur as a mixture of diastereoisomers but have a structure (12) in which the monomers have been joined in the reverse manner as that found for the coumarins (Fig. 1).

The origin of these dimeric coumarins is not an isolated phenomenon but is closely related to the biosynthesis of a series of dimeric 2-quinoline alkaloids,[9,10] indoles,[11,12] 2-phenethylamines[14,15] as well as some miscellaneous systems.[17] Moreover, these diverse materials are also found as racemates in every case.  An analysis of the dimeric alkaloid biogenesis is complicated by subsequent transformations following the Diels-Alder reaction.  The 4-hydroxy group of the 2-quinoline alkaloids is infrequently found methylated and consequently is more often involved in ring formation with adjacent isopentenyl groups.  Consideration of the coumarin dimers is less complicated because the orthohydroxy groups adjacent to the isopentenyl groups are always methylated preventing ring closures.

A number of questions arise regarding these Diels-Alder adducts.  Among these are:  what is the relative stereochemistry of the Diels-Alder adducts?  Do the Diels-Alder adducts arise by means of a concerted or non-concerted pathway?  Is the formation of the Diels-Alder adducts truly non-enzymatic?  Why have such diverse results been obtained on attempted synthesis of the dimers?

A recent paper by Schroeder and Stermitz[15] on isopentenyl substituted 2-phenethylamine alkaloids aids greatly in resolving some of these question.  These workers found a series of four racemic alkaloids in a Zanthoxylium species.  These alkaloids are all formal Diels-Alder adducts in which addition has occurred in a fashion (Fig. 2) entirely different from that found in the coumarin dimers (Fig. 1).

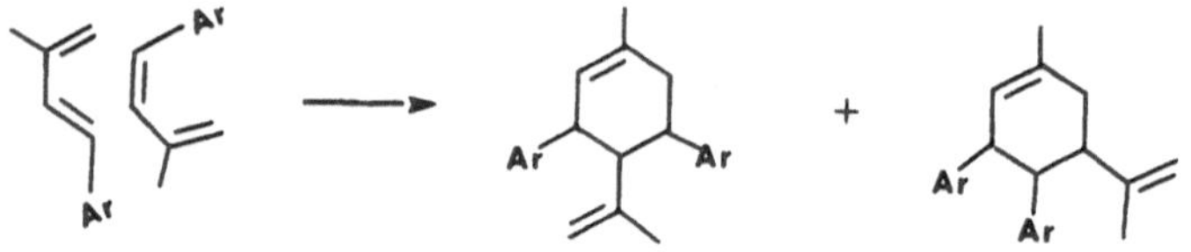

Fig. 2. Diels-Alder products from reaction with the in-chain double bond of the dienophile

Synthetic work[15] directed towards the preparation of these 2-phenethylamine alkaloids led to structural analogues of the coumarin dimers. Thus, when the isopentendienyl hordenine (15) was allowed to stand at room temperature, both diastereoisomers 16 and 17 were formed in a 9:1 ratio.

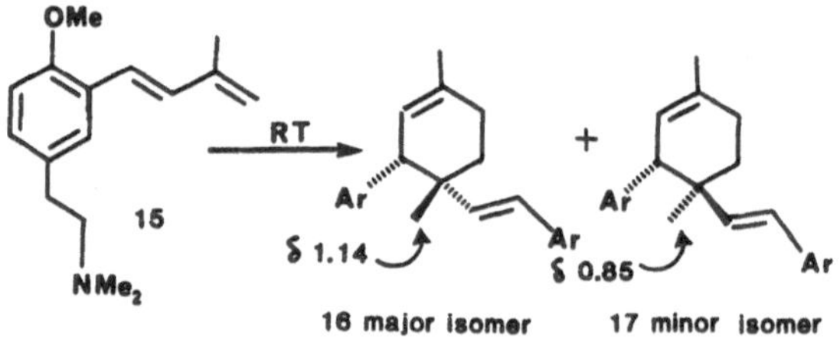

From the NMR data on both isomers, one can assign relative stereochemistry to thamnosin (8), toddasin (9) and isothamnosin A (11). The chemical shift of the tertiary C-methyl resonance in the two isomers, 16 and 17, differs substantially. In the major isomer, 16, the C-methyl resonance occurs at a normal value of $\delta$ 1.14 whereas in the minor isomer (17) it is shifted upfield to $\delta$ 0.85 due to the presence of an adjacent cis aromatic ring. The chemical shift of the tertiary C-methyl resonance in thamnosin (8) ($\delta$ 1.22),[3] toddasin (9) ($\delta$ 1.21)[5,6] and phebalin (10) ($\delta$ 1.23)[7] matches almost exactly the C-methyl signal of the major dimer (16) so that the aromatic ring must be trans to the C-methyl group in each case. These results are also supported by the X-ray structure determination of phebalin (10) thereby establishing independently the relative stereochemistry.[7]

If the formation of dimeric coumarins proceeds by a non-enzymatic Diels-Alder route, and from the work of Schroeder and Stermitz,[15] one would anticipate that minor amounts of diastereomeric adducts might be found in plants

from which coumarin dimers have already been isolated.  This
is probably the case with isothamnosin A and B which co-occur
with thamnosin in a <u>Ruta</u> species.[8]  The upfield position of
the C-methyl resonance reported for isothamnosin A at δ 0.58
suggests that, rather than structure <u>11</u>, it is the stereoisomer of thamnosin in which the C-methyl group is <u>cis</u> to
the aromatic ring.

The isopentadienylcoumarins, dehydroosthol (<u>18</u>)[18] from
<u>Choisya</u> <u>ternata</u> and avicennin (<u>19</u>)[19] from <u>Zanthoxylium</u>
<u>avicennae</u>, as well as 3-isopentadienyl-4-hydroxyacetophenone
<u>(20)</u>[20] and dehydroisoderricin (<u>21</u>)[21] have been reported as
naturally occurring, apparently unaccompanied by the corresponding Diels-Alder dimers.  Arthur and Lee[19] however report
that <u>19</u> forms a dimer of undetermined structure on heating.
A reexamination of these plants for the possible presence of
dimers related to the known monomers may be profitable.

DEHYDROOSTHOL (18)

AVICENNIN (19)

3-ISOPENTADIENYL-4-HYDROXY-
ACETOPHENONE (20)

DEHYDROISODERRICIN (21)

The synthetic work reported by Schroeder and Stermetz[15]
shows that dimerization of these isopentadienyl systems can
occur by two different routes.  One route is the classical
concerted Diels-Alder reaction (Fig. 1).  The second is an
acid catalyzed reaction of a carbocation generated from an
allyl alcohol or ether with a molecule of the diene.  This
reaction course follows that in Figure 2.  Since the latter
reaction pathway is non-concerted, exclusive <u>cis</u> addition in
the Diels-Alder reaction is not required and both <u>cis</u> and
<u>trans</u> products might result.  This non-concerted pathway is
illustrated in Figure 3.  These results explain the otherwise

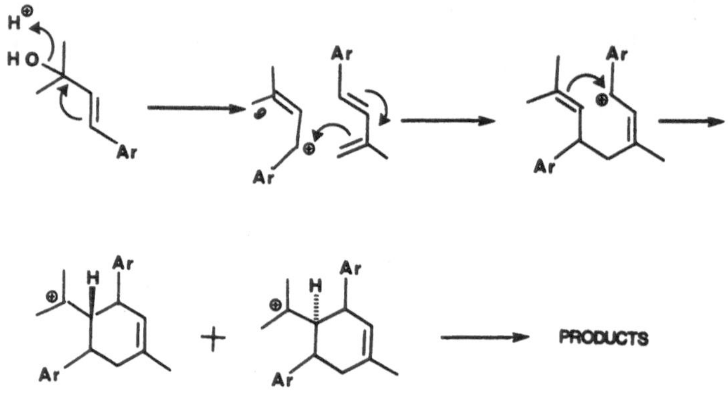

Fig. 3.   Mechanism of acid catalyzed formation of formal
Diels-Alder products.


unexpected <u>trans</u> ring junctures from an apparent Diels-Alder
reaction found among several sets of dimeric 2-quinolone
alkaloids which is illustrated for the case of vepridimerine
A (<u>cis</u>) and B (<u>trans</u>)[9] in Figure 4.

It is noteworthy that all of the nitrogen-containing
Diels-Alder adducts found in the Rutaceae arise by the reaction depicted in Figure 2 while dimerization of all the non-
nitrogen containing isopentadienyl substances have followed
the route illustrated in Figure 1.


STRUCTURE AND STEREOCHEMICAL IMPLICATIONS IN COUMARIN
BIOSYNTHESIS

The coumarin system in nature is frequently found
substituted either on carbon or oxygen by isopentenyl groups.
These isopentenyl groups often suffer successive oxidation
leading to a variety of epoxides, alcohols, ketones and
eventually to ring closure.  Degradation of ortho-C-
isopentenylphenols (<u>22</u>) through oxidation, ring closure and
fragmentation (illustrated for the linear case in Figure 5)
leads during biosynthesis to furanocoumarins, <u>6</u> and <u>7</u>.[22]
Many of the details of this scheme are supported by the
appropriate labelling studies.  Figure 5 also has many
features in common with that proposed for the biosynthesis
of furanoquinoline alkaloids.[22]

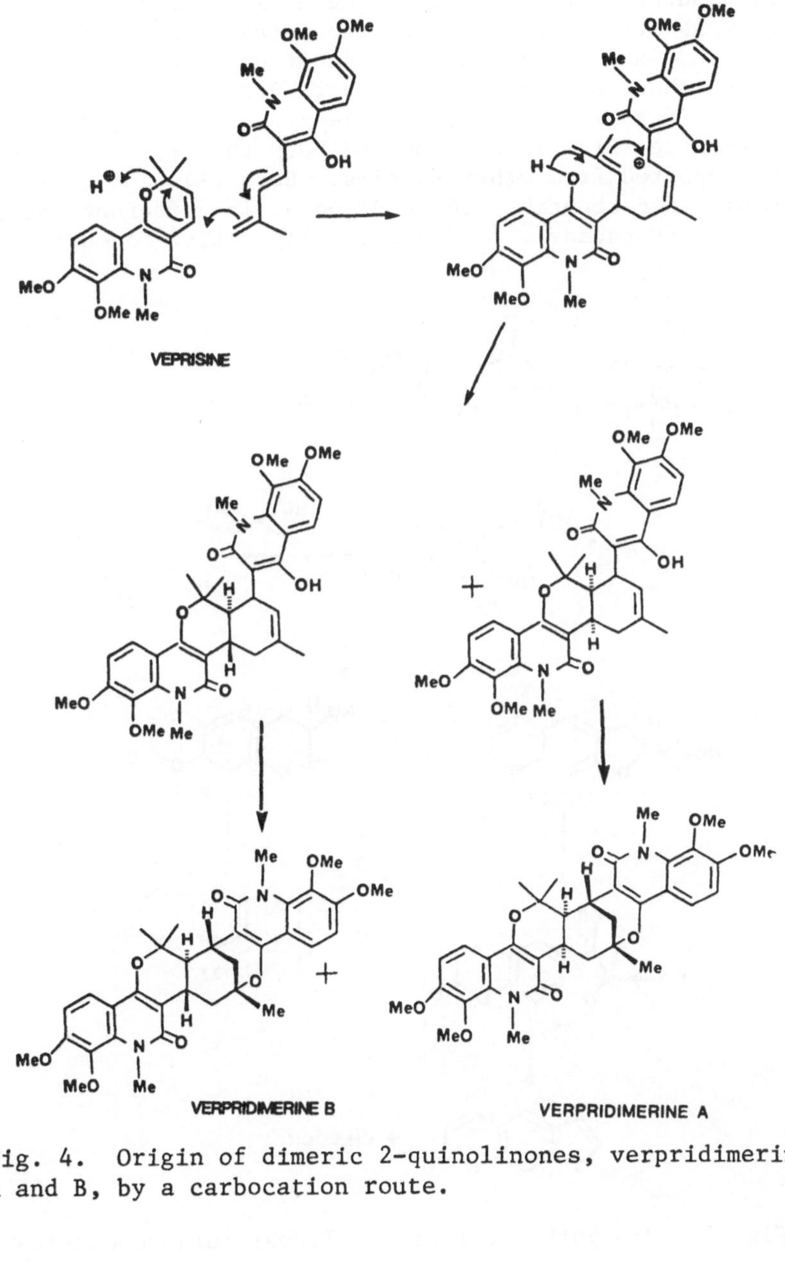

Fig. 4. Origin of dimeric 2-quinolinones, verpridimerine A and B, by a carbocation route.

Most of the intermediates in this sequence possess a single asymmetric carbon and are optically active. Nonetheless, a surprising fraction of these oxidization products of diverse structures occur as racemates[3] and several have been found in a partly racemic condition. Lemmich and co-workers[23] have summarized the distribution of oxypeucedanin in umbellifers and noted that (±)-oxypeucedanin (23) has been reported from 20 species and both the (+)- and (-)-forms reported from other species. Both (+)- and (-)-marmesin (nodakenetin) (26, R=H) occur in different species of Ptelea (Rutaceae).[24] The optically active coumarin

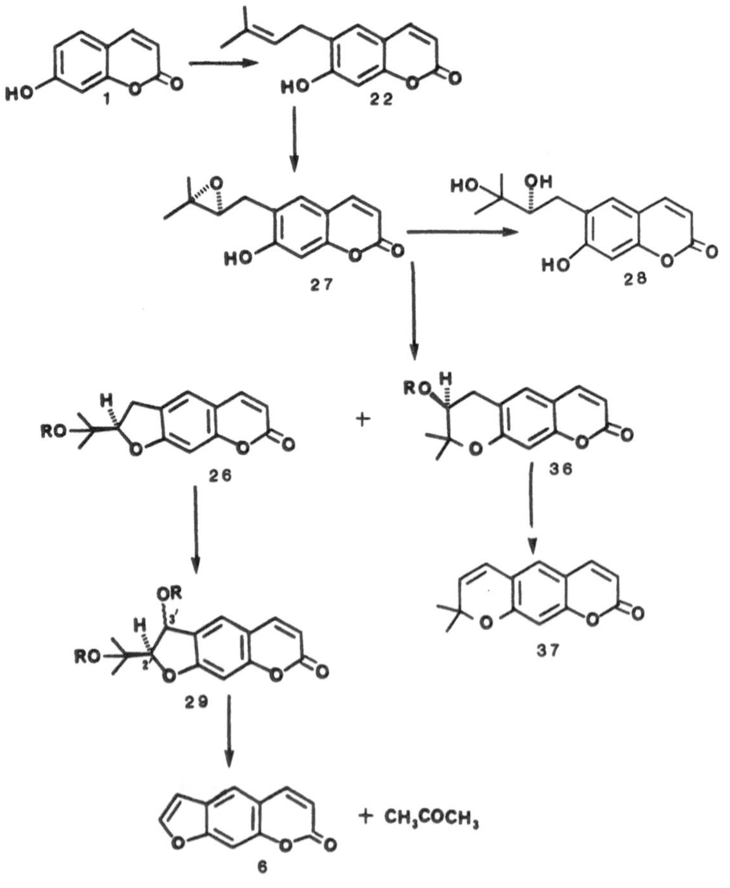

Fig. 5.  Biosynthetic route to linear furanocoumarins.

chalepin (24),[25] from Ruta, occurs as the racemate in
Helietta longifoliata.[26] (-)-Imperatorin oxide (25) has
been found partly racemic in Ptelea crenulata[24] while
Jefferies and co-workers[27] found (-)25 occurring in
Phebalium drummondii from one location but the racemic
mixture in plants at another location. Imperatorin oxide
in the (+)-form has been reported in Heracleium candicans
(Umbelliferae).[28]

Among the questions which arise when considering
absolute configuration of coumarins in relation to biosynthesis are: Do the changes in absolute configuration
correlate with the nature of the chemical transformations
in the various steps of coumarin biosynthesis? Does the
absolute configuration of oxidized isopentenyl substituted
coumarins correlate with their botanical distribution? Is
there a simple biosynthetic model which can accommodate the
occurrence of the different coumarin enantiomers as well as
racemates?

Although many different labelling studies have helped
support the pathway to furanocoumarins outlined in Figure
5, a few of the steps require further clarification. For
example, as yet there is no direct experimental evidence
showing that the initial ring closed 2-isopropanol derivative (26) arises directly from the epoxide (27) or via the
related diol (28). A displacement mechanism can accommodate
either 27 or 28 as possible intermediates. In either case,
in the absence of neighboring group effects, inversion of
configuration would be expected when proceeding from the
epoxide (27) to the dihydrofurans (26).

If the ortho-isopentenylphenol group (22) undergoes
methylation before oxidation and ring closure occurs, then
oxidized open chain isopentenyl species will accumulate in
the plant. If not, then furano (6 and 7) and pyranocoumarins (37) should be the main end product of the sequence.

Hydroxyisopropyl dihydrofuranocoumarins (26, R=H) are
well established as intermediates in the biosynthesis of
both angular and linear furanocoumarins.  This intermediate
requires the removal of the terminal three carbons of the
isopentenyl group (Fig. 5) to give the furanocoumarins, 6
or 7.  Current views on the method for removal of the
hydroxyisopropyl group involves a role for a 3'-hydroxy
derivative (29).  Indirect evidence supporting this view
comes from work on the parallel biosynthesis of furanoquinoline alkaloids.  Only one label is lost when 3',3'-doubly
labeled analogue of 26 is fed to the plant.[29]  This result
precludes an intermediary role for a 3'-keto marmesin.
These results suggest a fragmentation-elimination route by
way of a 3'-hydroxy or 3'-acyloxymarmesin (29) or, in the
angular case, a similar 3'-substituted columbianetin (3).
A large number of the required 3'-hydroxy or 3'-acyloxy
intermediates have been isolated; e.g. xanthoarnol (30)
(2',3'-trans),[30] vaginidiol (4) (2',3'-cis), smyrniorin (31)
(2',3'-cis).  The angular isomers are most frequently found

XANTHOARNOL (30)                SMYRANIORIN (31)

occurring in plants of the Umbelliferae.  With several
exceptions all of these compounds have the 2',3'-cis
configuration.

Reasonable mechanisms can be advanced leading to the
furanocoumarins (6 or 7) from either the cis or trans
precursor (29) (Fig. 6).  The trans configuration could
accommodate a classical trans elimination to generate the
furan double bond.  On the other hand, there is ample
precedent for cis elimination through a cyclic transition
state to yield a double bond (Fig. 6).  In either case
3'-acyloxyderivatives rather than the free hydroxy derivative might also be directly involved in elimination.

Isopentenyl and geranyl groups substituted either on
carbon or oxygen are widespread substituents in coumarins.
However, when occurring in the 3-position such terpene
groups are frequently found rearranged.  Grundon and

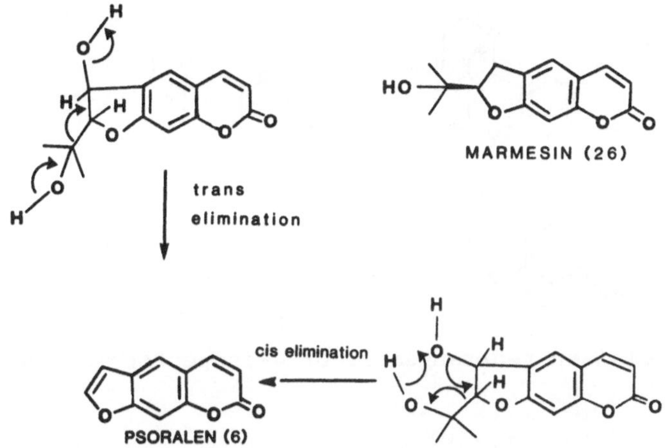

MARMESIN (26)

trans
elimination

cis elimination

PSORALEN (6)

Fig. 6. Possible fragmentation-elimination mechanism of
cis or trans-3'-hydroxymarmesin leading to furanocoumarins.

co-workers[31] showed that Ruta was able to convert labeled
umbelliferone (1), 4,7-dihydroxycoumarin (32) and
4-isopentenyloxy-7-hydroxycoumarin (33) to rutamarin (34).

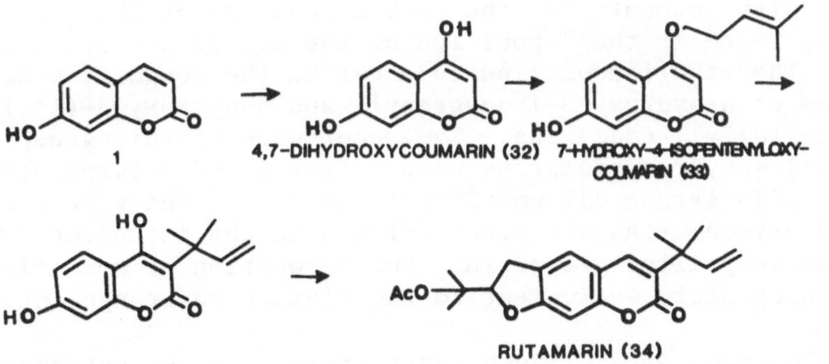

4,7-DIHYDROXYCOUMARIN (32)  7-HYDROXY-4-ISOPENTENYLOXY-
COUMARIN (33)

RUTAMARIN (34)

Unfortunately the 4-hydroxycoumarins, 32 and 33, have
not been demonstrated to occur naturally in Ruta. Nevertheless, the results suggest that 3-(1',1'-dimethylallyl)-
coumarins (e.g. 24 and 34) arise by a Claisen rearrangement
from the related 4-isopentenyloxycoumarin (33) (Fig. 7).
A Claisen rearrangement of the 4-isopentenyloxy group
accounts for the non-terminal point of attachment on the

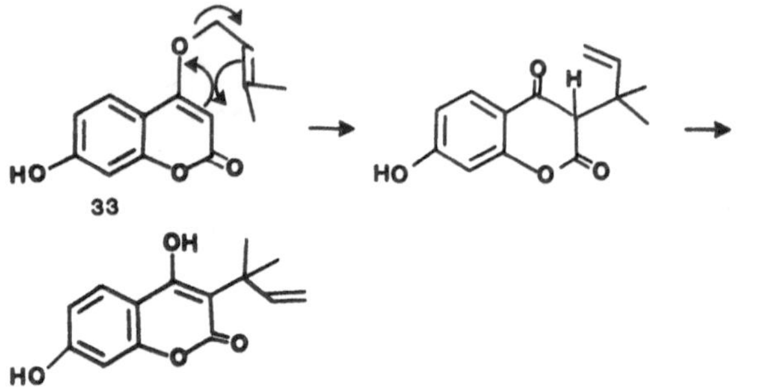

Fig. 7.   Claisen rearrangement of 4-isopentenyloxycoumarins.

isopentenyl group and the fact that 3-(1',1'-dimethylallyl)
groups occur twice as often as 3-isopentenyl groups in
coumarins.  This route to 3-terpenoid coumarins would
require the subsequent reduction of the 4-hydroxy group to
give 34.

A similar Claisen rearrangement of a 4-geranyloxycou-
marin also accounts for the carbon skeleton of the $C_{10}$
moiety found in the 3-position of the ethuliacoumarins (Fig.
8).[1]  The ethuliacoumarins co-occur in the Compositae with a
series of 4-hydroxy-3-isopentenyl- and 4-hydroxy-3-(1',1'-
dimethylallyl) coumarins.  The occurrence of this group of
natural 4-hydroxycoumarins with a rearranged 3-terpenoid
group adds strong circumstantial support to the view that
4-hydroxycoumarins are intermediates in the introduction of
1,1-dimethylallyl groups into the 3-position of coumarins
by a biogenetic equivalent to the Claisen rearrangement.

The idea of a Claisen rearrangement can be extended to
account for the structure of the terpenoid moiety in
ethuliconyzone (35)[32] which is obviously a degraded ethuliacoumarin.  If the initial Claisen rearrangement product
undergoes a further abnormal Claisen rearrangement, and this
is followed by ring closure, the appropriately substituted
cycloheptanone ring system found in ethuliconyzone (35) can
be generated (Fig. 9).  A similar rearrangement has previously been suggested to account for rearranged $C_5$ and $C_{10}$
terpenoid groups found in several 2-quinolinone alkaloids.[33]

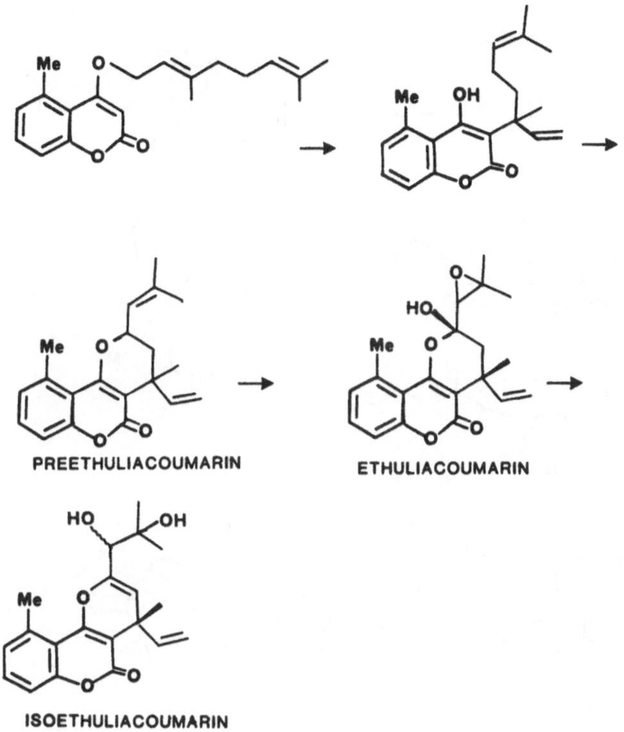

PREETHULIACOUMARIN     ETHULIACOUMARIN

ISOETHULIACOUMARIN

Fig. 8.   Probable biosynthetic route to the ethuliacoumarins.

Grundon and McColl[34] have discussed the relationship of
the absolute configuration of pairs of biosynthetically
related metabolites occurring in the same plant.  They point
out that the observed relationships are those expected from
Figure 5, although the number of examples are limited.
Thus, co-occurring epoxides (27) and their derived diols
(28) have the same absolute configuration since the latter
result from epoxide ring opening with retention of configuration.  Co-occurring hydroxyisopropyl dihydrofurans (26) and
hydroxy dimethylpyranocoumarins (36) have opposite absolute
configurations consistent with their origin from a common
oxide intermediate (27) by a displacement reaction.  The
former (26) arise by inversion at the asymmetric center while
36 arises by displacement at a non-asymmetric center and
therefore must have the same absolute configuration as the
starting epoxide (27).

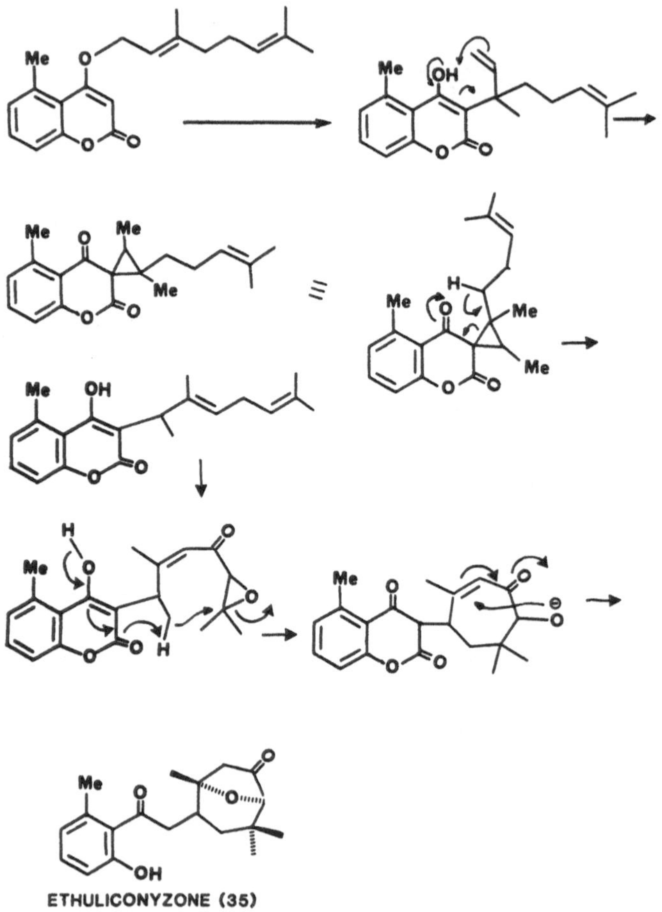

Fig. 9.   Probable biosynthetic route to ethuliconyzone.

Although the natural occurrence of both enantiomeric
forms in a single species is hardly confined to coumarins,
the occurrence of isopentenyl coumarins as racemates or in
both enantiomeric forms, depending on the botanical source,
has not been adequately explained.  Many coumarins reported
to occur as racemates or in both enantiomeric forms are
found compiled in Reference 3.  Since these examples are
not confined to any particular botanical group, there is no
chemotaxonomic correlation of one isomer or another with a
given taxa.

For any given case, the occurrence of two separate enzymes, each capable of making one of the two enantiomers, would require two rather unique and unlikely evolutionary events. The problem may be resolved for cases involving a relatively simple asymmetric reaction (e.g. epoxidation of an alkene) by assuming that the enzyme active site is capable, in many cases, of accommodating both arrangements of the substrate. If the active site could be modified by participation of a second regulatory molecule, which might perhaps also be a natural product, the substrate might be restricted to only one orientation on the active site and only one enantiomeric product would be formed. Laboratory analogies exist for such a model.[35] The occurrence of such regulatory substances may vary during the growing cycle leading to different enantiomeric composition at different stages of plant growth. Since most research on natural products is done on mature plants, little is known regarding the variation of optical activity with plant maturity.

The relative rates of the various biosynthetic steps may determine to a great degree which substances will occur in a given plant species. This may be a result of the degree of compartmentalization and transfer of substrate across membranes.[36] Variation in this transfer rate depends on the state of the plant and explains differences in composition occasionally reported for some species or in tissue culture.

## REFERENCES

1. BOHLMANN, F., C. ZDERO. 1982. Glaucolides and other constituents from South African Vernonia species. Phytochemistry 21: 2263–2267 and preceding papers in this series.
   SHUKLA, V.S., S.C. DUTTA, R.N. BARUAH, R.P. SHARMA, G. THYAGARAJAN, W. HERZ, N. LUMAR, K. WATANABE, J.G. BLOUNT. 1982. New 5-methylcoumarins from Ethulia conyzoides. Phytochemistry 21: 1725–1731.
2. MURRAY, R.D.H., Z.D. JORGE. 1984. Claisen rearragements. XV. Structure revision of the coumarin, celerin, by synthesis. Tetrahedron 40: 5229–5233 and preceding papers in this series.
3. MURRAY, R.D.H., J. MENDEZ, S.A. BROWN. 1982. The Natural Coumarins, Wiley and Sons, New York, New York, 702 pp.

4.  GUISE, G.B., E. RITCHIE, R.G. SENIOR, W.C. TAYLOR.
    1967.  The chemical constituents of Australian
    Zanthoxylum species.  IV. Two new coumarins from
    Zanthoxylum suberosum C. T. White (Syn. Zanthoxylum
    dominianum Merr. & Perry; Zanthoxylum ovalifolium
    Wight).  Aust. J. Chem. 20: 2429-2439.
    KUTNEY, J.P., T. INABA, D.L. DREYER.  1968.  The
    structure of thamnosin.  A novel dimeric coumarin
    system.  J. Amer. Chem. Soc. 90: 813-814.
    KUTNEY, J.P., T. INABA, D.L. DREYER.  1970.  Further
    studies on the constituents of Thamnosma montana
    Torr. and Frem.  The structure of thamnosin, a
    novel dimeric coumarin system.  Tetrahedron 26:
    3171-3184.
5.  SHARMA, P.N., A. SHOEB, R.S. KAPIL, S.P. POPLI.  1980.
    Toddasin, a new dimeric coumarin from Toddalia
    asiatica.  Phytochemistry 19: 1258-1260.
6.  CHAKRABORTY, D.P., S. ROY, A. CHAKRABORTY, A.K. MANDAL,
    B.K. CHOWDHURY.  1980.  Structure and synthesis of
    mexolide; A new antibiotic dicoumarin from Murraya
    exotica Linn. [Syn. Murraya paniculata (L.) Jack.].
    Tetrahedron 36: 3563-3564.
7.  BROWN, K., R.C. CAMBIE, D. HALL.  1971.  Structure of
    phebalin: a biscoumarin derivative.  Chem. and
    Ind., 1020.
    BROWN, K.L., A.I.R. BURFITT, R.C. CAMBIE, D. HALL,
    K.P. MATHAI.  1975.  Constituents of Phebalium nudum.
    III. Structures of Phebalin and Phebalarin.  Aust.
    J. Chem. 28: 1327-1337.
8.  GONZALEZ, A.G., R.J. CARDONA, E. DIAZ CHICO, H. LOPEZ
    DORTA, F. RODRIGUEZ LUIS.  1977.  New sources of
    natural coumarins.  XXXIII. Bicoumarins of Ruta
    sps. Tene. 29662.  An. Quim. 73: 1510-1511; 1978.
    Chem. Abstr. 89: 160105.
9.  NGADJUI, T.B., J.F. AYAFOR, B.L. SONDENGAM, J.D.
    CONNOLLY, D.S. RYCROFT, S.A. KHALID, P.G. WATERMAN,
    N.M.D. BROWN, M.F. GRUNDON, V.N. RAMACHANDRAN.
    1982.  The structures of vepridimerines A-D, four
    new dimeric prenylated quinolone alkaloids from
    Vepris louisii and Oricia renieri (Rutaceae).
    Tetrahedron Lett., 2041-2044.
10. JURD, L., R.Y. WONG.  1981.  New quinolinone alkaloids
    from the heartwood of Euxylophora paraensis.  Aust.
    J. Chem. 34: 1625-1632.
    JURD, L., R.Y. WONG, M. BENSON.  1982.  The structures
    of paraensidimerin A and C, two bisquinolinone

alkaloids from Euxylophora paraensis.  Aust. J.
Chem. 35: 2505-2517.

JURD, L., M. BENSON, R.Y. WONG.  1983.  New quinolinone
and bis-quinolinone alkaloids from Euxylophora
paraensis.  Aust. J. Chem. 36: 759-786.

11.  TILLEQUIN, F., M. KOCH, M. BERT, T. SENENET.  1979.
Plantes de Nouvelle Caledonie.  LV. Isoborreverine
et borreverine, alcaloides bis-indoliques de
Flindersia fournieri.  J. Nat. Products 42: 92-95.

POUSSET, J.L., A. CAVE, A. CHIARONI, C. RICHE.  1977.
A novel bis-indole alkaloid.  X-ray crystal structure determination of borreverine and its rearrangement product on diacetylation.  Chem. Comm. 261-
262.

TILLEQUIN, F., M. KOCH.  1979.  Trois nouveaux
alcaloides bis-indoliques de Flindersia fournieri.
Phytochemistry 18: 1559-1561.

TILLEQUIN, F., M. KOCH.  1979.  Deux nouveaux
alcaloides bis-indoliques de Flindersia fournieri.
Phytochemistry 18: 2066-2067.

12.  KONG, Y., K. CHENG, R.C. CAMBIE, P.G. WATERMAN.  1985.
Yuehchukene: a novel indole alkaloid with anti-
implantation activity.  Chem. Comm. 47: 47-48.

CHENG, K., Y. KONG, T. CHAN.  1985.  Biometric
synthesis of yeuhchukene.  Chem. Comm., 48-49.

13.  SUBBA RAO, G.S.R., B. RAVINDRANATH, V.P. SASHI KUMAR.
1984.  Volatile constituents of Clausena
willdenovii; Structures of the furanoterpenes
a-clausenan, diclausenan A and diclausenan B.
Phytochemistry 23: 399-401.

14.  CAOLO, M.A., F.R. STERMITZ.  1979.  A new zanthoxylum
alkaloid structurally related to tetrahydrocanna-
binol.  Tetrahedron 35: 1487-1492.

SWINEHART, J.A., F.R. STERMITZ.  1980.  Bishordeninyl
terpene alkaloids and other constituents of
Zanthoxylum culantrillo and Zanthoxylum coriaceum.
Phytochemistry 19: 1219-1223.

15.  SCHROEDER, D.R., F.R. STERMITZ.  1985.  Isolation and
synthesis of bishordeninyl terpene alkaloids.  Some
experiments relating to the natural occurrence of
formal Diels-Alder adducts.  Tetrahedron 41:
4309-4320.

16.  MOCK, J.R., R.C. SENIOR, W.C. TAYLOR.  1980.  The
chemical constituents of Australian Zanthoxylum
species.  VII. Some transformations of suberosin.
Aust. J. Chem. 33: 395-411.

        BARNES, C.S., M.I. STRONG, J.L. OCCOLOWITZ. 1963.
        The 2,2-dimethylchromen dimer 2,2-diphenylchromen
        and related compounds. Tetrahedron 19: 839-847.
17.   BOHLMANN, F., R. MATHUR, J. JAKUPOVIC, R.K. GUPTA,
        R.M. KING, H. ROBINSON. 1982. Furanoheliangolides
        and other compounds from Celea hymenolepis.
        Phytochemistry 21: 2045-2048.
18.   BOHLMANN, F., H. FRANK, C. ZDERO. 1972. Natuerlich
        vorkommende cumarin-derivate. An. Quim. 68: 765-
        767.
19.   ARTHUR, H.C., C.M. LEE. 1960. An examination of the
        Rutaceae of Hong Kong. Part V. A new coumarin,
        avicennin from the bark of Zanthoxylum avicennae.
        J. Chem. Soc., 4654-4657.
        ARTHUR, H.C., W.D. OLLIS. 1963. A revised structure
        for avicennin. J. Chem. Soc., 3910-3912.
20.   BOHLMANN, F., M. GRENZ. 1970. Neue isopentenyl-
        acetophenon-derivative aus Helianthella uniflora.
        Chem. Ber. 103: 90-96.
21.   CUCA SUAREZ, L.E., F. DELLA MONACHE, G.B. MARINI
        BETTOLO, F. MENICHINI. 1980. Tephrosieae.
        X. Three new prenylated flavanones from Tephrosia
        sp. Farmaco Ed. Sci. 35: 796-800; 1980. Chem.
        Abstr. 93: 182824.
22.   GRUNDON, M.F. 1978. The biosynthesis of aromatic
        hemiterpenes. Phytochemistry 34: 143-161.
23.   LEMMICH, J., P.A. PEDERSEN, B.E. NIELSEN. 1971.
        Coumarins and terpenoids from the fruits of
        Ligusticum sequieri. Phytochemistry 10: 3333-
        3334 and references cited therein.
24.   DREYER, D.L. 1969. Coumarins and alkaloids of the
        genus Ptelea. Phytochemistry 8: 1013-1020.
25.   BROOKER, R.M., J.N. EBLE, N.A. STARKOVSKY. 1967.
        Chalepensin, chalepin, and chalepin acetate,
        three novel furocoumarins from Ruta chalepensis.
        Lloydia 30: 73-77.
26.   POZZI, H., E. SANCHEZ, J. COMIN. 1967. Studies on
        Argentine plants. XXII. Heliettin, a new furo-
        coumarin from Helietta longifoliata Britt.
        Tetrahedron 23: 1129-1137.
27.   CHOW, P.W., A.M. DUFFIELD, P.R. JEFFERIES. 1966. The
        chemistry of the Western Australian Rutaceae.
        V. The isolation of some coumarins and the struc-
        ture of phebalosin. Aust. J. Chem. 19: 483-488.
28.   SHARMA, Y.N., A. ZAMAN, A.R. KIDWAI. 1964. Chemical
        examination of Heraceleum candicans. I.

Isolation and structure of a new furocoumarins-heraclenin. Tetrahedron 20: 87-90.

29. GRUNDON, M.F., D.M. HARRISON, C.G. SPYROPOULOS. 1975. Biosynthesis of aromatic isoprenoids. Part III. Mechanism of formation of the furan ring and origin of the 4-methoxy group in the biosynthesis of furoquinoline alkaloids. J. Chem. Soc. Perkin I, 302-304.

30. ISHII, H., T. ISHIKAWA, H. SEKIGUCHI, K. HOSAYA. 1973. Xanthoarnol. New dihydrofuranocoumarin. Chem. Pharm. Bull. 21: 2346-2348; 1974. Chem. Abstr. 80: 45639.

31. DONNELLY, W.J., M.F. GRUNDON, V.N. RAMACHANDRAN. 1977. Biosynthesis of aromatic isoprenoids. Part IV. Origin of the 1,1-dimethylallyl group of the coumarin, rutamarin. Proc. Roy. Irish Acad. Sect. B 77B: 443-447; 1978. Chem. Abstr. 89: 103824.

32. BALBAA, S.I., A.F. HALIM, F.T. HALAWEISH, F. BOHLMANN. 1980. New 5-methylcoumarin derivatives from Ethulia conyzoides. Phytochemistry 19: 1519-1522.

33. GRUNDON, M.F. 1979. Quinoline alkaloids related to anthranilic acid. In The Alkaloids. (R.H.F. Manske, R. Rodrigo, Eds.), Vol. XVII, Academic Press, New York, New York, pp. 105-191.

34. GRUNDON, M.F., I.S. McCOLL. 1975. Stereochemical aspects of the biological oxidation of aryl isoprenoids. The asymmetric synthesis and absolute configuration of meranzin and of meranzin hydrate. Phytochemistry 14: 143-150.
    STECK, W., S.A. BROWN. 1970. Biosynthesis of angular furanocoumarins. Can. J. Biochem. 48: 872-880.

35. KATSUKI, T., K.B. SHARPLESS. 1980. The first practical method for asymmetric epoxidation. J. Amer. Chem. Soc. 102: 5974-5976.
    SHARPLESS, K.B., C.H. BEHRENS, T. KATSUKI, A.W.M. LEE, V.S. MARTUB, M. TAKATANI, S.M. VITI, F.J. WALKER, S.S. WOODARD. 1983. Stereo and regioselective openings of chiral 2,3-epoxy alcohols. Versatile route to optically pure natural products and drugs. Unusual kinetic resolutions. Pure and Applied Chem. 55: 589-604.

36. LUCKNER, M., B. DIETTRICH, W. LERBS. 1980. Progress in Phytochemistry. (L. Reinhold, J.B. Harborne, T. Swain, Eds.), Vol. 6, Pergamon Press, Oxford, pp. 103-142.